Springer-AAS Acoustics Series

This series publishes peer reviewed high quality monographs and contributed volumes on all topics in acoustics. Books in this series range from those focused on a particular aspect of acoustics and vibration to practical handbooks covering a range of topics. The advantage for authors is that the inclusion of a book as part of this series will demonstrate high quality content of the book. While this series encourages authors and topics that are relevant to the Australasian region, proposals for contributions to the series are not be restricted to this region.

Martin Ochmann · Rafael Piscoya

Moving Acoustic Sources

Theory, Applications and Localization

Australian Acoustical Society

Springer

Martin Ochmann
Emeritus of BHT Berliner Hochschule für Technik
Berlin, Germany

Rafael Piscoya
wax GmbH
Berlin, Germany

ISSN 2948-2062 ISSN 2948-2070 (electronic)
Springer-AAS Acoustics Series
ISBN 978-981-96-0870-6 ISBN 978-981-96-0871-3 (eBook)
https://doi.org/10.1007/978-981-96-0871-3

This Springer imprint is published by the registered company Springer Nature Singapore Pte Ltd.
The registered company address is: 152 Beach Road, #21-01/04 Gateway East, Singapore 189721, Singapore

Preface

We live in a society where, as part of globalization, travel by land, water and air, as well as exchanging goods, plays an increasingly significant role. As a result, in recent years and decades, traffic flows have increased enormously, leading to both advantages and disadvantages. Almost all parts of the world have become accessible to us within a few hours, but at the same time, energy consumption and noise pollution from traffic have increased immensely. Therefore, it is an important task to minimize the unwanted noise generated by traffic. Of course, traffic sources are mostly moving sources. Thus, noise reduction will be more successful the more accurately one can localize such sound sources and understand and calculate the sound field emitted by them.

The research project "*LION–Localization and identification of moving noise sources*" (launched in 2019), which was funded by the German Research Foundation (DFG), served precisely this purpose. This book contains the results that the two authors of the book were able to achieve within the project.

Large parts of the book are published here for the first time, offering suggestions for subsequent research work. For this reason, the book is intended primarily for researchers in the field of theoretical and numerical acoustics who want to address the theory and applications of moving sound sources.

Of course, we realize that no one is safe from making mistakes, so we would be grateful to readers who bring errors to our attention whenever they find them.

In the epilogue of our first jointly authored book *Theory and Application of Acoustic Sources Using Complex Analysis*, published in 2021, we pointed out that "*we have only considered stationary complex sources, moving sources are not treated in the book*" and our hope of being able to deal with this exciting topic has now been realized.

The present book is again a joint work of both authors, based on a long and fruitful cooperation in many research projects. Nevertheless, we would like to mention that the extensive Chap. 2, covering nine different themes on methods in the frequency range, was written by Martin Ochmann. Chapters 3–5 mainly develop methods in the time domain and were authored by Rafael Piscoya. Therefore, it should not surprise or even irritate the reader that the style of these two parts of the book differs slightly.

Berlin, Germany
Spring 2024

Martin Ochmann
Rafael Piscoya

Acknowledgements

The research project "*LION–Localization and identification of moving noise sources*" started in 2019 and was funded by the Deutsche Forschungsgemeinschaft (DFG, German Research Foundation)—Project number 422372636.

Thus, the authors gratefully acknowledge the financial support of the DFG for this research project.

We would also like to thank Dr. Marcus Maeder from the Technical University of Munich for his careful proofreading, which has contributed to the improved readability of the book.

Additionally, we express our gratitude to Hon. Assoc. Prof. Marion Burgess, and to Dr. Loyola D'Silva, the Senior Editor at Springer Nature in Singapore along with his team, for their invaluable support.

Contents

Chapter 1
Introduction

1.1 Motivation

In 2016, the first author had the idea of initiating a joint project in which researchers from Germany, Austria and Switzerland would work together on a current topic in the field of acoustics. The identification and localization of moving sound sources crystallized as an exciting topic, as the study of moving sound sources is of great importance, since, for example, airplanes, trains and cars generate significant noise, which should be minimized as far as possible. The application to the German Research Foundation, which was submitted jointly by the Berlin University of Applied Sciences (BHT), the Technical University of Berlin, the Acoustic Research Institute of the Austrian Academy of Sciences in Vienna and the Swiss Federal Laboratories for Materials Science and Technology was successful. Therefore, the so-called project *LION–Localization and identification of moving noise sources* could start in 2019. The two authors of this book are responsible for the BHT part of the project.

Many questions were raised at the beginning of the project, including the following important issues:

(1) In what way do the time and frequency domains interact in the localization and identification of moving sound sources and which of these two domains can be better used for localization?
(2) What influence do ground effects have on the accuracy of the results?
(3) Is it possible to use the well-known equivalent source method (ESM) to localize sound sources, and does it deliver good results even at low frequencies, in contrast to known methods such as beamforming?

These questions motivated our research and the results form the content of this book.

Regarding the first question, the first author observed that the scientific literature on moving sources essentially refers to the time domain, where the Lorentz transformation is often used for rectilinear motion. Studies in the frequency domain

M. Ochmann and R. Piscoya, *Moving Acoustic Sources*,
Springer-AAS Acoustics Series, https://doi.org/10.1007/978-981-96-0871-3_1

are rarely found. This is, of course, because we are dealing with two different time scales: The oscillation of the source and its time-dependent motion through space. However, especially in the frequency domain, very useful results and closed-form expressions for the spectrum can be found. Therefore, a large part of the book is devoted to studying the frequency spectrum of moving point sources.

Concerning the second question, we can consider the influence of the ground in a rigorous way by considering half-space solution. To this end, we study the sound field of sources moving over infinite planes.

Both authors have long been concerned with the ESM for calculating the sound radiation from ***stationary*** sources. Therefore, this topic is explored in the third part of the book, in particular how to formulate the ESM for moving sources and whether it can be used efficiently for source localization.

Moving sources also play an important role in fields other than acoustics. Here we give two selected examples: In electromagnetic theory, the field of sources moving on straight and circular orbits is studied in Chap. VII of [1], where Clemmow treats the problem by using the plane wave spectrum representation. In soil dynamics, the propagation of surface vibrations due to high-speed moving harmonic rectangular loads is analyzed in [2].

1.2 Overview About the Content of the Book and Related Literature

The extensive Chap. 2 is dedicated to examining the frequency spectrum of harmonically radiating point sources and of extended sources moving along various trajectories. The restriction to point sources is not that strong, as more complex sources can be constructed by superimposing several such sources. However, when whole vibrating bodies, such as vehicles, move through the medium, integral equation methods must be applied. This leads to the famous Ffowcs Williams and Hawkings (FW-H) equation, which is formulated in the time domain and can handle moving boundaries and foreign bodies in a very general way [3]. Similar approaches for vibrating bodies in the time domain can be found, for example, in [4–6].

For a linear, uniform motion, the FW-H equation can be transformed into the frequency domain after a Galilean transformation. This was shown by Lockard for the two-dimensional [7] and also for the three-dimensional case [8]. Using the Galilean transformation has the disadvantage that the distance between the sound source and the receiver remains constant, and, as a result, the Doppler effect does not occur at all. However, we can demonstrate that the FW-H equation for general orbits and extended bodies can be transformed into the frequency domain, yielding a boundary integral equation.

Most of the studies on moving sources that can be found in the literature have been conducted mainly in the time domain (see, e.g., [9, Chap. 11], [10, Chap. 9], [11, Chap. 14], [12, Chap. N], [13], [14, Chap. 9] and others). One reason for this is that two different time scales have to be considered: the temporal oscillation behavior of the source and its time-dependent motion through space. Direct calculations of the frequency spectrum of moving sources are rarely found in the scientific literature and are typically limited to specific cases. However, with the help of a temporal Fourier transform, the dependence on time can be removed, so that we obtain an inhomogeneous Helmholtz equation. In the subsections of Chap. 2, we solve this equation for different trajectories of the source in free space and half-spaces over ground, thereby obtaining the radiated spectrum.

In Sect. 2.1, we start with the description of the general problem and the solution method in the frequency domain. The starting point is an inhomogeneous wave equation for the acoustic velocity potential in which the right-hand side describes the moving point source with the aid of Dirac delta functions. In the following, we always assume that the moving source emits a mono-frequent harmonic signal with angular frequency Ω. The temporal Fourier transformation leads to a Helmholtz equation, where the right-hand side is now given by a Fourier integral over Dirac delta functions. This integral can be solved explicitly for a large class of trajectories of the source.

In Sect. 2.2, we start with the simplest motion and consider a source moving along a straight line in one, two and three dimensions with constant positive velocity. In all three cases, the spectrum can be described by a simple analytical formula. A similar short derivation in three dimensions can be found in [9, pp. 728–729]. In the literature, on the other hand, the Lorentz transformation is often used to solve the problem in the time domain. This allows to transform the wave equation for a moving source into a wave equation for a stationary source (see, e.g., [9], Chap. 11.2 or [15, 16] for sources moving in half-spaces). Thus, the Lorentz transformation eliminates the motion of the source. For example, for a moving and mono-frequently oscillating source in physical space, one obtains a steady and single-frequency oscillating source, but now in Lorentz space. A detailed and recommendable work on this subject—but only available in German—was carried out by Ollendorf (see [17–19]). Unfortunately, this technique can only be applied to rectilinear motions with constant speed, which is a very strong limitation. A generalization to arbitrary trajectories is not possible by means of the Lorentz transformation.

In Sect. 2.3 we generalize the method in three ways: First, we show that the frequency spectra for sources traveling along an explicitly given arbitrary trajectory can be expressed by an explicit integral representation in Cartesian coordinates and is henceforth referred to as the *Cartesian convolution integral* (CCI) since we use the free-field Green's function in Cartesian coordinates as a convolution kernel. In the specific case of a source moving along a straight line from minus infinity to plus infinity, the CCI can be explicitly solved (see Sect. 2.2) and aligns with the solution known from existing literature. Secondly, we demonstrate that we can handle finite or semi-infinite trajectories by performing the integration within the CCI only over a finite or semi-infinite interval. This is not only more practical since in real experiments only finite times and thus finite trajectories are measured. It is also numerically much more stable since the infinite convolution integral contains terms that oscillate strongly as the limits of the integral approach infinity. The integral with a limited integration range, on the other hand, can be solved without any problems. As an application example, we consider sound sources moving on a finite parabolic curve. Third, we study damping effects and investigate to what extent they affect the spectrum. To this end, we explore the damping mechanism by considering a slightly viscous fluid by adding an additional term to the wave equation. However, the damping effect in air is only relevant when large distances between the source and the receiver are considered.

So far, all investigations have been carried out in Cartesian coordinates. But rotating sound sources, for example, are also of importance. Therefore, in Sect. 2.4 we start from polar coordinates and derive for this two-dimensional case the spectrum of a line source moving on a circular path and also the spectrum of a source on an Archimedean helix. For the circular path, we obtain a line spectrum in which the amplitudes are products of Bessel and Hankel functions. The corresponding solution in the time domain is easy to achieve and is also given by Poletti [20]. The formula for the spectrum of an Archimedean helix is new and more complicated, but can be expressed in closed analytical form if the spectral pressure at the origin is considered.

In Sect. 2.5, we consider rotating sources in three dimensions. Practical applications include the calculation of noise generated by propellers or wind turbines. We use three-dimensional cylindrical coordinates to derive first the frequency spectrum of a source moving along a circular trajectory and then that of a source moving along a circular helix. Surprisingly and fortunately, the spectrum of the helix is even simpler than that of the pure circular orbit and can be expressed by an infinite series of cylindrical functions. Comparisons between this series solution and the results obtained by the CCI show excellent agreement between the two methods. Such spectral representations appear to have been published for the first time in [21].

In Sect. 2.6, we now turn to spherical coordinates to obtain an integral-free representation for the spectrum of a source on a circular path in three-dimensional space. The final solution is a series of spherical functions and agrees with the results found in the literature by Poletti [22] and Mao et al. [23]. Only the derivation is done in a different way.

If we raise the Archimedean spiral into the third dimension, we get the conical helix. The spectrum for a source moving along such a conical helix is derived in spherical coordinates and was also studied in [21]. The result is validated with the help of the CCI. Here, again, there is an excellent agreement between the two methods.

The case studied in Sect. 2.7 concerns a radiating monopole moving along an elliptical helix. Given the complexity of solving the wave equation in elliptical cylindrical coordinates (see, e.g., [24] or [25]), we refrain from presenting a series representation with Mathieu functions. The reason for this can be seen in the following points. Firstly, Green's function in elliptical coordinates has a complicated structure. Secondly, the Mathieu functions and the circular functions resulting from the normal temporal Fourier transform are not orthogonal. Therefore, such a derivation is not performed here and is not used for further numerical analysis. Instead, we use the CCI to calculate the frequency spectra and compare the results for the elliptical helix with those obtained for circular helices inside and outside the elliptical helix.

In Sect. 2.8, we turn to the spectrum of sources in half-spaces, specifically in spaces over impedance grounds. For this purpose, we combine the very general superposition principle formulated and proved in [26] with the spectral solutions for sources moving in the free field. In this way, we obtain closed-form spectral formulas for moving sources over grounds with arbitrary impedance boundary conditions. The inverse Fourier transform of these formulas leads to the time domain solutions derived in [15] and [16]. Here, the significant advantage of the formulation in the frequency domain becomes apparent, namely the simple form of the impedance boundary condition in contrast to its representation in the time domain as a convolution integral. This advantage is particularly visible in three dimensions: The solutions in the time domain look very different for different impedances, but in the frequency domain, one obtains a single integral representation, whose inverse Fourier transformation leads to these differently constructed time-domain solutions.

It is worth noting that the influence of an impedance plane can be represented by a line source below the plane in non-physical space [27]. Quite similarly, a moving source can be described by the superposition of stationary point sources along the flight path (i.e., a line source, too, corresponding to the trajectory), showing the compatibility of the superposition integral with the motion of the source. Both effects can be described by a suitable superposition of point sources.

In the last Sect. 2.9 of the second chapter, we study a vibrating and non-deformable body with a finite volume not equal to zero as a real source instead of point sources. For this reason, we start with a reduced form of the Ffowcs Williams and Hawkings (FW-H) equation. The body moves along an arbitrary path in the same way as described in Sect. 2.3.1 for a point source. The Fourier transform leads to an inhomogeneous Helmholtz equation with a relatively complex right-hand side containing the so-called thickness and loading noise. By convolution with Cartesian Green's function, we obtain an integral representation of the sound field outside the body, which has the same structure as the well-known boundary element formula for stationary sources. Only the kernel function within the integrals must be replaced by another core function. We call this representation the "*BE solution*" and apply it to a pulsating sphere traveling in a straight line. It is possible to show that the BE solution can be converted into a known analytical solution using three approximations. Furthermore, the numerical evaluation of the BE solution agrees well with an analytical solution derived for a sphere moving at a low Mach number.

The study of moving sources in half-spaces is further extended and deepened in Chap. 3, where the focus shifts toward solutions in the time domain. While we continue to work with line and point sources, we now also accommodate a broader range of time dependencies, moving away from exclusive mono-frequent vibrations. In addition, we consider more general ground properties.

In Chaps. 3, 4 and 5, we move from frequency to time domain. In Chap. 3, we deduce formulas for the sound field of sources moving above an infinite flat interface between two fluid media, radiating sound with a general time dependence, including sound of finite duration. We show that these formulas can be obtained through the convolution of Green's function of the stationary case and the source function that includes the motion of the source. We apply this approach to sources in two and three dimensions that have Green's function that can be expressed in closed form and then validate them, comparing the results with other formulas available in the literature. For the two-dimensional case, the second semi-infinite medium can be locally or non-locally reacting, while for the three-dimensional case, only a locally reacting medium is considered. Chapter 4 is dedicated to the conventional ESM for simulating radiation and scattering. We start by reviewing the basic concepts of the ESM in the frequency domain for stationary sources and give some examples of application. Then, we move to the time domain. The time domain formulation follows the usual iterative approach that finds the solution at each new time step from the results of the previous steps. Such a method is unstable and requires some type of regularization. We discuss different regularization methods and present several examples. We consider two cases, stationary bodies and moving bodies. For the latter, the local fluid perturbation due to the motion of the body is neglected. In Chap. 5, we study the application of the time domain ESM deduced in Chap. 4 to the source localization of moving sources. Now, not only does the sound field produced by the moving body need to be reproduced, but also the equivalent source distribution must reflect the position of the real sources with accuracy. We found that the iterative approach is not able to provide an accurate localization of the sources. Therefore, a new approach that solves the strengths of the equivalent sources at all time steps in one calculation and enforces the sparsity of the solution is suggested. The examples show that the new approach gives good results but also has some disadvantages, which are discussed.

In Chap. 2, *the imaginary unit is denoted by* i *with* $i^2 = -1$, *and the time convention* $e^{-i\omega t}$ *is used.*

In Chaps. 3, 4 and 5, *the imaginary unit is denoted by* j *instead, and the time convention* $e^{+j\omega t}$ *is adopted.*

References

1. P.C. Clemmow, *The Plane Wave Spectrum Representation of Electromagnetic Fields*. Oxford University Press, Oxford (reissue 1996)
2. G. Lefeuve-Mesgouez, A.T. Peplow, D. Le Houédec, Surface vibration due to a sequence of high speed moving harmonic rectangular loads. Soil Dyn. Earthq. Eng. **22**, 459–473 (2002)
3. J.E. Ffowcs Williams, D.L. Hawkings, Sound generation by turbulence and surfaces in arbitrary motion. Philos. Trans. R. Soc. Lond., Ser. A, Math. Phys. Sci. **264**, 321–342 (1969)
4. X. Wu, A. Akay, Sound radiation from vibrating bodies in motion. J. Acoust. Soc. Am. **91**(5), 2544–2555 (1992)
5. S.F. Wu, Z. Wang, Extended Kirchhoff integral formulations for sound radiation from vibrating cylinders in motion. Trans. ASME **115**, 324–331 (1993)
6. S.F. Wu, Z. Zhou, Simulation of vehicle pass-by noise radiation. J. Vib. Acoust. **121**, 197–203 (1999)
7. D.P. Lockard, An efficient, two-dimensional implementation of the Ffowcs Williams and Hawkings equation. J. Sound Vib. **229**(4), 897–911 (2000)
8. D.P. Lockard, A comparison of Ffowcs Williams-Hawkings solvers for airframe noise applications, in *8th AIAA/CEAS Aeroacoustics Conference*, 1–11, 2002
9. P.M. Morse, K.U. Ingard, *Theoretical Acoustics* (Princeton University Press, 1986)

10. A.P. Dowling, J.E. Ffowcs Williams, *Sound and Sources of Sound* (Ellis Horwood Ltd., Chicester, West Sussex, 1983)
11. D.G. Crighton, A.P. Dowling, J.E. Ffowcs Williams, M. Heckl, F.G. Leppington, *Modern Methods in Analytical Acoustics*, Lecture Notes (Springer, Berlin Heidelberg GmbH, 1992)
12. P. Költzsch, Flow acoustics, in *Formulas of Acoustics*, 2nd edn, ed. by F.P. Mechel (Springer, Berlin, 2008)
13. J. Delfs, *Basics of Aeroacoustics*, Lecture notes, WS 2024/2025 at Technische Universität Braunschweig (2023). https://elib.dlr.de/202954/1/Notes_Basics_of_Aeroacoustics_Delfs.pdf
14. S.W. Rienstra, A. Hirschberg, An Introduction to Acoustics, Eindhoven University of Technology, 021. https://www.win.tue.nl/~sjoerdr/papers/boek.pdf
15. D. Dragna, P. Blanc-Benon, Sound radiation by a moving line source above an impedance plane with frequency-dependent properties. J. Sound Vib. **349**, 259–275 (2015)
16. M. Ochmann, Exact solutions for sound radiation from a moving monopole above an impedance plane. J. Acoust. Soc. Am. **133**(4), 1911–1921 (2013)
17. F. Ollendorf, Beiträge zur Theorie des Doppler-Effektes (Contribution to the theory of the Doppler-Effect). ACUSTICA **49**, 1–12 (1981)
18. F. Ollendorf, Versuch einer relativistischen Schalltheorie bewegter Körper (A tentative formulation of relativistic acoustics). ACUSTICA **50**, 83–98 (1982)
19. F. Ollendorf, Das Frequenzspektrum des Doppler-Effektes (The frequency spectrum of the Doppler effect). ACUSTICA **50**, 99–108 (1982)
20. M.A. Poletti, Cylindrical harmonic expansion of the sound field due to a rotating line source, in *Proceedings of 20th International Congress on Acoustics, ICA 2010*, Sydney, Australia, 2010
21. M. Ochmann, The frequency spectrum of a harmonically radiating monopole moving along arbitrary trajectories—with application to helical orbits. J. Theor. Comp. Acoust. **31**(1), 2350010-1–2350010-19, open access, 2023
22. M.A. Poletti, Series expansions of rotating two and three dimensional sound fields. J. Acoust. Soc. Am. **128**, 3363–3374 (2010)
23. Y. Mao, Y. Gu, D. Qi, H. Tang, An exact frequency-domain solution of the sound radiated from the rotating dipole point source. J. Acoust. Soc. **132**(3), 1294–1302 (2012)
24. R. Barakat, Diffraction of plane waves by an elliptic cylinder. J. Acoust. Soc. **35**(12), 1990–1996 (1963)
25. F.P. Mechel, *Mathieu Functions—Formulas, Generation Use* (S. Hirzel Verlag, Stuttgart—Leipzig, 1997)
26. M. Ochmann, R. Piscoya, *Theory and Application of Acoustic Sources Using Complex Analysis*, 1st ed. (Springer Nature, Singapore, 2021)
27. M. Ochmann, The complex equivalent source method for sound propagation over an impedance plane. J. Acoust. Soc. Am. **116**(6), 3304–3311 (2004)

Chapter 2
The Frequency Spectrum of Moving Sources

Abstract The sound field of harmonically radiating point sources moving along arbitrary trajectories in one, two and three dimensions is studied in the frequency domain. First, we analyze sources moving rectilinearly with constant velocity over an infinite time interval. Then, the method is generalized threefold by taking into account arbitrary trajectories, introducing damping effects, and considering finite observation times of the source's motion. The so-called "Cartesian convolution integral" is derived and applied to sources that can move on generally formulated trajectories, such as circular, conical and elliptical helices. For purely circular trajectories and helical orbits, an alternative approach in cylindrical or spherical coordinates leads to closed-form expressions in the form of infinite series for the frequency spectra. Comparing the results of the Cartesian convolution integral with those of the series expansion shows excellent agreement. We also investigate moving sources above an impedance plane in one, two, and three dimensions. Finally, we study the frequency spectrum of "real sources" with a finite, non-zero volume based on the Ffowcs Williams and Hawkings equation. For this reason, we derive a frequency domain formulation of the Ffowcs Williams and Hawkings equation.

2.1 Description of the Problem and Method of Solution

The starting point of the investigations in this chapter is an acoustic monopole that moves along an arbitrary spatial trajectory $\vec{x}_0(t) = (x_0(t), y_0(t), z_0(t))^T$ with the corresponding time-dependent velocity $\vec{u}(t) = (u_1(t), u_2(t), u_3(t))^T = \frac{\mathrm{d}}{\mathrm{d}t}(x_0(t), y_0(t), z_0(t))^T$ in free space and emits an arbitrary time signal described by the function $q(t)$, where t denotes the time and the superscript T means transposition. This is sketched in Fig. 2.1.

The resulting sound pressure p is the causal solution of the inhomogeneous time-dependent wave equation

$$\Delta p - \frac{1}{c^2}\frac{\partial^2 p}{\partial t^2} = -\frac{\partial}{\partial t}\left[q(t)\delta(\vec{x} - \vec{x}_0(t))\right], \tag{2.1}$$

M. Ochmann and R. Piscoya, *Moving Acoustic Sources*,
Springer-AAS Acoustics Series, https://doi.org/10.1007/978-981-96-0871-3_2

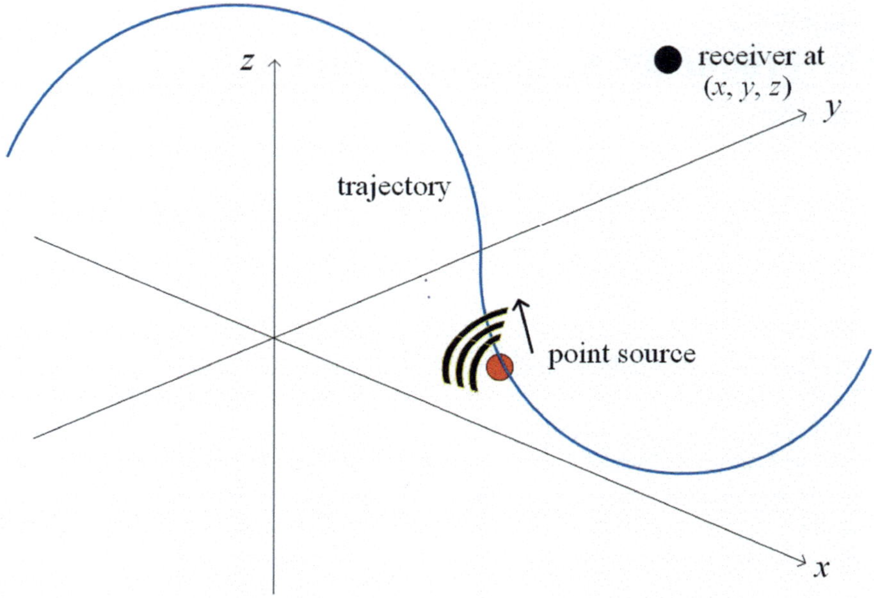

Fig. 2.1 An acoustical source moving along an arbitrary trajectory in free space

where $\Delta = \nabla^2$ is the Laplace operator and c is the speed of sound in air. The function $q(t)$ has the dimension of a mass flow: $[q] = \text{kg/s}$. It can easily be shown that $q(t)$ is the mass flow rate of a small point source (see Mechel [1, p. 34] or Morse and Ingard [2, pp. 721 and 728]). Note however, that the dimension of $q(t)$ depends on the dimension of the space considered. As usual, δ is the three-dimensional Dirac delta function. Using the acoustic velocity potential defined by $p = -\rho \frac{\partial \varphi}{\partial t}$ with air density ρ, the wave equation reads

$$\Delta\varphi - \frac{1}{c^2}\frac{\partial^2\varphi}{\partial t^2} = \frac{q(t)}{\rho}\delta(\vec{x} - \vec{x}_0(t)). \tag{2.2}$$

In the following chapters, we consider a source that radiates at a fixed angular frequency $\Omega > 0$

$$q(t) = Q_1 \exp(-i\Omega t) \text{ with } i^2 = -1. \tag{2.3}$$

The task is to calculate the frequency spectrum of the pressure p or, in other words, the temporal Fourier transform

$$\hat{p}(\omega) := F_t[p(t)] := \int_{-\infty}^{\infty} p(t)e^{i\omega t}\mathrm{d}t. \tag{2.4}$$

Clearly, $\hat{p}(\omega)$ is nothing but the complex sound pressure and has the dimension Pa·s or Pa/Hz, i.e., the pressure spectrum is, more precisely, a pressure per frequency. The acoustic potential in the frequency domain $\hat{\varphi} = \int_{-\infty}^{\infty} \varphi(t)e^{i\omega t}dt$ is given by

$$\hat{p}(\omega) = i\omega\rho\hat{\varphi}. \tag{2.5}$$

Thus, in the frequency domain, pressure and potential only differ by the constant factor $i\omega\rho$ and are, therefore, essentially the same quantity.

Applying the temporal Fourier transform to the wave Eq. (2.1) with source term (2.3), we obtain the inhomogeneous Helmholtz equation for the pressure $\hat{p}(\omega) = i\omega\rho\hat{\varphi}$

$$\begin{aligned}\Delta\hat{p} + \left(\frac{\omega}{c}\right)^2 \hat{p} &= \int_{-\infty}^{\infty} i\omega q(t)\delta(\vec{x} - \vec{x}_0(t))e^{i\omega t}dt \\ &= -Q\int_{-\infty}^{\infty} e^{i(\omega-\Omega)t}\delta(\vec{x} - \vec{x}_0(t))dt \end{aligned} \tag{2.6}$$

with $Q = -i\omega Q_1$.

While the wave Eq. (2.1) has two different time scales, expressed in the acoustical radiation factor $\exp(-i\Omega t)$ and in the movement of the source with velocity $\vec{u}(t)$, the Helmholtz Eq. (2.6) only depends on the two frequencies ω and Ω. It seems obvious, therefore, that a solution in the frequency domain is much easier to obtain than in the time domain. The Fourier integral on the right-hand side of Eq. (2.6)

$$I_F = \int_{-\infty}^{\infty} e^{i(\omega-\Omega)t}\delta(\vec{x} - \vec{x}_0(t))dt. \tag{2.7}$$

can be solved analytically for certain cases. For example, in Cartesian coordinates, the Dirac delta function in three dimensions (3D) can be split into three one-dimensional (1D) Dirac delta functions $\delta(\vec{x} - \vec{x}(t)) = \delta(x - x_0(t))\delta(y - y_0(t))\delta(z - z_0(t))$.

Here, we do not differentiate between Dirac delta functions in one, two or three dimensions since this is evident from the context. Now, the property of the Dirac delta function is well-known (see for example Delfs [3, pp. 41–42]) that

$$\delta(f(t)) = \sum_j \delta(t - t_j)/\left|\frac{\partial f}{\partial t}\right|_{t=t_j}, \tag{2.8}$$

where the t_j are all solutions of the generally nonlinear equation $f(t_j) = 0$.

Only if we can find these zeros explicitly, it is possible to obtain an analytical solution for the integral (2.7); if not, the integral has to be solved numerically. For example, consider the partial Dirac delta function $\delta(x - x_0(t))$ on the right-hand side of Eq. (2.6). Due to

$$\frac{\mathrm{d}}{\mathrm{d}t}f(t) = \frac{\mathrm{d}}{\mathrm{d}t}(x - x_0(t)) = -u_1(t),$$

we obtain directly

$$\delta(x - x_0(t)) = \sum_j \frac{\delta(t - t_j)}{|u_1(t)|_{t=t_j}}, \tag{2.9}$$

and thus Eqs. (2.7), (2.8) and (2.9) lead to

$$I_F = \sum_j \frac{1}{|u_1(t)|_{t=t_j}} e^{i(\omega-\Omega)t_j} \delta(y - y_0(t_j))\delta(z - z_0(t_j)). \tag{2.10}$$

After having obtained an explicit expression of the Fourier integral, the corresponding inhomogeneous Helmholtz Eq. (2.6) has to be solved to get the desired complex pressure $\hat{p}(\omega)$. The method is first demonstrated for the derivation of the frequency spectrum of a point source in 3D, which moves along an infinite straight line at a constant velocity.

2.2 Radiating Monopole Moving Along a Straight Line at Constant Speed

2.2.1 *Three-Dimensional Case (3D Case)*

2.2.1.1 Derivation of the Spectral Pressure of a Moving Point Source

Without loss of generality, we assume that the source moves along the x-axis with the constant positive speed $u > 0$ as shown in Fig. 2.2.

If the source moves along a different direction $\vec{d}$, we rotate the coordinate system so that the new x-axis points in the direction $\vec{d}$. It follows that $\vec{x}_0(t) = (ut, 0, 0)^T$ and that the wave Eq. (2.2) together with (2.3) results in

$$\Delta\varphi - \frac{1}{c^2}\frac{\partial^2}{\partial t^2}\varphi = \frac{Q_1}{\rho}e^{-i\Omega t}\delta(x - ut)\delta(y)\delta(z). \tag{2.11}$$

The corresponding inhomogeneous Helmholtz equation for the pressure $\hat{p}(\omega) = i\omega\rho\hat{\varphi}$ is now

$$\Delta\hat{p} + \left(\frac{\omega}{c}\right)^2\hat{p} = -Q\delta(y)\delta(z)\int_{-\infty}^{\infty} e^{i(\omega-\Omega)t}\delta(x - ut)\mathrm{d}t \tag{2.12}$$

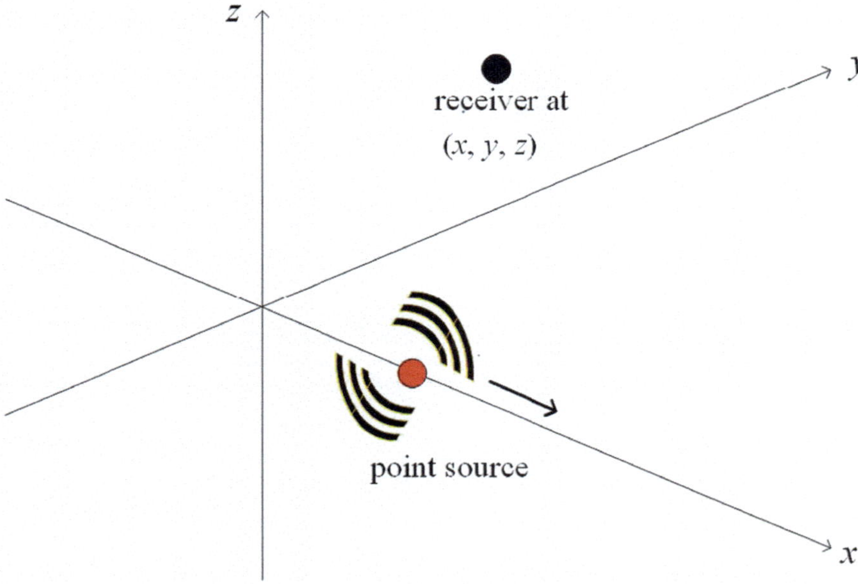

Fig. 2.2 An acoustical source moving along the x-axis

with $Q = -i\omega Q_1$. From the property (2.9), it follows that $\delta(x - ut) = \delta(t - x/u)/u$. Thus, the sifting property of the delta function enables us to solve the Fourier integral easily leading to

$$\Delta\hat{p} + \left(\frac{\omega}{c}\right)^2 \hat{p} = -\frac{Q}{u} e^{i(\omega-\Omega)\frac{x}{u}} \delta(y)\delta(z). \tag{2.13}$$

After having obtained an explicit expression for the right-hand side of the Helmholtz equation, the second step now consists of finding the solution to Eq. (2.13).

We describe two different methods of solution: A method using cylindrical coordinates and the Green's function method (GFM). The latter is more generally applicable and is also used for other trajectories of the source.

For the application of the ***method in cylindrical coordinates***, we use the coordinates $x,\ r,\ \phi$ with

$$x = x,\quad y = r\cos\phi,\quad z = r\sin\phi,\ r = \sqrt{y^2 + z^2}. \tag{2.14}$$

Due to the symmetry of the radiation problem (see Fig. 2.2), the sound pressure is independent of the polar angle ϕ around the x-axis. Thus, writing Eq. (2.13) in such cylindrical coordinates and considering that $\delta(y, z) = \delta(r)/(2\pi r)$, we obtain

$$\frac{\partial^2\hat{p}}{\partial x^2} + \frac{1}{r}\frac{\partial}{\partial r}\left(r\frac{\partial\hat{p}}{\partial r}\right) + \left(\frac{\omega}{c}\right)^2 \hat{p} = -\frac{Q}{u} e^{i(\omega-\Omega)\frac{x}{u}} \frac{\delta(r)}{2\pi r}. \tag{2.15}$$

The approach

$$\hat{p} = \frac{Q}{u} e^{i(\omega-\Omega)\frac{x}{u}} \tilde{p}(r) \tag{2.16}$$

leads to the ordinary differential equation

$$\frac{1}{r}\frac{\partial}{\partial r}\left(r\frac{\partial \tilde{p}}{\partial r}\right) + \kappa^2 \tilde{p} = -\frac{\delta(r)}{2\pi r} \text{ with } \kappa^2 = \left(\frac{\omega}{c}\right)^2 - \left(\frac{\omega-\Omega}{u}\right)^2 \tag{2.17}$$

with known outwardly propagating solution (see Duffy [4, p. 277])

$$\tilde{p}(r) = \frac{i}{4} H_0^{(1)}(\kappa r), \tag{2.18}$$

which is the Green's function in 2D. Equations (2.16) and (2.18) give

$$\hat{p}(\omega) = \frac{iQ}{4u} e^{i\left(\frac{\omega-\Omega}{u}\right)x} H_0^{(1)}\left(r\sqrt{\left(\frac{\omega}{c}\right)^2 - \left(\frac{\omega-\Omega}{u}\right)^2}\right), \tag{2.19}$$

where $H_0^{(1)}$ is the Hankel function of first kind and order zero. This solution approach was also used by Morse and Ingard [2], pp. 728–729 and their solution is compared with Eq. (2.19) a bit further down in the text.

The corresponding acoustic velocity potential in the frequency domain can be written

$$\hat{\varphi}(\omega) = \frac{-iQ_1}{4u\rho} e^{i\left(\frac{\omega-\Omega}{u}\right)x} H_0^{(1)}\left(r\sqrt{\left(\frac{\omega}{c}\right)^2 - \left(\frac{\omega-\Omega}{u}\right)^2}\right), \tag{2.20}$$

since $\hat{\varphi}(\omega) = \hat{p}(\omega)/(i\omega\rho)$ and $Q = -i\omega Q_1$.

We secondly apply the ***Green's function method (GFM)***for solving the inhomogeneous Helmholtz Eq. (2.13). The 3D Green's function $\hat{g}$ is defined by

$$\Delta\hat{g} + \left(\frac{\omega}{c}\right)^2 \hat{g} = -\delta(x)\delta(y)\delta(z) \tag{2.21}$$

and has the well-known expression

$$\hat{g} = \frac{1}{4\pi\sqrt{x^2+y^2+z^2}} e^{i\frac{\omega}{c}\sqrt{x^2+y^2+z^2}} \tag{2.22}$$

(see, for example, Duffy [4, p. 276]). By convoluting the Green's function with the negative right-hand side (i.e., the excitation spectrum) of the inhomogeneous

Helmholtz Eq. (2.13)

$$R(x, y, z) = \frac{Q}{u} e^{i(\omega-\Omega)\frac{x}{u}} \delta(y)\delta(z), \tag{2.23}$$

we obtain

$$\begin{aligned}\hat{p} &= \int_{-\infty}^{\infty}\int_{-\infty}^{\infty}\int_{-\infty}^{\infty} \hat{g}(x-\xi, y-\eta, z-\zeta) R(\xi, \eta, \zeta) \mathrm{d}\xi \mathrm{d}\eta \mathrm{d}\zeta \\ &= \frac{Q}{4\pi u} \int_{-\infty}^{\infty} \frac{e^{i\frac{\omega}{c}\sqrt{(x-\xi)^2+y^2+z^2}}}{\sqrt{(x-\xi)^2+y^2+z^2}} e^{i(\omega-\Omega)\frac{\xi}{u}} \mathrm{d}\xi .\end{aligned} \tag{2.24}$$

By changing the variable of integration to $\alpha = x - \xi$, we obtain

$$\hat{p} = \frac{Q}{4\pi u} e^{i(\omega-\Omega)\frac{x}{u}} \int_{-\infty}^{\infty} \frac{e^{i\frac{\omega}{c}\sqrt{\alpha^2+y^2+z^2}}}{\sqrt{\alpha^2+y^2+z^2}} e^{-i(\omega-\Omega)\frac{\alpha}{u}} \mathrm{d}\alpha. \tag{2.25}$$

Since

$$\int_{-\infty}^{\infty} \frac{e^{i\frac{\omega}{c}\sqrt{\alpha^2+y^2+z^2}}}{\sqrt{\alpha^2+y^2+z^2}} e^{-i(\frac{\omega-\Omega}{u})\alpha} \mathrm{d}\alpha = \pi i H_0^{(1)}\left(r\sqrt{\left(\frac{\omega}{c}\right)^2 - \left(\frac{\omega-\Omega}{u}\right)^2} \right), \tag{2.26}$$

we have obtained solution (2.19) again. Only the proof of the integral representation (2.26) must be added:

Proof of Eq. (2.26) Equation (2.26) can be easily recognized by comparing the Green's functions in 2D and in 3D: The Fourier transform of the 3D Helmholtz equation (2.22) with respect to x

$$\tilde{\hat{g}} := F_x[\hat{g}] := \int_{-\infty}^{\infty} \hat{g} e^{-ik_x x} \mathrm{d}x \tag{2.27}$$

leads to

$$\Delta\tilde{\hat{g}} + (k^2 - k_x^2)\tilde{\hat{g}} = -\delta(y)\delta(z), \quad k = \frac{\omega}{c}. \tag{2.28}$$

Thus, on the one hand, $\tilde{\hat{g}}$ is the well-known 2D-Green's function (cf. Eq. (2.18))

$$\tilde{\hat{g}} = \frac{i}{4} H_0^{(1)}\left(r\sqrt{k^2 - k_x^2} \right) \tag{2.29}$$

and on the other hand, it is the Fourier transform of the solution of Eq. (2.21), i.e., the Fourier transform of the 3D Green's function

$$\tilde{\tilde{g}} = \frac{1}{4\pi} \int_{-\infty}^{\infty} \frac{e^{i\frac{\omega}{c}\sqrt{x^2+y^2+z^2}}}{\sqrt{x^2+y^2+z^2}} e^{-ik_x x} \mathrm{d}x. \tag{2.30}$$

By comparing the last two equations and setting $k_x = \frac{\omega-\Omega}{u}$, Eq. (2.26) follows immediately.

2.2.1.2 Verifying the Frequency Spectrum and Comparison with Other Solutions

We can also verify the correctness of this spectral solution as follows. The solution of Eq. (2.11) in the time domain can be obtained by using a modified Lorentz transformation[1] (see Ochmann [5])

$$x_L = \gamma^2(x - Mct), \;\; y_L = \gamma y, \;\; z_L = \gamma z, \;\; t_L = \gamma^2(t - Mx/c), \tag{2.31}$$

and

$$x = x_L + Mct_L, \;\; y = \gamma^{-1} y_L, \;\; z = \gamma^{-1} z_L, \;\; t = t_L + Mx_L/c, \tag{2.32}$$

where $M = u/c$ is the Mach number and $\gamma = 1/\sqrt{1 - M^2}$. By applying this transformation to the wave equation (2.11), we obtain the new equation

$$\Delta_L \varphi_L - \frac{1}{c^2}\frac{\partial^2}{\partial t_L^2}\varphi_L = \gamma^2 \frac{Q_1}{\rho} e^{-i\Omega t_L} \delta(x_L)\delta(y_L)\delta(z_L), \tag{2.33}$$

where the acoustic potential φ_L in Lorentz space is now a solution of the wave equation with a stationary monopole source at the origin and Δ_L is the Laplace operator written in Lorentz coordinates.

On the right-hand side of (2.33), we have taken advantage of the property of the Dirac delta function $\delta(t - a)f(t) = \delta(t - a)f(a)$, which together with (2.32) yields

$$\begin{aligned} \delta(x_L)e^{-i\Omega t} &= \delta(x_L)e^{-i\Omega(t_L + Mx_L/c)} = e^{-i\Omega t_L}\delta(x_L)e^{-i\Omega M x_L/c} \\ &= e^{-i\Omega t_L}\delta(x_L)e^0 = e^{-i\Omega t_L}\delta(x_L) \end{aligned}$$

(cf. Morse and Ingard [2, p. 722]).

The solution of Eq. (2.33) is the 3D Green's function

[1] Because of the factor γ^2 one speaks of the *modified* Lorentz transformation.

$$\varphi_L = \frac{-\gamma^2 Q_1}{4\pi r_L \rho} e^{ik_0 r_L} e^{-i\Omega t_L} \tag{2.34}$$

with $r_L = \sqrt{x_L^2 + y_L^2 + z_L^2}$ and wave number $k_0 = \Omega / c$.

Substituting (2.31) into (2.34) gives the solution for a moving monopole source in physical space (cf. Ochmann [5, Eq. (21b)] or Morse and Ingard [2, p. 723])

$$\varphi = \frac{-Q_1}{4\pi R_1 \rho} e^{ik_0\gamma^2 R_1} e^{-ik_0\gamma^2(ct - Mx)} \text{ with } R_1 = \sqrt{(x-ut)^2 + \left(y^2 + z^2\right)/\gamma^2}. \tag{2.35}$$

In Appendix A it is shown that the temporal Fourier transform of the potential φ correctly leads to the expected expression for the potential in the frequency domain (2.20)

$$\begin{aligned} F_t[\varphi(t)] &= F_t\left[\frac{-Q_1}{4\pi R_1 \rho} e^{ik_0\gamma^2 R_1} e^{-ik_0\gamma^2(ct - Mx)}\right] \\ &= \frac{-iQ_1}{4u\rho} e^{i\left(\frac{\omega-\Omega}{u}\right)x} H_0^{(1)}\left(r\sqrt{\left(\frac{\omega}{c}\right)^2 - \left(\frac{\omega-\Omega}{u}\right)^2}\right) = \hat{\varphi}(\omega). \end{aligned} \tag{2.36}$$

Ollendorf [6] has taken a similar path of derivation via the temporal Fourier transform. However, it is clear from the calculation in Appendix B and from Ollendorf's work that the derivation starting directly from the Helmholtz Eq. (2.13) in the frequency domain combined with the Green's function method is much easier to perform.

Also, in Sect. 11.2 "*Sound Emission from Moving Sources*" of the textbook "Theoretical Acoustics" by Morse and Ingard [2], we find under the heading "*Power spectrum*" on p. 728 the same approach for a source moving with uniform velocity in a straight line. Morse and Ingard noticed that "*it is instructive to start directly from the wave equation ... and take its Fourier transform*" (quoted from [2, p. 728]). They also solved the Helmholtz equation using the radially symmetrical approach (2.16) and obtained the solution (see [2, p. 729, below Eq. (11.2.23)])—written in our notation

$$p^+ = \frac{\omega q_0}{16\pi u} e^{i\left(\frac{\omega-\Omega}{u}\right)x} H_0^{(1)}\left(r\sqrt{\left(\frac{\omega}{c}\right)^2 - \left(\frac{\omega-\Omega}{u}\right)^2}\right). \tag{2.37}$$

If we compare our Eq. (2.13) with the Eq. (11.2.21) of Morse and Ingard [2], we see that we have to choose

$Q = -i\omega q_0/(4\pi)$ and thus obtain identical results

$p^+ = \hat{p}(\omega)$, i.e., (2.19) = (2.37).

Morse and Ingard used the spectral representation to consider also the supersonic case with $M > 1$, as the spectral method is not restricted to the subsonic case with $M < 1$. We will compare the spectra of both cases in the next section. In addition, Morse and Feshbach calculated the acoustic power of the moving source (see pp. 731–732 in [2]). For that reason, they used the spectral representation (2.37) and the asymptotic form of the Hankel function for large arguments. The acoustic power is then determined by integrating the acoustic intensity "*over the surface of an infinitely long cylinder enclosing the line of motion of the source*" (quoted from [2, p. 730]). Their result is that the acoustic power of the moving source

$$P_{\text{mov}} = P_{\text{stat}}/\left(1 - M^2\right)^2 \tag{2.38}$$

is greater than the acoustic power of the stationary source

$$P_{stat} = \frac{|q_0\Omega|^2}{8\pi\rho c} \tag{2.39}$$

by a factor of $\gamma^4 = \left(1 - M^2\right)^{-2}$ (see Eq. (11.2.30) in [2]).

2.2.1.3 Discussion and Evaluation of the Spectral Pressure

(a) **Subsonic case** $M < 1$

The expression

$$\hat{p}(\omega) = \frac{iQ}{4u} e^{i\left(\frac{\omega-\Omega}{u}\right)x} H_0^{(1)}\left(r\sqrt{\left(\frac{\omega}{c}\right)^2 - \left(\frac{\omega-\Omega}{u}\right)^2}\right) \tag{2.40}$$

for the frequency spectrum of the acoustic pressure can be easily calculated (e. g., with Matlab). For example, if the source radiates at a frequency of 400 Hz and moves at a speed of 34.3 m/s, we obtain the following spectrum shown in Fig. 2.3.

The Doppler shift shows the two frequency peaks at about $f_u \approx 364$ Hz and at $f_o \approx 444$ Hz in agreement with the well-known estimates

$$f_- = \frac{\omega_-}{2\pi} = \frac{\Omega}{2\pi(1+M)} \approx 364 \text{ Hz and } f_+ = \frac{\omega_+}{2\pi} = \frac{\Omega}{2\pi(1-M)} \approx 444 \text{ Hz.} \tag{2.41}$$

It should be noted that the argument of the Hankel function in Eq. (2.19) becomes zero at both Doppler frequencies, and thus, the corresponding amplitudes are infinite, as damping is not included. Therefore, the height of the peaks in Fig. 2.3 and in similar figures cannot be displayed correctly and depends on the numerical frequency resolution.

In the "Doppler regime" $\Delta_{\text{Dopp}} = (\omega_-, \omega_+)$ the frequency spectrum is essentially different from zero. Outside the interval Δ_{Dopp}, the pressure amplitudes converge very quickly toward zero. This can be seen from the argument of the Hankel function in Eq. (2.19). A small calculation shows that the square root—we choose the positive

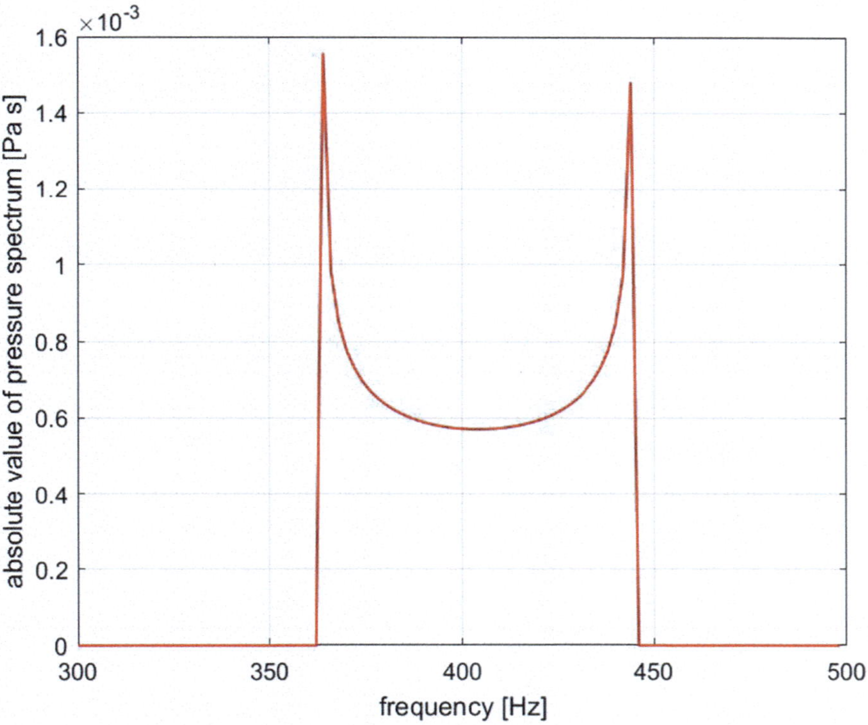

Fig. 2.3 Absolute value of the spectral pressure with parameters $(x, y, z) = (10\text{ m}, 10\text{ m}, 10\text{ m})$, $\Omega = 2\pi\, 400\, 1/\text{s}$, $M = 0.1 (u = 34.3\text{ m/s})$, $Q = 1\text{ kg/s}^2$

branch of the square root to obtain radiating solutions or solutions that decay at infinity—can be expressed by using the Doppler frequencies $\omega_- = \Omega/(1+M)$ and $\omega_+ = \Omega/(1-M)$ as follows:

$$\kappa = \begin{cases} +\sqrt{\left(\frac{\omega}{c}\right)^2 - \left(\frac{\omega-\Omega}{u}\right)^2} = +\sqrt{(1-M^2)(\omega-\omega_-)(\omega_+-\omega)}/u \text{ if } \omega \in \Delta_{\text{Dopp}} \\ +i\sqrt{\left|\left(\frac{\omega}{c}\right)^2 - \left(\frac{\omega-\Omega}{u}\right)^2\right|} = +i\sqrt{(1-M^2)(\omega-\omega_-)(\omega-\omega_+)}/u, \text{ if } \omega \notin \Delta_{\text{Dopp}} \end{cases} . \tag{2.42}$$

For $\omega \in \Delta_{\text{Dopp}} = (\omega_-, \omega_+)$ κ is, therefore, a real number in the subsonic case $M < 1$. The argument of the Hankel function in Eq. (2.19) becomes real, too, and the pressure amplitudes are significantly different from zero in the Doppler regime.

On the other hand, the square root becomes purely imaginary for $\omega \notin \Delta_{\text{Dopp}} = (\omega_-, \omega_+)$. Thus, the argument of the Hankel function in Eq. (2.19) becomes imaginary outside the Doppler interval, and the pressure amplitudes quickly tend toward zero. This can be seen as follows: Due to Abramowitz and Stegun [7, p. 375, Eq. 9.6.4], the Hankel function becomes the modified Bessel function $K_0(x)$

$$H_0^{(1)}(ix) = \frac{2}{\pi i} K_0(x). \tag{2.43}$$

for real x. Thus, we obtain from Eq. (2.19) outside of the Doppler regime

$$\hat{p}(\omega) = \frac{iQ}{4u} e^{i\left(\frac{\omega-\Omega}{u}\right)x} H_0^{(1)}(i|\kappa|r) = \frac{Q}{2\pi u} e^{i\left(\frac{\omega-\Omega}{u}\right)x} K_0(|\kappa|r), \tag{2.44}$$

and K_0 goes exponentially toward zero for growing arguments (see [7, p. 378, Eq. 9.7.2]).

Formula (2.44) corresponds to Eq. (II.2.13) for the "*supercritical range of high-frequency phase rates*" and to Eq. (II.2.19) for the "*subcritical range of low-frequencies*" according to Ollendorff's description in [6, p. 108]. This drastic drop in the pressure is shown in Fig. 2.4 for frequencies below the lower Doppler frequency:

(b) **Supersonic case** $M > 1$

Since the square root κ can be expressed by

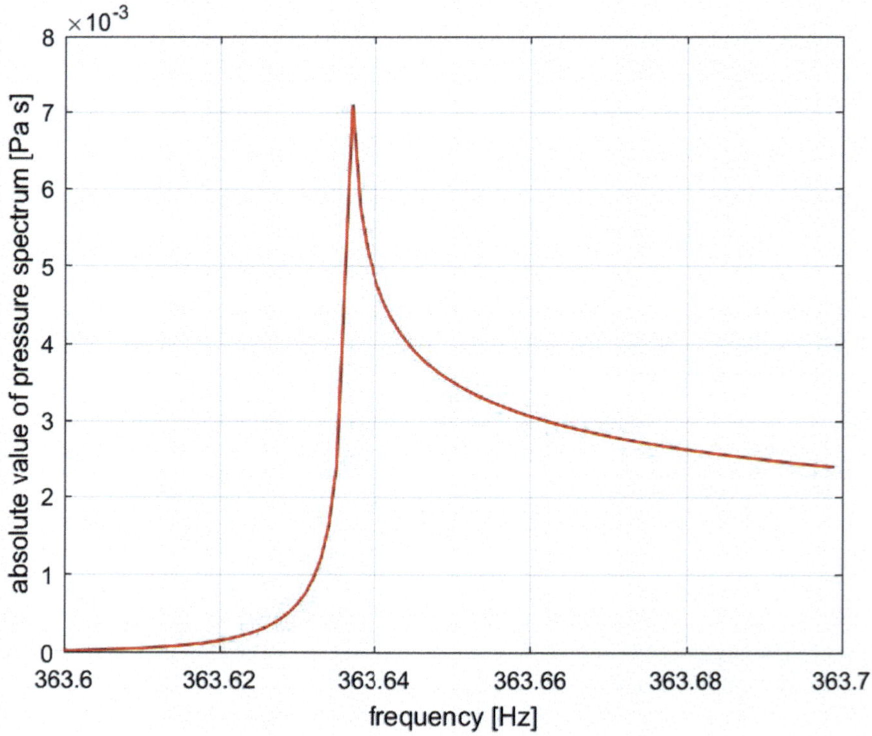

Fig. 2.4 Absolute value of the spectral pressure in the neighborhood of the lower Doppler frequency $f_- = 363.64$ Hz; parameters are $(x, y, z) = (10\text{ m}, 10\text{ m}, 10\text{ m})$, $\Omega = 2\pi\, 400$ 1/s, $M = 0.1$ ($u = 34.3$ m/s), $Q = 1$ kg/s^2

$$\kappa = +\sqrt{\left(\frac{\omega}{c}\right)^2 - \left(\frac{\omega - \Omega}{u}\right)^2} = \sqrt{(M^2 - 1)\left(\omega + \frac{\Omega}{M-1}\right)\left(\omega - \frac{\Omega}{M+1}\right)}/u,$$

it is easy to see that κ is a real number in the supersonic case $M > 1$ only if (cf. [2, p. 730])

$$-\infty < \omega < -\frac{\Omega}{M-1} \text{ or } \frac{\Omega}{M+1} < \omega < \infty.$$

Only in these two intervals, the argument of the Hankel function in Eq. (2.19) becomes real, and the pressure amplitudes differ substantially from zero.

Morse and Ingard noticed that the appearance of negative frequencies in the spectrum "*expresses the fact that the signals that were emitted last from the source arrive first at the point of observation*" (quoted from [2, p. 730]).

As an example, we choose $M = 5$ and a radiating frequency of 600 Hz. As expected, the numerical evaluation of expression (2.19) in Fig. 2.5 shows two peaks, one at $f = -\Omega/(2\pi(M-1)) = -150$ Hz and the other at $f = \Omega/(2\pi(M+1)) = 100$ Hz, and two separate intervals where the spectrum deviates significantly from zero, see Fig. 2.5.

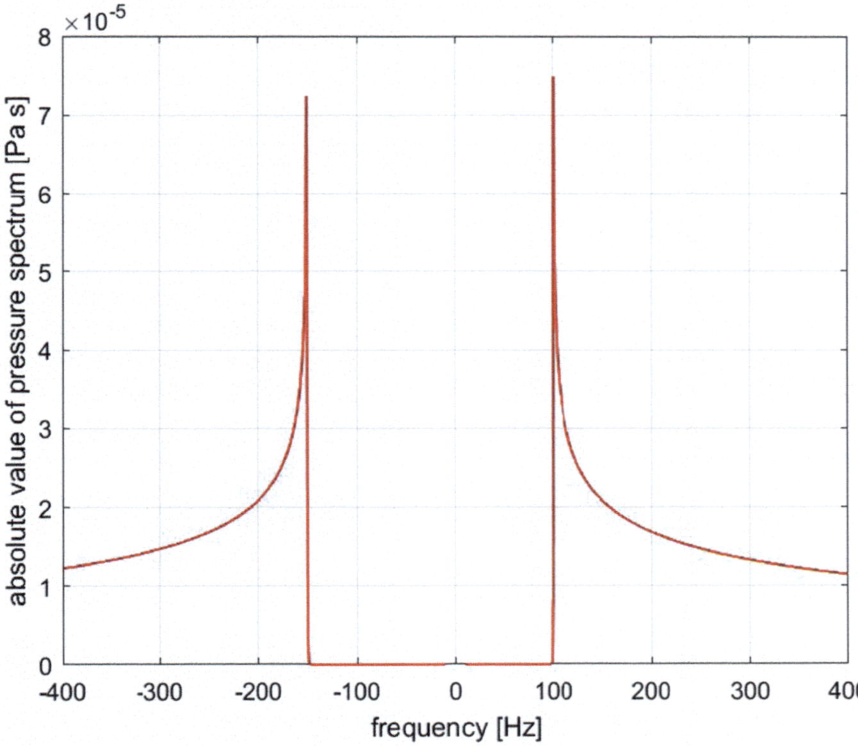

Fig. 2.5 Absolute value of the spectral pressure with parameters $(x, y, z) = (10 \text{ m}, 10 \text{ m}, 10 \text{ m})$, $\Omega = 2\pi 600\,1/\text{s}$, $M = 5$, $Q = 1 \text{ kg/s}^2$

The supersonic case is not of great importance in most practical applications and, because of the occurrence of shock waves, falls into the field of nonlinear acoustics, which is not covered in this book. Therefore, the supersonic case will not be discussed further. More details can be found in the relevant literature, e.g., in Morse and Ingard [2, Sect. 11.2] or in Dowling and Ffowcs Williams [8, Chap. 9].

2.2.2 *Two-Dimensional Case (2D Case)*

2.2.2.1 Derivation of the Spectral Pressure of a Moving Line Source

A line source pulsating with the angular frequency Ω and moving with constant velocity u in the x-direction can be treated analogously to the three-dimensional case in the previous Sect. 2.2.1.

The 2D wave equation for the acoustic potential of a function of x and y becomes

$$\Delta_2\varphi_2 - \frac{1}{c^2}\frac{\partial^2\varphi_2}{\partial t^2} = \frac{Q_1}{\rho}e^{-i\Omega t}\delta(x-ut)\delta(y), \tag{2.45}$$

where $\Delta_2 := \frac{\partial^2}{\partial x^2} + \frac{\partial^2}{\partial y^2}$ is the two-dimensional Laplace operator.

The corresponding inhomogeneous Helmholtz equation for the pressure $\hat{p}_2(\omega) = i\omega\rho\hat{\varphi}_2$ in the frequency domain is given by

$$\begin{aligned}\Delta_2\hat{p}_2 + \left(\frac{\omega}{c}\right)^2\hat{p}_2 &= -Q\delta(y)\int_{-\infty}^{\infty} e^{i(\omega-\Omega)t}\delta(x-ut)\mathrm{d}t\\ &= -\frac{Q}{u}e^{i(\omega-\Omega)\frac{x}{u}}\delta(y)\end{aligned} \tag{2.46}$$

with $Q = -i\omega Q_1$. Please note that the source amplitude Q in 2D now has the dimension of a mass flow per time and length: $[Q] = kg/(ms^2)$. We apply the Green's function method for solving this inhomogeneous Helmholtz equation. The Green's function $\hat{g}_2$ in 2D is defined by the equation

$$\Delta_2\hat{g}_2 + \left(\frac{\omega}{c}\right)^2\hat{g}_2 = -\delta(x)\delta(y) \tag{2.47}$$

and has the well-known expression (cf. Eq. 2.18)

$$\hat{g}_2 = \frac{i}{4}H_0^{(1)}\left(\frac{\omega}{c}\sqrt{x^2+y^2}\right). \tag{2.48}$$

By convoluting $\hat{g}_2$ with the negative right-hand side of the Helmholtz Eq. (2.46)

$$R(x,y) = \frac{Q}{u}e^{i(\omega-\Omega)\frac{x}{u}}\delta(y), \tag{2.49}$$

we obtain

$$\hat{p}_2 = \int_{-\infty}^{\infty}\int_{-\infty}^{\infty} \hat{g}_2(x-\xi, y-\eta)R(\xi,\eta)\mathrm{d}\xi\mathrm{d}\eta$$
$$= \frac{Qi}{4u}\int_{-\infty}^{\infty} H_0^{(1)}\left(\frac{\omega}{c}\sqrt{(x-\xi)^2+y^2}\right)e^{i(\omega-\Omega)\frac{\xi}{u}}\mathrm{d}\xi. \tag{2.50}$$

Similar to the 3D case, the integral can be solved explicitly with the result

$$\hat{p}_2 = \frac{Qi}{2u}e^{i(\omega-\Omega)\frac{x}{u}}\frac{e^{i\kappa|y|}}{\kappa}, \quad \kappa = \sqrt{\left(\frac{\omega}{c}\right)^2 - \left(\frac{\omega-\Omega}{u}\right)^2}. \tag{2.51}$$

Proof of Eq. (2.51): By changing the variable of integration in Eq. (2.50) to $\alpha = x - \xi$, we obtain

$$\hat{p}_2 = \frac{Qi}{4u}e^{i(\omega-\Omega)\frac{x}{u}}\int_{-\infty}^{\infty} H_0^{(1)}\left(\frac{\omega}{c}\sqrt{\alpha^2+y^2}\right)e^{-i\left(\frac{\omega-\Omega}{u}\right)\alpha}\mathrm{d}\alpha. \tag{2.52}$$

To obtain formula (2.51), we still have to prove that

$$\int_{-\infty}^{\infty} H_0^{(1)}\left(\frac{\omega}{c}\sqrt{\alpha^2+y^2}\right)e^{-i\left(\frac{\omega-\Omega}{u}\right)\alpha}\mathrm{d}\alpha = \frac{2e^{i\sqrt{\left(\frac{\omega}{c}\right)^2-\left(\frac{\omega-\Omega}{u}\right)^2}|y|}}{\sqrt{\left(\frac{\omega}{c}\right)^2-\left(\frac{\omega-\Omega}{u}\right)^2}}. \tag{2.53}$$

For this reason, we compare the Green's functions in 1D and in 2D: The Fourier transform of the 2D Helmholtz Eq. (2.47) with respect to x

$$\tilde{\hat{g}}_2 := F_x[\hat{g}_2] := \int_{-\infty}^{\infty} \hat{g}_2 e^{-ik_x x}\mathrm{d}x$$

leads to

$$\Delta\tilde{\hat{g}}_2 + \left(k^2 - k_x^2\right)\tilde{\hat{g}}_2 = -\delta(y),\ k = \frac{\omega}{c}. \tag{2.54}$$

Thus, on the one hand, $\tilde{\hat{g}}_2$ is the well-known 1D-Green's function (see Duffy [4, p. 275])

$$\tilde{\hat{g}}_2 = \frac{i}{2\sqrt{k^2-k_x^2}}e^{i\sqrt{k^2-k_x^2}|y|} \tag{2.55}$$

and on the other hand, it is the Fourier transform of the solution of Eq. (2.47), i.e., the Fourier transform with respect to x of the 2D-Green's function (2.48)

$$\tilde{\hat{g}}_2 = \frac{i}{4} \int\limits_{-\infty}^{\infty} H_0^{(1)}\left(\frac{\omega}{c}\sqrt{x^2+y^2}\right) e^{-ik_x x} \mathrm{d}x. \tag{2.56}$$

Comparing the last two equations, we get

$$\int\limits_{-\infty}^{\infty} H_0^{(1)}\left(\frac{\omega}{c}\sqrt{x^2+y^2}\right) e^{-ik_x x} \mathrm{d}x = \frac{2}{\sqrt{k^2-k_x^2}} e^{i\sqrt{k^2-k_x^2}|y|}. \tag{2.57}$$

By choosing $k_x = \frac{\omega-\Omega}{u}$, Eq. (2.53) is proven.

In addition, the pressure in the time domain should be briefly derived. The use of the inverse temporal Fourier transform of $\hat{p}_2(\omega)$ is too cumbersome. It is easier to apply the modified Lorentz transformation similarly as in three dimensions (see Eqs. (2.31 and 2.32)) or in one dimension, as carried out later in Sect. 2.2.3. Hence, by applying $x_L = \gamma^2(x - Mct)$, $y_L = \gamma y$, $t_L = \gamma^2(t - Mx/c)$ to the 2D wave equation (2.45), we obtain

$$\Delta_{2L}\varphi_{2L} - \frac{1}{c^2}\frac{\partial^2}{\partial t_L^2}\varphi_{2L} = \gamma\frac{Q_1}{\rho}e^{-i\Omega t_L}\delta(x_L)\delta(y_L), \tag{2.58}$$

where the potential φ_{2L} in Lorentz space is now a solution of the 2D wave equation with a stationary line source at the origin with strength $-\gamma Q_1/\rho$.[2] Here, Δ_{2L} is the 2D Laplace operator written in the modified Lorentz coordinates.

Equation (2.58) is easily solved (cf. Eq. (2.48)) with the result

$$\varphi_{2L} = \frac{-iQ_1\gamma}{4\rho} H_0^{(1)}\left(k_0\sqrt{x_L^2+y_L^2}\right) e^{-i\Omega t_L}, \; k_0 = \Omega/c. \tag{2.59}$$

By performing the inverse Lorentz transformation, the solution for a moving line source in physical space is obtained

$$\varphi_2 = -\frac{iQ_1\gamma}{4\rho} H_0^{(1)}\left(k_0\sqrt{\left[\gamma^2(x-Mct)\right]^2 + (\gamma y)^2}\right) e^{-i\Omega\gamma^2(t-Mx/c)}. \tag{2.60}$$

This solution is well-known and can often be found in the relevant literature, see, e.g., Dragna and Blanc-Benon [9, Eq. (29)], Piscoya and Ochmann [10, Eq. (18)], and [11, Eq. (35)].

[2] In 3D the source strength is $-\gamma^2 Q_1/\rho$.

2.2.2.2 Discussion and Evaluation of the Spectral Pressure of a Moving Line Source

Clearly, the simple formula of the frequency spectrum (2.51) can be easily evaluated. If we use the same parameters as shown in Fig. 2.3 for the 3D case, then we get the following Fig. 2.6.

The comparison with the 3D case in the previous section shows that the Doppler frequencies remain the same in 2D and 3D. However, the spectral amplitudes are higher in 2D than in 3D, which is evident since the line source can be viewed as a composite of many single point sources.

Just as in 3D, the argument of the exponential function in Eq. (2.51) becomes imaginary outside the Doppler range, and the pressure amplitudes quickly approach zero. This drastic decrease is shown in Fig. 2.7 for frequencies below the lower Doppler frequency $f_{-} = 363.6$ Hz.

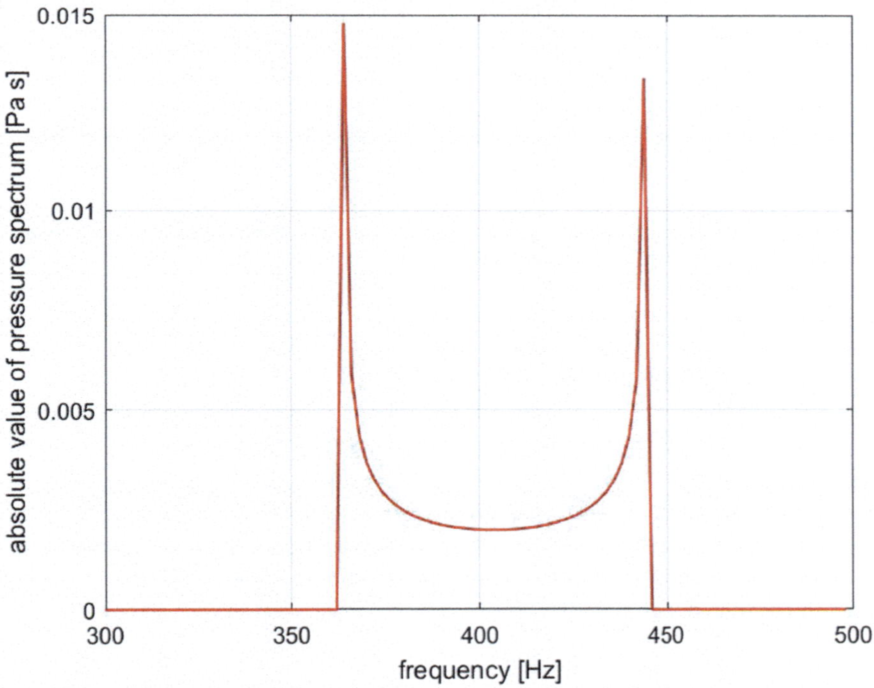

Fig. 2.6 Absolute value $|\hat{p}_2|$ of the spectral pressure with parameters $(x, y) = (10\text{ m}, 10\text{ m})$, $\Omega = 2\pi 400\,1/\text{s}$, $M = 0.1(u = 34.3\text{ m/s})$, $Q = 1\text{ kg}/(\text{m s}^2)$

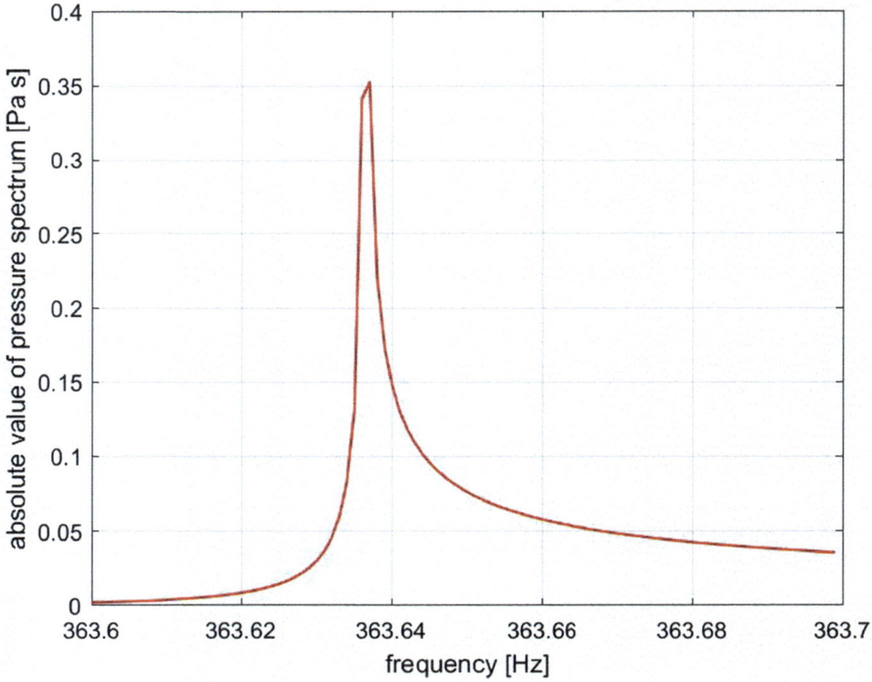

Fig. 2.7 Absolute value $|\hat{p}_2|$ of the spectral pressure in the neighborhood of the lower Doppler frequency $f_- = 363.6$ Hz; parameters are $(x, y) = (10$ m, 10 m$)$, $\Omega = 2\pi 400\,1/\text{s}$, $M = 0.1(u = 34.3$ m/s$)$, $Q = 1$ kg$/(\text{m s}^2)$

2.2.3 *One-Dimensional Case (1D Case)*

Obviously, a radiating and moving point source in 1D is not of great practical importance, but the mathematical treatment we will provide offers some interesting mathematical and physical insights, particularly with regard to the convergence of the infinite convolution integral and the transition from the frequency to the time domain. Such a source can be thought of as a pulsating infinite yz-plane in 3D space, moving in a straight line along the x-axis (Fig. 2.8).

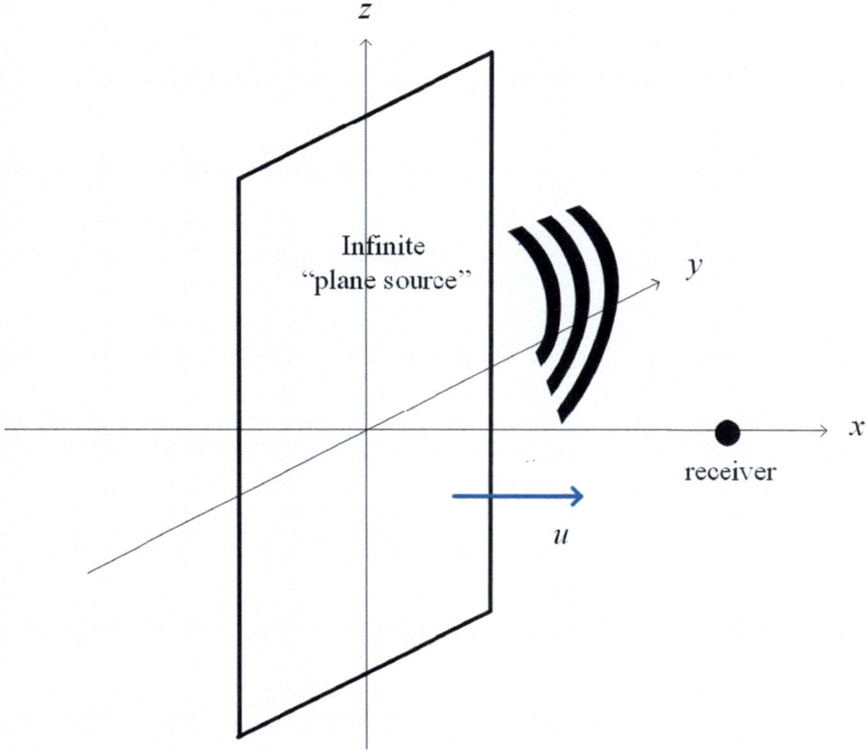

Fig. 2.8 An acoustical "plane source" in the *yz*-plane moving along the *x*-axis

2.2.3.1 Derivation of the Spectral Pressure of the Moving Source in 1D

We start with the 1D wave equation for the potential φ_1

$$\frac{\partial^2 \varphi_1}{\partial x^2} - \frac{1}{c^2}\frac{\partial^2 \varphi_1}{\partial t^2} = \frac{Q_1}{\rho} e^{-i\Omega t}\delta(x - ut). \tag{2.61}$$

The temporal Fourier transform leads to the inhomogeneous one-dimensional vibration equation for the spectral pressure $\hat{p}_1(\omega) = i\omega\rho\hat{\varphi}_1$

$$\frac{\partial^2 \hat{p}_1}{\partial x^2} + \left(\frac{\omega}{c}\right)^2 \hat{p}_1 = -\frac{Q}{u} e^{i\left(\frac{\omega-\Omega}{u}\right)x},\ Q = -i\omega Q_1, \tag{2.62}$$

which can be easily solved with the ansatz $\hat{p} = Ce^{i\zeta x}$, $\zeta = (\omega - \Omega)/u$ if the resonance case $(\omega - \Omega)/u = \pm\omega/c \Leftrightarrow \omega_{\mp} = \frac{\Omega}{1 \pm M}$ at both Doppler frequencies is excluded. The ansatz leads to the particular solution

$$C = -\frac{Q}{u}\frac{1}{\kappa^2} \text{ with } \kappa^2 = \left(\frac{\omega}{c}\right)^2 - \left(\frac{\omega - \Omega}{u}\right)^2, \tag{2.63}$$

and therefore, the spectral pressure $\hat{p}_1$ in 1D is given by

$$\hat{p}_1 = -\frac{Q}{u}\frac{e^{i\left(\frac{\omega-\Omega}{u}\right)x}}{\kappa^2} + A\cos\left(\frac{\omega}{c}x\right) + B\sin\left(\frac{\omega}{c}x\right). \tag{2.64}$$

Assuming that there are no free oscillations, we set $A = B = 0$, which leads to

$$\hat{p}_1 = -\frac{Q}{u}\frac{e^{i\left(\frac{\omega-\Omega}{u}\right)x}}{\kappa^2} \text{ for } \omega \neq \omega_\pm. \tag{2.65}$$

Of course, the poles of this particular solution are the Dopplers frequencies ω_- and ω_+.

It is now instructive to derive this solution using the Green's function method since the convolution integral can be solved analytically, and thus its behavior at infinity can be studied. The 1D Green's function $\hat{g}_1$ in the frequency domain is defined by the equation

$$\frac{\mathrm{d}^2\hat{g}_1}{\mathrm{d}x^2} + \left(\frac{\omega}{c}\right)^2\hat{g}_1 = -\delta(x) \tag{2.66}$$

and has the well-known expression (see Duffy [4, p. 275])

$$\hat{g}_1(x) = \frac{i}{2k}e^{ik|x|},\ k = \omega/c \tag{2.67}$$

By convoluting $\hat{g}_1$ with the negative right-hand side of Eq. (2.62)

$$R(x) = \frac{Q}{u}e^{i\left(\frac{\omega-\Omega}{u}\right)x},$$

we obtain

$$\begin{aligned}\hat{p}_1 &= \int_{-\infty}^{\infty}\hat{g}_1(x-\xi)R(\xi)\mathrm{d}\xi \\ &= \int_{-\infty}^{\infty}\hat{g}_1(\xi)R(x-\xi)\mathrm{d}\xi \\ &= \frac{i}{2k}\frac{Q}{u}e^{i\left(\frac{\omega-\Omega}{u}\right)x}\int_{-\infty}^{\infty}e^{ik|\xi|}e^{-i\left(\frac{\omega-\Omega}{u}\right)\xi}\mathrm{d}\xi.\end{aligned} \tag{2.68}$$

The integral $I := \int_{-\infty}^{\infty} e^{ik|\xi|}e^{-i\left(\frac{\omega-\Omega}{u}\right)\xi}\mathrm{d}\xi$ can be solved analytically with the result

$$I = \int_{-\infty}^{0} e^{-i\left(k+\frac{\omega-\Omega}{u}\right)\xi} \mathrm{d}\xi + \int_{0}^{\infty} e^{i\left(k-\frac{\omega-\Omega}{u}\right)\xi} \mathrm{d}\xi$$
$$= \left.\frac{e^{-i\left(k+\frac{\omega-\Omega}{u}\right)\xi}}{-i\left(k+\frac{\omega-\Omega}{u}\right)}\right|_{-\infty}^{0} + \left.\frac{e^{i\left(k-\frac{\omega-\Omega}{u}\right)\xi}}{i\left(k-\frac{\omega-\Omega}{u}\right)}\right|_{0}^{\infty}. \tag{2.69}$$

Thus.
$I = I_0 + I_\infty$ with

$$I_0 = \frac{1}{-i\left(k+\frac{\omega-\Omega}{u}\right)} - \frac{1}{i\left(k-\frac{\omega-\Omega}{u}\right)} \tag{2.70}$$

and

$$I_\infty = \lim_{\xi\to-\infty} \frac{e^{-i\left(k+\frac{\omega-\Omega}{u}\right)\xi}}{i\left(k+\frac{\omega-\Omega}{u}\right)} + \lim_{\xi\to\infty} \frac{e^{i\left(k-\frac{\omega-\Omega}{u}\right)\xi}}{i\left(k-\frac{\omega-\Omega}{u}\right)}. \tag{2.71}$$

The first term I_0 leads to the particular solution (2.65)

$$\frac{i}{2k}\frac{Q}{u} e^{i\left(\frac{\omega-\Omega}{u}\right)x} I_0 = \frac{i}{2k}\frac{Q}{u}\frac{i2k}{k^2-\left(\frac{\omega-\Omega}{u}\right)^2} e^{i\left(\frac{\omega-\Omega}{u}\right)x}$$
$$= -\frac{Q}{u}\frac{e^{i\left(\frac{\omega-\Omega}{u}\right)x}}{\kappa^2} = \hat{p}_1.$$

However, the contribution of the integral at infinity I_∞ only tends toward zero if we introduce a certain damping by adding a small imaginary part to the wave number. Then we get

$$I_\infty = -i \lim_{\xi\to\infty} \left\{ e^{ik\xi} \left[\frac{e^{i\left(\frac{\omega-\Omega}{u}\right)\xi}}{\left(k+\frac{\omega-\Omega}{u}\right)} + \frac{e^{-i\left(\frac{\omega-\Omega}{u}\right)\xi}}{\left(k-\frac{\omega-\Omega}{u}\right)} \right] \right\} = 0 \tag{2.72}$$

if $k = \frac{\omega}{c} + i\varepsilon$, $\varepsilon > 0$. This convergence behavior is investigated by a numerical calculation of the spectral pressure for different values of the damping ε. For this purpose, we use the Matlab function *integral* and compare the result with the analytical solution (2.65). As an example, we choose $f = 500$ Hz and $M = 0.2$. The results are summarized in Table 2.1 and in the corresponding Figs. 2.9, 2.10, 2.11, and 2.12. "Convergence = no" means that the Matlab function *integral* does not find a convergent solution if the integration interval is taken to be $(-\infty, \infty)$.

In the last two figures, no difference can be seen between the blue and red curves. For example, the maximum difference in the last figure is smaller than 10^{-12} Pa s. We observe that a damping of five per thousand is sufficient to ensure the convergence of the infinite integral. For smaller values of ε, we can obtain a very good agreement between the numerical and the analytical solution, although the integral does not

Table 2.1 The effect of damping on the spectral pressure

Dampingε	0.0002	0.0005	0.001	0.005
Convergence	No	No	No	Yes
Agreement with analytical solution	No	Nearly	Yes	Yes
Figures	2.9	2.10	2.11	2.12

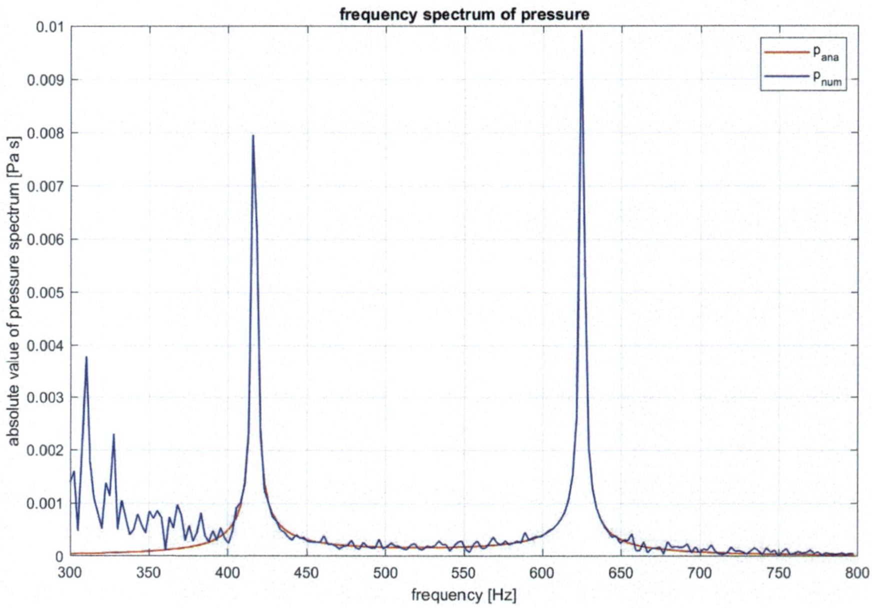

Fig. 2.9 $\varepsilon = 0.0002$: absolute value of the 1D frequency spectrum of the pressure: p_{ana} = analytical expression (2.65), p_{num} = numerical integral expression (2.68); $\Omega = 2\pi 500$ 1/s, $M = 0.2$, $Q = 1$ kg/(ms)^2; interval of integration $(-\infty, \infty)$

converge. In this case, Matlab displays a warning, for example.: "*Reached the limit on the maximum number of intervals in use. Approximate bound on error is 1.3e-06. The integral may not exist, or it may be difficult to approximate*".

In Sect. 2.3.3, the effect of damping is examined in more detail. There, it is shown that, for physical reasons, attenuation must be introduced in the wave number as well as in the amplitude, and its quantity depends on the value of the frequency ω, while the small parameter ε here only serves to ensure the convergence of the integral.

In conclusion, we see that the convolution integral *only* converges numerically if some kind of damping is added to the wavenumber. Then, the integral can be extended from minus infinity to plus infinity. Another possibility is to truncate the interval of integration to a finite size, which can also be considered as a finite measurement time. This procedure is analyzed in Sect. 2.3.2.

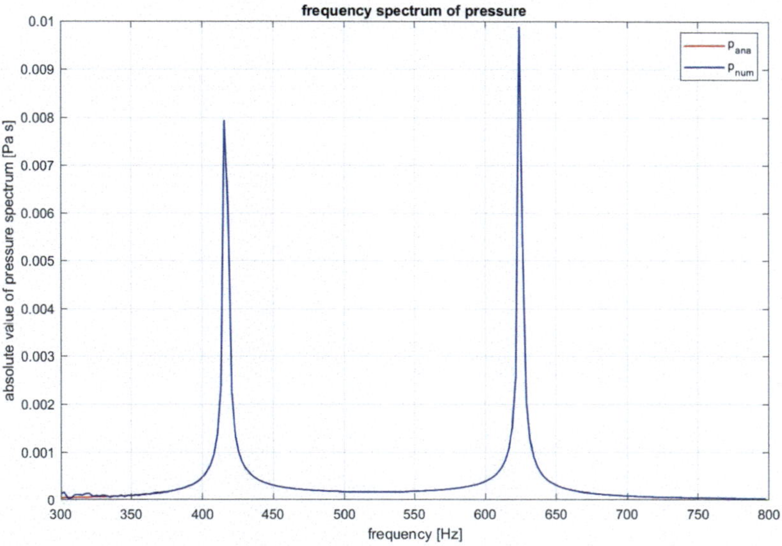

Fig. 2.10 $\varepsilon = 0.0005$: absolute value of the 1D frequency spectrum of the pressure: p_{ana} = analytical expression (2.65), p_{num} = numerical integral expression (2.68); $\Omega = 2\pi 500$ 1/s, $M = 0.2$, $Q = 1$ kg/(ms)^2; interval of integration $(-\infty, \infty)$

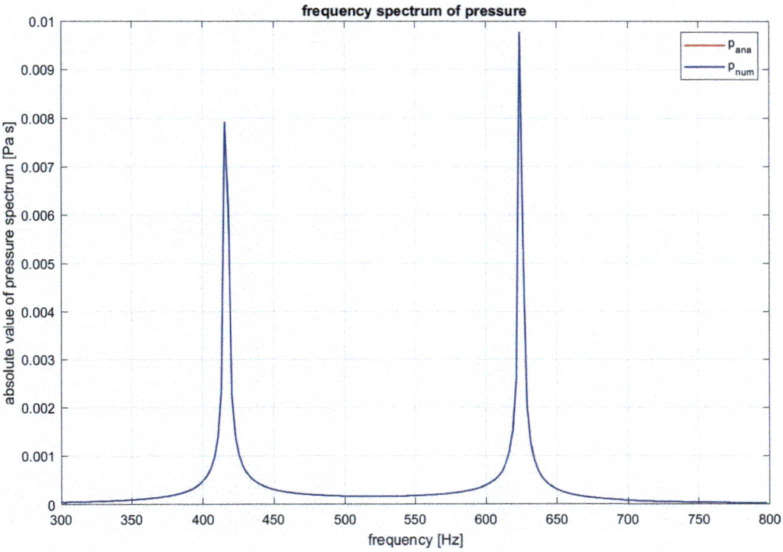

Fig. 2.11 $\varepsilon = 0.001$: absolute value of the 1D frequency spectrum of the pressure: p_{ana} = analytical expression (2.65), p_{num} = numerical integral expression (2.68); $\Omega = 2\pi 500\,1/\text{s}$, $M = 0.2$, $Q = 1$ kg/(ms)^2; interval of integration $(-\infty, \infty)$

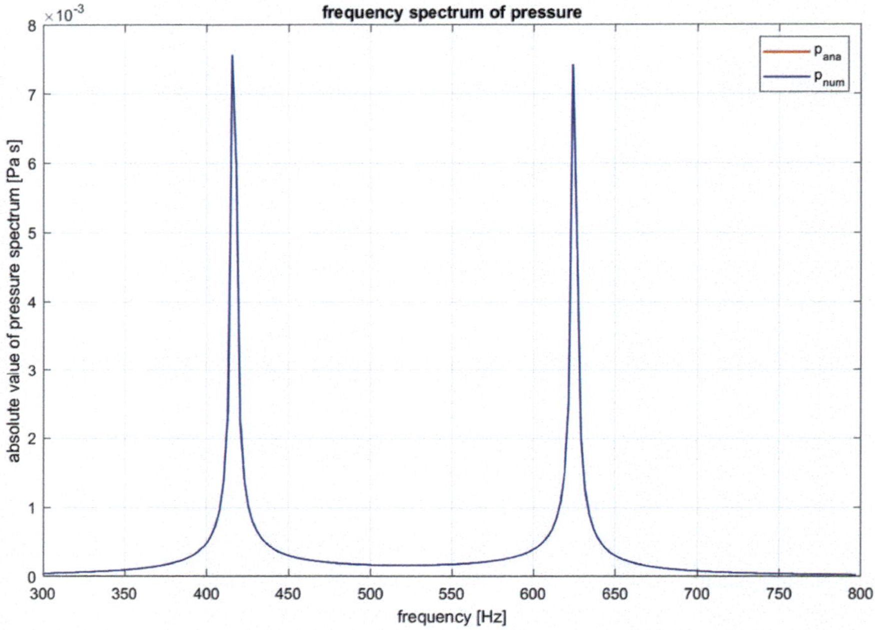

Fig. 2.12 $\varepsilon = 0.005$: absolute value of the 1D frequency spectrum of the pressure: p_{ana} = analytical expression (2.65), p_{num} = numerical integral expression (2.68); $\Omega = 2\pi 500\,1/\text{s}$, $M = 0.2$, $Q = 1\ \text{kg}/(\text{ms})^2$; interval of integration $(-\infty, \infty)$

2.2.3.2 Time Domain Acoustic Pressure or Potential of a Moving Source in 1D

The inverse Fourier transform of the spectral pressure $\hat{p}_1$ gives the time domain pressure

$$p(t) = \frac{1}{2\pi u} \int_{-\infty}^{\infty} i\omega Q_1 \frac{e^{i\left(\frac{\omega-\Omega}{u}\right)x} e^{-i\omega t}}{\kappa^2(\omega)} \mathrm{d}\omega = \frac{-Q_1}{2\pi u} \frac{\partial}{\partial t} \int_{-\infty}^{\infty} \frac{e^{i\left(\frac{\omega-\Omega}{u}\right)x} e^{-i\omega t}}{\kappa^2(\omega)} \mathrm{d}\omega \tag{2.73}$$

with

$$\kappa^2(\omega) = \left(\frac{\omega}{c}\right)^2 - \left(\frac{\omega-\Omega}{u}\right)^2 = \frac{(M^2-1)}{u^2}(\omega-\omega_-)(\omega-\omega_+), \tag{2.74}$$

where $\omega_- = \Omega/(1+M)$ and $\omega_+ = \Omega/(1-M)$.

Note that $Q = -i\omega Q_1$ cannot be drawn in front of the integral. Since $p = -\rho\frac{\partial\varphi}{\partial t}$, the acoustic potential in the time domain is given by

$$\varphi(t) = \frac{Q_1}{2\pi\rho u} \int_{-\infty}^{\infty} \frac{e^{i\left(\frac{\omega-\Omega}{u}\right)x} e^{-i\omega t}}{\kappa^2(\omega)} \mathrm{d}\omega \tag{2.75}$$

This Fourier integral can be solved in three different ways. We start with an application of Cauchy's residue theorem.

(a) **Derivation of the time domain potential by means of the residue theorem**

The Doppler frequencies $\omega_- = \Omega/(1+M)$ and $\omega_+ = \Omega/(1-M)$ are the poles of the integrand in Eq. (2.75). Since no damping is included, they lie on the real axis, and the question arises as to how the integration path in the complex plane around the poles can be chosen to obtain the physically correct solution to the radiation problem. Before investigating this, we first calculate the residues (see [12])

$$R(\omega_\pm) = \mathrm{Res}(F(\omega), \omega_\pm) = \lim_{\omega \to \omega \pm} (\omega - \omega_\pm) F(\omega), \tag{2.76}$$

where (see Eq. (2.74))

$$F(\omega) = \frac{-e^{i\left(\frac{\omega-\Omega}{u}\right)x} e^{-i\omega t}}{u\kappa^2(\omega)} = \frac{u e^{i\left(\frac{\omega-\Omega}{u}\right)x} e^{-i\omega t}}{\left(1-M^2\right)(\omega-\omega_-)(\omega-\omega_+)} \tag{2.77}$$

and therefore

$$\varphi(t) = \frac{Q_1}{2\pi\rho} \int_{-\infty}^{\infty} \frac{e^{-i\Omega\frac{x}{u}} e^{i\frac{\omega}{u}(x-ut)}}{u\kappa^2(\omega)} \mathrm{d}\omega = \frac{-Q_1}{2\pi\rho} \int_{-\infty}^{\infty} F(\omega) \mathrm{d}\omega. \tag{2.78}$$

The residues of $F(\omega)$ are given by

$$R(\omega_-) = \frac{-1}{2k_0} e^{i\frac{\omega_-}{c}(-x-ct)} = \frac{-1}{2k_0} e^{-i\frac{\omega_-}{c}(x-ut)} e^{-i\Omega t}, \; k_0 = \frac{\Omega}{c} \tag{2.79}$$

and

$$R(\omega_+) = \frac{1}{2k_0} e^{i\frac{\omega_+}{c}(x-ct)} = \frac{1}{2k_0} e^{i\frac{\omega_+}{c}(x-ut)} e^{-i\Omega t}, \tag{2.80}$$

where we have used the conversions $\left(\frac{\Omega}{c}\right)\frac{1}{1\pm M}(x \pm ct) = \left(\frac{\Omega}{c}\right)\frac{1}{1\pm M}(x-ut) \pm \Omega t$.

Both poles lie on the real axis, and we try to use the integration path suggested by Duffy in Fig. 5.1.1 [4, p. 275], where Duffy's $\pm k_0$ corresponds to our poles $\omega_\pm$.

Thus, we solve the integral (2.78) by closing the path of integration in the upper half-plane for $x - ut > 0$ and obtain by means of the residue theorem (see Fig. 2.13: red curve)

$$\varphi(t) = \frac{-Q_1}{2\pi\rho}\{2\pi i R(\omega_+)\} = \frac{-iQ_1}{2k_0\rho} e^{i\frac{\omega_+}{c}(x-ut)} e^{-i\Omega t} \text{ for } x - ut > 0. \tag{2.81}$$

By closing the path of integration in the lower half-plane for $x - ut < 0$, we find that (see Fig. 2.13: green curve)

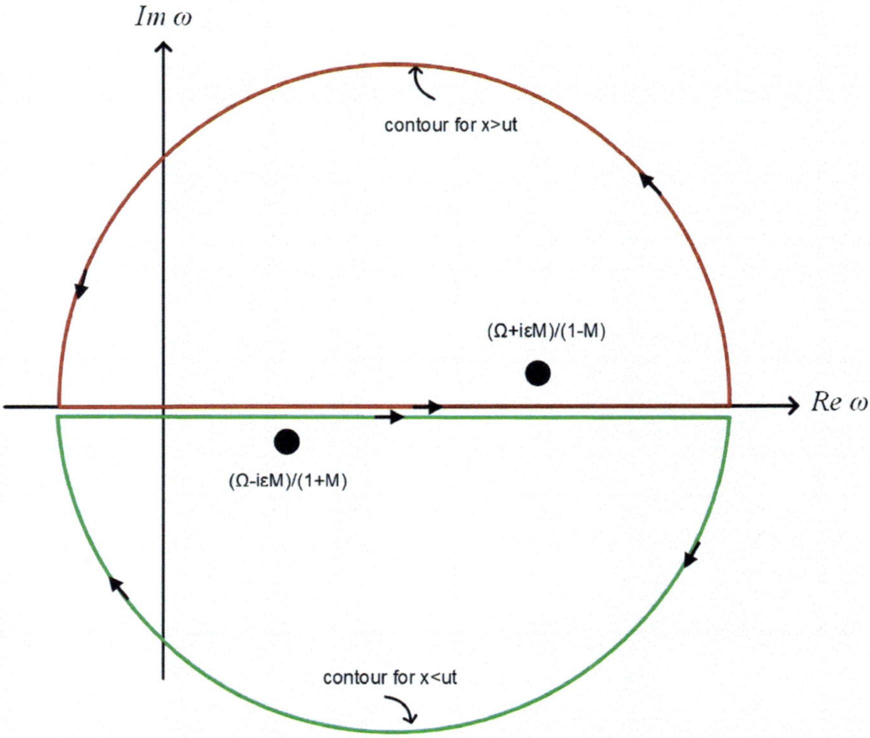

Fig. 2.13 Path of integration leading to Eqs. (2.81) and (2.82)

$$\varphi(t) = \frac{-Q_1}{2\pi\rho}\{-2\pi iR(\omega_-)\} = \frac{-iQ_1}{2k_0\rho}e^{-i\frac{\omega_-}{c}(x-ut)}e^{-i\Omega t} \quad \text{for } x-ut<0. \tag{2.82}$$

In Eqs. (2.81) and (2.82), we have applied Jordan's Lemma (see Duffy [4, p. 393] or Skudrzyk [13, p. 47]), and the minus sign in front of the factor $2\pi i$ results from the clockwise direction of the green contour in Fig. 2.13.

In summary, the 1D potential in the time domain can be written as

$$\varphi(t) = \frac{-Q_1 i}{2k_0\rho}e^{i\left(\frac{\Omega}{c}\right)\frac{1}{1-\tilde{M}}|x-ut|}e^{-i\Omega t} \text{ with } \tilde{M} := M\,\mathrm{sgn}(x-ut), \tag{2.83}$$

where

$$\mathrm{sgn}(z) = \begin{cases} 1 & \text{for } z>0 \\ -1 & \text{for } z<0 \end{cases}$$

is the signum function. Expression (2.83) can also be converted into the representation commonly used in the literature as follows: We expand the exponent in (2.81) by $1+M$ and in (2.82) by $1-M$ and obtain

$$\varphi(t) = \frac{-iQ_1}{2k_0\rho} e^{ik_0 E} \text{ with } E = \begin{cases} \frac{(1+M)(x-ut)}{1-M^2} - ct, x - ut > 0 \\ \frac{-(1-M)(x-ut)}{1-M^2} - ct, x - ut < 0 \end{cases}. \tag{2.84}$$

Since $M(x - ut) - ct(1 - M^2) = -(ct - Mx)$, we get

$$E = \frac{1}{1 - M^2}\{|x - ut| - (ct - Mx)\}$$

and finally

$$\varphi = \frac{-iQ_1}{2k_0\rho} e^{\frac{ik_0}{1-M^2}\{|x-ut|-(ct-Mx)\}}. \tag{2.85}$$

If we let u (and, therefore, also M) go to zero, we obtain the well-known solution

$$\varphi(t) = \frac{-iQ_1}{2k_0\rho} e^{ik_0|x|} e^{-i\Omega t} \text{ if } u \to 0 \tag{2.86}$$

for a source at rest (cf. Duffy [4, p. 275, Eq. (5.1.8)]).

The path of integration chosen above can also be physically justified. We have seen in the previous section that the Green's function method only provides the correct particular solution if the wavenumber has a small positive imaginary part $k^* = \frac{\omega}{c} + \frac{i\tilde{\varepsilon}}{c}$, $\tilde{\varepsilon} > 0$ (see Eq. 2.72). If we use such a complex k^* instead of the real one, we get

$$\kappa *^2(\omega) = \left(\frac{\omega}{c} + \frac{i\tilde{\varepsilon}}{c}\right)^2 - \left(\frac{\omega - \Omega}{u}\right)^2 = \frac{(M^2 - 1)}{u^2}(\omega - \omega_-^*)(\omega - \omega_+^*)$$

with $\omega_-^* = \frac{1}{1+M}(\Omega - i\tilde{\varepsilon}M)$ and $\omega_+^* = \frac{1}{1-M}(\Omega + i\tilde{\varepsilon}M)$. Thus, the smaller residue ω_-^* is slightly below the real axis, and the larger residue ω_+^* is slightly above it. Now the integration path in the upper half-plane for $x > ut$ contains only the residue $R(\omega_+)$ leading to (2.81), and the integration path in the lower half-plane for $x < ut$ contains the residue $R(\omega_-)$ leading to (2.82). A similar argumentation can be found by Skudrzyk in his Sect. 27.9.2 [13]. The contours are sketched in Fig. 2.13.

(b) **Derivation of the time domain pressure by means of the modified Lorentz transform**

The Lorentz transform makes it possible to quickly derive the pressure in the time domain generated by a rectilinearly moving source. The 1D wave Eq. (2.61) reads in the modified Lorentz coordinates $x_L = \gamma^2(x - Mct)$, $t_L = \gamma^2(t - Mx/c)$—see the analogous derivation in 3D given in [2, Sect. 11.2] or in Ochmann [5]

$$\frac{\partial^2 \varphi_L}{\partial x_L^2} - \frac{1}{c^2}\frac{\partial^2 \varphi_L}{\partial t_L^2} = \frac{Q_1}{\rho} e^{-i\Omega t_L}\delta(x_L). \tag{2.87}$$

The solution is the well-known Green's function (cf. Eq. (2.67))

$$\varphi_L(t_L) = \frac{-iQ_1}{2\rho k_0} e^{ik_0|x_L|} e^{-i\Omega t_L}, \; k_0 = \frac{\Omega}{c}, \tag{2.88}$$

which can be expressed through back substitution into the physical coordinates

$$\varphi(t) = \frac{-iQ_1}{2\rho k_0} e^{ik_0\left|\gamma^2(x-Mct)\right|} e^{-i\Omega\gamma^2(t-Mx/c)} \tag{2.89}$$

or

$$\varphi(t) = \frac{-iQ_1}{2\rho k_0} e^{ik_0 \frac{1}{1-M^2}\{|x-ut|-(ct-Mx)\}}. \tag{2.90}$$

As expected, we once again obtain the result (2.85).

(c) **Derivation of the time domain pressure by means of the Green's function method**

Clearly, the Green's function method can also be applied in the time domain. For this reason we consider the 1D wave equation

$$\frac{\partial^2 g}{\partial x^2} - \frac{1}{c^2}\frac{\partial^2 g}{\partial t^2} = -\delta(t-\tau)\delta(x-\xi) \tag{2.91}$$

which has the Green's function

$$g(x,t;\xi,\tau) = \frac{c}{2}H(t-\tau-|x-\xi|/c) = \frac{c}{2}H(t-\tau-(x-\xi)/c)H(t-\tau+(x-\xi)/c) \tag{2.92}$$

as solution (see [4, p. 79]), where

$$H(x) = \begin{cases} 1, & x > 0 \\ 0, & x < 0 \end{cases} \tag{2.93}$$

is the Heaviside unit step function. The 1D wave Eq. (2.61) of a moving source can be solved by convoluting its negative right-hand side $\tilde{Q}$ with the 1D Green's function

$$\begin{aligned}
\varphi(x,t) &= \int_{-\infty}^{\infty}\int_{-\infty}^{\infty} \tilde{Q}(\xi,\tau)g(x,t;\xi,\tau)\,\mathrm{d}\xi\,\mathrm{d}\tau \\
&= \frac{-c}{2}\int_{-\infty}^{\infty}\int_{-\infty}^{\infty} \frac{Q_1}{\rho}e^{-i\Omega\tau}\delta(\xi-u\tau)H\left(t-\tau-\frac{x-\xi}{c}\right) \\
&\quad\times H\left(t-\tau+\frac{x-\xi}{c}\right)\mathrm{d}\xi\,\mathrm{d}\tau \\
&= \frac{-c}{2}\frac{Q_1}{\rho}\int_{-\infty}^{\infty} e^{-i\Omega\tau}H\left(t-\tau-\frac{x-u\tau}{c}\right)H\left(t-\tau+\frac{x-u\tau}{c}\right)\mathrm{d}\tau.
\end{aligned} \tag{2.94}$$

The argument of the first and of the second Heaviside unit step function is greater than zero for

$$\tau < \tau_1 = \frac{t - x/c}{1 - M} \text{ and for } \tau < \tau_2 = \frac{t + x/c}{1 + M}. \tag{2.95}$$

Hence, the integration in (2.94) need only be extended over the interval $I = (-\infty, \min\{\tau_1, \tau_2\})$

$$\varphi(x, t) = -\frac{c}{2}\frac{Q_1}{\rho} \int\limits_{-\infty}^{\min\{\tau_1, \tau_2\}} e^{-i\Omega\tau} \mathrm{d}\tau. \tag{2.96}$$

Since $\tau_1 < \tau_2 \Leftrightarrow x - ut > 0$, we obtain

$$\varphi(x, t) = -\frac{c}{2}\frac{Q_1}{\rho} \int\limits_{-\infty}^{\tau_1} e^{-i\Omega\tau} \mathrm{d}\tau = \frac{-iQ_1}{2\Omega\rho/c}\left[e^{-i\Omega\tau_1} - \lim_{\tau\to-\infty} e^{-i\Omega\tau}\right] \text{for } x - ut > 0. \tag{2.97}$$

Assuming that the radiated frequency has a small damping component $\Omega \to \Omega + i\varepsilon$ so that the limit vanishes, we obtain

$$\varphi(x, t) = \frac{-iQ_1}{2\rho k_0} e^{i\frac{k_0}{1-M}(x-ct)} = \frac{-iQ_1}{2\rho k_0} e^{i\frac{k_0}{1-M}(x-ut)} e^{-i\Omega t} \text{ for } x - ut > 0. \tag{2.98}$$

In the same way we get

$$\varphi(x, t) = \frac{-iQ_1}{2\rho k_0} e^{-i\frac{k_0}{1+M}(x-ut)} e^{-i\Omega t} \text{ for } \tau_1 > \tau_2 \Leftrightarrow x - ut < 0. \tag{2.99}$$

Equations (2.98) and (2.99) agree with the solutions (2.81)/(2.82) previously obtained.

In summary, all three different mathematical solution techniques—the complex contour integration, the modified Lorentz transform, and the Green's function method—lead to the same result as expected.

2.2.4 *Comparison of the Results for One, Two and Three Dimensions*

Table 2.2 shows the formulas for the acoustic velocity potential in the time domain and for the pressure in the frequency domain in one, two and three dimensions.

This comparison reveals an interesting feature: In the frequency domain, the x-dependence is the same for all dimensions and is given by the function

Table 2.2 Compilation of the results for the pressure of a source moving in a straight line

	Frequency domain	Time domain
Def	$X(x) := e^{i\left(\frac{\omega-\Omega}{u}\right)x}$; $Q = -i\omega Q_1$ $\kappa^2 := \left(\frac{\omega}{c}\right)^2 - \left(\frac{\omega-\Omega}{u}\right)^2$	$T(t) := e^{-ik_0\gamma^2(ct-Mx)}$
1D	$\hat{p}_1 = \frac{-Q}{u}\frac{X(x)}{\kappa^2}$ (2.65)	$\varphi = \frac{-iQ_1}{2\rho k_0} e^{ik_0\gamma^2\lvert x-ut\rvert} T(t)$ (2.85)
2D	$\hat{p}_2 = \frac{iQ}{2u} X(x) \frac{e^{i\kappa\lvert y\rvert}}{\kappa}$ (2.51)	$\varphi_2 = \frac{-iQ_1\gamma}{4\rho} H_0^{(1)}\left(k_0\sqrt{\left[\gamma^2(x-Mct)\right]^2 + (\gamma y)^2}\right) \cdot T(t)$ (2.60)
3D	$\hat{p} = \frac{iQ}{4u} X(x) H_0^{(1)}(\kappa r)$ (2.19)	$\varphi = \frac{-Q_1}{4\pi\rho R_1} e^{ik_0\gamma^2 R_1} T(t)$ (2.35)

$X(x) := e^{i\left(\frac{\omega-\Omega}{u}\right)x}$. The complexity of the solution increases from dimension to dimension. In the time domain, there is a common factor $T(t) = e^{-ik_0\gamma^2(ct-Mx)}$.

2.3 Generalizations of the Spectral Calculation in Cartesian Coordinates

We generalize the methods derived in Cartesian coordinates and presented in Sect. 2.2 in three additional points: (1) The trajectory of the moving source no longer has to be straight, but can assume an arbitrary curve. (2) We consider finite times of overflight of the source. (3) Damping effects can be included.

2.3.1 Radiating Monopole Moving Along an Explicitly Given Arbitrary Trajectory

We first consider the ***two-dimensional case*** as in Sect. 2.2.2 and assume that the trajectory of the moving source can be described by a function explicitly given in the *xy*-plane by

$$y_0 = \Phi(x_0) \tag{2.100}$$

or as a curve parameterized by time t

$$\begin{pmatrix} x_0(t) \\ y_0(t) \end{pmatrix} = \begin{pmatrix} ut \\ \Phi(t) \end{pmatrix} \text{ for } t \in (-\infty, \infty). \tag{2.101}$$

The 2D wave equation in x and y becomes

$$\Delta_2\varphi_2 - \frac{1}{c^2}\frac{\partial^2\varphi_2}{\partial t^2} = \frac{Q_1}{\rho}e^{-i\Omega t}\delta(x-ut)\delta(y-\Phi(t)). \tag{2.102}$$

By applying the temporal Fourier transform, we obtain the Helmholtz equation for the pressure $\hat{p}_2(\omega) = i\omega\rho\hat{\varphi}_2$

$$\Delta_2\hat{p}_2 + \left(\frac{\omega}{c}\right)^2\hat{p}_2 = -Q\int_{-\infty}^{\infty} e^{-i(\Omega-\omega)t}\delta(x-ut)\delta(y-\Phi(t))dt$$
$$= -\frac{Q}{u}e^{i\left(\frac{\omega-\Omega}{u}\right)x}\delta(y-\Phi(x/u)). \tag{2.103}$$

with $Q = -i\omega Q_1$.

Similar to what was performed in Sect. 2.2.2.1, the Green's function method leads to

$$\hat{p}_2 = \frac{Qi}{4u}\int_{-\infty}^{\infty} H_0^{(1)}\left(\frac{\omega}{c}\sqrt{(x-\xi)^2+(y-\Phi(\xi/u))^2}\right)e^{i(\omega-\Omega)\frac{\xi}{u}}d\xi, \tag{2.104}$$

which is easily obtained by substituting y through $y-\Phi(\xi/u)$ in Eq. (2.50). In what follows, we refer to this representation and the associated 3D representations of the frequency spectrum as the *Cartesian convolution integral* (CCI).

In ***three dimensions***, the trajectory can even be an arbitrary curve—implicit or explicit—in the yz-plane associated with any x-position $x_0(t) = ut$:

$$\begin{pmatrix} x_0(t) \\ y_0(t) \\ z_0(t) \end{pmatrix} = \begin{pmatrix} ut \\ \Phi(t) \\ \psi(t) \end{pmatrix} \text{ for } t \in (-\infty, \infty). \tag{2.105}$$

The Green's function method applied to the wave equation in 3D

$$\Delta\varphi - \frac{1}{c^2}\frac{\partial^2\varphi}{\partial t^2} = \frac{Q_1}{\rho}e^{-i\Omega t}\delta(x-ut)\delta(y-\Phi(t))\delta(z-\psi(t)) \tag{2.106}$$

yields the CCI

$$\hat{p}_3 = \frac{Q}{4\pi u}\int_{-\infty}^{\infty}\frac{e^{i\frac{\omega}{c}\sqrt{(x-\xi)^2+(y-\Phi(\xi/u))^2+(z-\psi(\xi/u))^2}}}{\sqrt{(x-\xi)^2+(y-\Phi(\xi/u))^2+(z-\psi(\xi/u))^2}}e^{i(\omega-\Omega)\frac{\xi}{u}}d\xi, \tag{2.107}$$

where we have made the substitutions $y \to y-\Phi(\xi/u)$ and $z \to z-\psi(\xi/u)$ in Eq. (2.24).

In general, Eqs. (2.104) and (2.107) have to be solved numerically. Only for very special chosen functions Φ and ψ the CCI can be solved in closed form. For

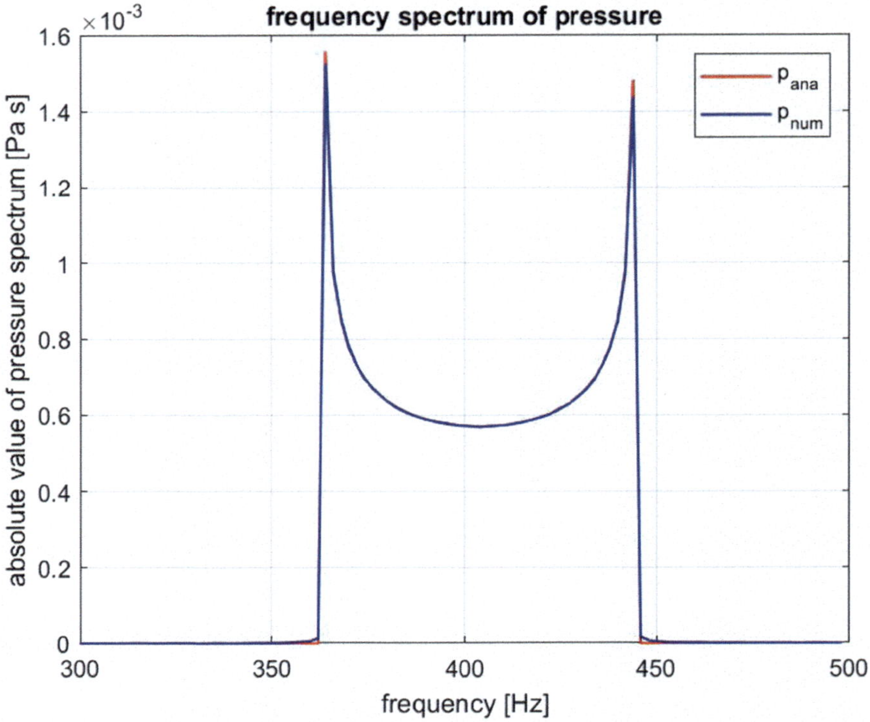

Fig. 2.14 Absolute value of the 3D spectral pressure: p_{ana} = analytical expression (2.19), p_{num} = CCI (2.107) for $\Phi(t) = \psi(t) = 0$; parameters $(x, y, z) = (10\ \mathrm{m}, 10\ \mathrm{m}, 10\ \mathrm{m})$, $\Omega = 2\pi 400$ 1/s, $M = 0.1$, $Q = 1\ \mathrm{kg/s^2}$; interval of integration [−500 m, 500 m]

example, the case $\Phi(t) = \psi(t) = constant$ leads to the analytical solutions for infinite rectilinear trajectories given in Chap. 2.2. The integrals in (2.104) and (2.107) contain terms that oscillate rapidly for $\xi \to \pm\infty$ and require special numerical treatment. However, this problem only occurs when the trajectory and the corresponding observation time of the source are infinite. In most practical cases, the observation time, and thus the integration interval can be restricted, as will be shown in the next section. Nevertheless, we will first address the case of an infinitely long trajectory. Of course, there is a lot of literature on how to calculate numerically such highly oscillatory integrals, and many references can be found in the book of Deaño et al. "*Computing Highly Oscillatory Integrals*" [14]. However, since the integral tends to zero for $\xi \to \pm\infty$, we expect a relatively good solution by integration over a sufficiently large but finite interval. This can be verified by calculating the integrals for a rectilinear trajectory and comparing the result with the known analytical solutions (2.19) and (2.51) in 3D and in 2D, respectively. For instance, by computing the 3D spectrum depicted in Fig. 2.3 using the CCI (2.107) for $\Phi(t) = \psi(t) = 0$ and comparing it with the analytical solution (2.19), we obtain Fig. 2.14, where we have used the Matlab function *integral* over the integration interval $[-500m, 500m]$.

For this integration interval, both curves exhibit excellent agreement.

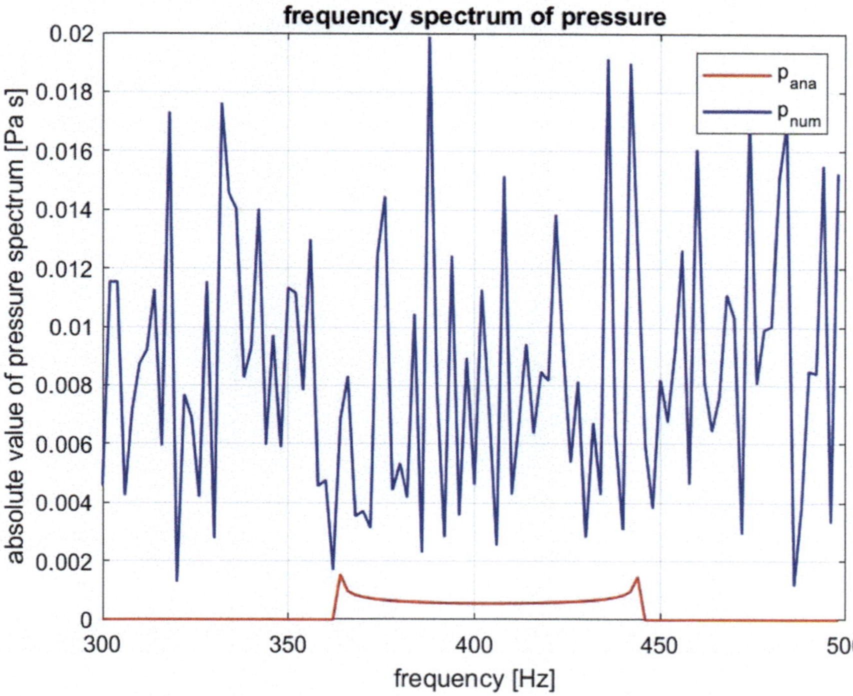

Fig. 2.15 Absolute value of the 3D spectral pressure: p_{ana} = analytical expression (2.19), p_{num} = CCI (2.107) for $\Phi(t) = \psi(t) = 0$; parameters $(x, y, z) = (10\text{ m}, 10\text{ m}, 10\text{ m})$, $\Omega = 2\pi 400$ 1/s, $M = 0.1$, $Q = 1\text{ kg/s}^2$; interval of integration $(-\infty, \infty)$

If the integration is extended to the interval "$(-\infty, \infty)$" by choosing the interval (−Inf, Inf) in the Matlab function *integral*, an incorrect numerical result is obtained due to the strong oscillations of the integrand, as shown in Fig. 2.15.

Now we choose $\Phi(t) = ut$ and $\psi(t) = 0$, which means that the source moves along the diagonal in the xy-plane. Therefore, the speed of the source is equal to $u_\Sigma = \sqrt{2}\,u$. The frequency spectrum, compared to the previous one where the source is moving along the x-axis with speed u, is shown in Fig. 2.16.

The main difference between the two curves is the broadening of the Doppler range. The Doppler shift now shows the two frequency peaks at about $f_u \approx 352$ Hz and at $f_o \approx 464$ Hz—again in relatively good agreement with the well-known estimates

$$f_- = \frac{\Omega}{2\pi(1 + u_\Sigma/c)} = 350.4 \text{ Hz and } f_+ = \frac{\Omega}{2\pi(1 - u_\Sigma/c)} = 465.9 \text{ Hz}. \quad (2.108)$$

Thus, the numerical evaluation of the CCI (2.107) leads to reliable results for known analytical solutions. For nonlinear trajectories, however, analytical solutions are rarely available for comparison, and the question arises of what it means to

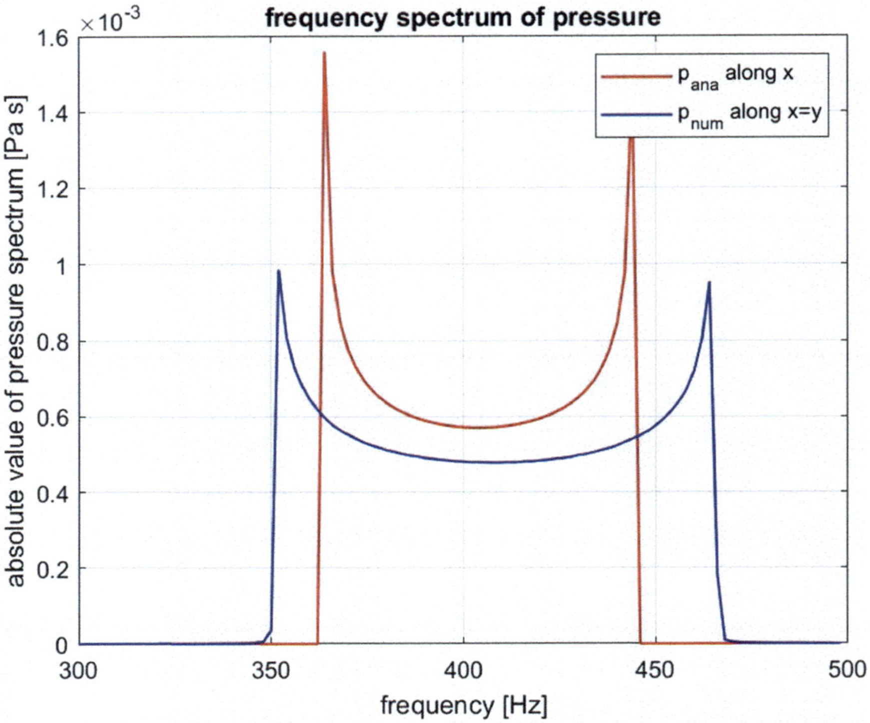

Fig. 2.16 Absolute value of the 3D spectral pressure: p_{ana} = analytical expression (2.19) moving along the x-axis with speed u, p_{num} = numerical integral expression (2.107) for $\Phi(t) = ut$ and $\psi(t) = 0$ moving along the diagonal $x = y$ with speed $u_\Sigma = \sqrt{2}\,u$; parameters $(x, y, z) = (10\text{ m}, 10\text{ m}, 10\text{ m})$, $\Omega = 2\pi 400$ 1/s, $u = 34.3$ m/s, $Q = 1$ kg/s^2; interval of integration $[-500\text{ m}, 500\text{ m}]$

restrict the integration domain to a finite interval and how the resulting error is to be interpreted. This question will be fully clarified in the next section.

We have only performed numerical calculations for the 3D case. If we evaluate the integral (2.104) for the 2D case instead, we obtain results that are very similar to those of the 3D spectrum and are, therefore, not presented here.

2.3.2 *Finite Observation Times*

When measuring the acoustic signal emitted by a moving source, the measurement time is limited to a finite time interval $[t_0, t_1]$ for practical reasons. For $t \notin [t_0, t_1]$, the signal is assumed to be zero. Thus, the right-hand side of the wave Eq. (2.106) must be multiplied by the characteristic function

$$\chi_{[t_0,t_1]}(t) := \begin{cases} 1, & t \in [t_0, t_1] \\ 0, & t \notin [t_0, t_1] \end{cases}.$$

Then the 3D wave equation is given by

$$\Delta\varphi - \frac{1}{c^2}\frac{\partial^2\varphi}{\partial t^2} = \frac{Q_1}{\rho}e^{-i\Omega t}\delta(x - ut)\delta(y - \Phi(t))\delta(z - \psi(t))\chi_{[t_0,t_1]}(t). \quad (2.109)$$

As usual, the Fourier transform yields the Helmholtz equation for the pressure $\hat{p}_3(\omega) = i\omega\rho\hat{\varphi}$

$$\Delta\hat{p}_3 + \left(\frac{\omega}{c}\right)^2\hat{p}_3 = -\frac{Q}{u}e^{-i(\Omega-\omega)\frac{x}{u}}\delta(y - \Phi(x/u))\delta(z - \psi(x/u))\chi_{[x_0,x_1]}(x). \quad (2.110)$$

Note that $\chi_{[t_0,t_1]}(t)\big|_{t=x/u} = H\left(\frac{x-x_0}{u}\right) - H\left(\frac{x-x_1}{u}\right) = \chi_{[x_0,x_1]}(x)$ with $x_0 = ut_0$, $x_1 = ut_1$.

The subsequent convolution with the Green's function leads to the CCI

$$\hat{p}_3 = \frac{Q}{4\pi u}\int\limits_{x_0=ut_0}^{x_1=ut_1}\frac{e^{i\frac{\omega}{c}\sqrt{(x-\xi)^2+(y-\Phi(\xi/u))^2+(z-\psi(\xi/u))^2}}}{\sqrt{(x-\xi)^2+(y-\Phi(\xi/u))^2+(z-\psi(\xi/u))^2}}e^{i(\omega-\Omega)\frac{\xi}{u}}\,d\xi. \quad (2.111)$$

We have thus rigorously proven that integration over a finite interval has the physical meaning that the signal is only recorded over a certain time interval. Accordingly, only a finite part of the—possibly infinite—trajectory is considered. For example, in Fig. 2.14 from the last section, the interval is equal to $I = [-500\text{ m}, 500\text{ m}]$, and therefore, the signal contributes to the spectrum only when the source moves from -500 m until 500 m. Outside of I, it is assumed that the signal is very small and negligible. Thus, there is hardly any difference between the analytical solution for the infinitely long trajectory and the 1000 m long one.

But now we are able to calculate the spectra for finite trajectories of different lengths and study the deviation from the theoretical infinitely long flight path. Let us again consider the simplest case of linear motion, i.e., the source moves along the x-axis so that $\Phi(t) = \psi(t) = 0$.

The comparison is shown in Fig. 2.17, where the source moves over a distance of 60 m, 8 m and 1.6 m. The analytical solution is again related to the full interval $(-\infty, \infty)$.

The figure shows that as the observation interval becomes shorter, the two Doppler frequencies contract to the original emission frequency, the amplitude decreases, and the pattern of a sinc function is formed.

As another example of the application of the CCI (2.111), let us study the effect of *the curvature of the trajectory* on the spectrum by passing from a straight trajectory to a parabolic curve, as shown in Fig. 2.18, where the source travels from left to right.

The blue and green trajectories of the source are described by

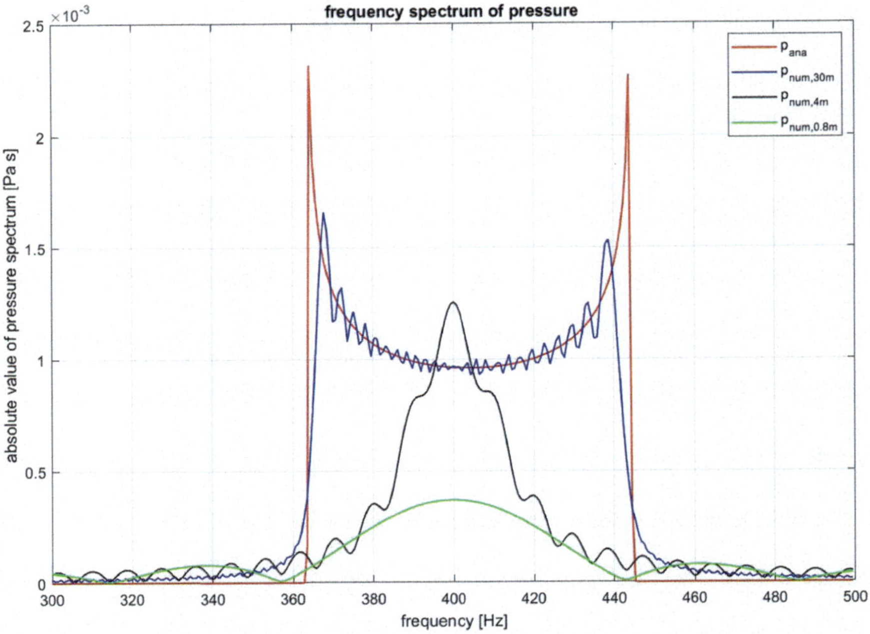

Fig. 2.17 Absolute value of the 3D frequency spectrum of the pressure: p_{ana} = analytical expression (2.19), $p_{\text{num,Xm}}$ = CCI (2.111) over a trajectory of length $[-X/2, X/2]$ m; parameters: $(x, y, z) = (0\text{ m}, 5\text{ m}, 0\text{ m})$, $\Omega = 2\pi 400$ 1/s, $M = 0.1$, $Q = 1$ kg/s^2

$$\begin{pmatrix} x_0(t) \\ y_0(t) \\ z_0(t) \end{pmatrix} = \begin{pmatrix} ut \\ \Phi(t) \\ \psi(t) \end{pmatrix} = \begin{pmatrix} ut \\ at^2 \\ 0 \end{pmatrix} \text{ for } t \in (-1, 1)\text{s}, \tag{2.112}$$

where the acceleration is $a = u$ in m/s^2 for the blue parabola and $a = 3 \cdot u$ in m/s^2 for the green parabola with $u = 34.3$ m/s. For comparison, the black finite straight line is also considered as a trajectory. By introducing the curve (2.112) into the CCI (2.111), we obtain

$$\hat{p}_3 = \frac{Q}{4\pi u} \int_{x_0=ut_0}^{x_1=ut_1} \frac{e^{i\frac{\omega}{c}\sqrt{(x-\xi)^2+\left(y-a(\xi/u)^2\right)^2+z^2}}}{\sqrt{(x-\xi)^2+\left(y-a(\xi/u)^2\right)^2+z^2}} e^{i(\omega-\Omega)\frac{\xi}{u}} \mathrm{d}\xi. \tag{2.113}$$

The spectra of the pressure are shown in Fig. 2.19, where the colors of the spectra correspond to the colors of the trajectories in Fig. 2.18.

We see that the difference between the solution for the infinite straight line (red) and for the finite straight line (black) is relatively small. When curvature comes into play, the two Doppler frequencies move closer together (blue curve) and finally contract to the original emission frequency (green curve) since the parabola winds more around the receiving point.

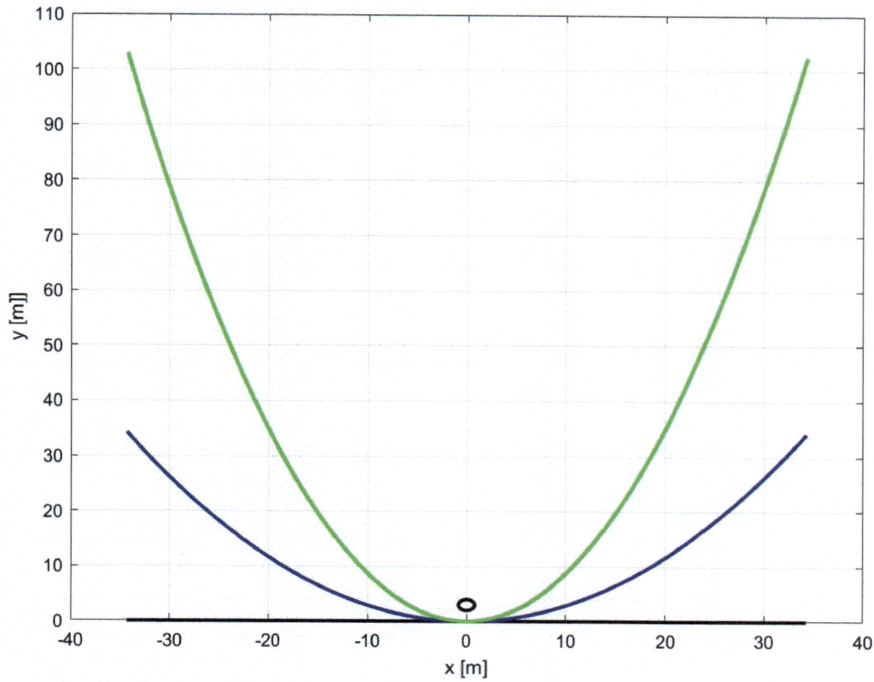

Fig. 2.18 Trajectories and position of receiver at $(x, y, z) = (0\text{ m}, 3\text{ m}, 0\text{ m})$, (little black circle) in the xy-plane

As a third interesting example, we will consider the following curve in Sect. 2.5.2

$$\begin{pmatrix} x_0(t) \\ y_0(t) \\ z_0(t) \end{pmatrix} = \begin{pmatrix} r_0 \cos(wt) \\ r_0 \sin(wt) \\ v_z t \end{pmatrix} \text{ for } t \in [t_0, t_1]. \tag{2.114}$$

The trajectory of the radiating source is a circular helix of radius r_0 and pitch $2\pi v_z/w$ winding around the z-axis with z-velocity v_z. Now the source coordinate $z_0(t)$—instead of $x_0(t)$—depends linearly on time, so that the CCI has the form

$$\hat{p}_3 = \frac{Q}{4\pi v_z} \int\limits_{z_0=v_z t_0}^{z_1=v_z t_1} \frac{e^{i\frac{\omega}{c}\sqrt{(x-\Phi(\zeta/v_z))^2+(y-\psi(\zeta/v_z))^2+(z-\zeta)^2}}}{\sqrt{(x-\Phi(\zeta/v_z))^2+(y-\psi(\zeta/v_z))^2+(z-\zeta)^2}} e^{i(\omega-\Omega)\frac{\zeta}{v_z}} \mathrm{d}\zeta, \tag{2.115}$$

where $\Phi(t) = r_0 \cos(wt)$ and $\psi(t) = r_0 \sin(wt)$.

Thus, we obtain the solution

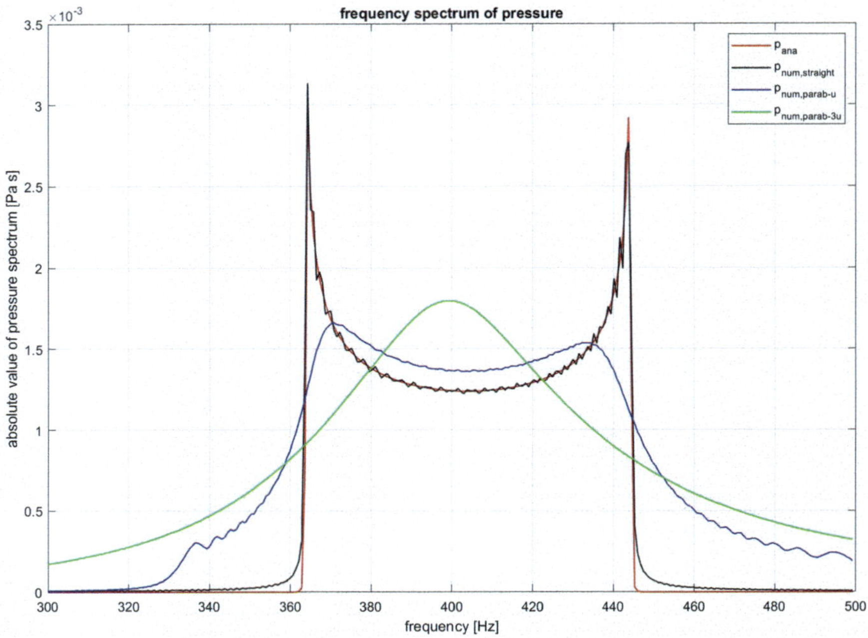

Fig. 2.19 Absolute value of the 3D spectral pressure: p_{ana} = analytical expression (2.19), $p_{\text{num,straight}}$ = CCI (2.111) for a trajectory of length $[-34.3, 34.3]$ m, $p_{\text{num,parab-u}}$ for a parabola with $a = 34.3$ m/s^2; $p_{\text{num,parab-3u}}$ for a parabola with $a = 3 \cdot 34.3$ m/s^2; parameters: receiver at $(x, y, z) = (0\text{ m}, 3\text{ m}, 0\text{ m})$, $\Omega = 2\pi 400$ 1/s, $M = 0.1$, $Q = 1$ kg/s^2

$$\hat{p}_3 = \frac{Q}{4\pi v_z} \int\limits_{z_0 = v_z t_0}^{z_1 = v_z t_1} \frac{e^{i\frac{\omega}{c}\sqrt{(x - r_0\cos(w\zeta/v_z))^2 + (y - r_0\sin(w\zeta/v_z))^2 + (z-\zeta)^2}}}{\sqrt{(x - r_0\cos(w\zeta/v_z))^2 + (y - r_0\sin(w\zeta/v_z))^2 + (z-\zeta)^2}} e^{i(\omega - \Omega)\frac{\zeta}{v_z}} \,\mathrm{d}\zeta\,. \tag{2.116}$$

In Sect. 2.5.2, we examine this case in detail, derive an infinite series in cylindrical coordinates as an analytical representation of the spectral pressure, and show that the CCI (2.116) perfectly agrees with the results of the series.

Obviously, the same procedure in *two dimensions* leads to the Cartesian convolution integral (cf. Eq. 2.104)

$$\hat{p}_2 = \frac{Qi}{4u} \int\limits_{x_0 = ut_0}^{x_1 = ut_1} H_0^{(1)}\left(\frac{\omega}{c}\sqrt{(x-\xi)^2 + (y - \Phi(\xi/u))^2}\right) e^{i(\omega-\Omega)\frac{\xi}{u}} \,\mathrm{d}\xi\,, \tag{2.117}$$

if the acoustic signal emitted by a moving line source is constrained to the finite time interval $[t_0, t_1]$ with $x_0 = ut_0$, $x_1 = ut_1$. The only difference compared to the 3D case is that we have to replace the 3D Green's function in Eq. (2.111) with the 2D Green's function given by the Hankel function.

2.3.3 Radiating Source Moving in a Viscous Fluid

In this section, we take a look at the influence of damping on the spectrum of a moving source. In Sect. 2.2.3.1, damping was introduced as an artificial factor ε in order to ensure the convergence of the infinite integrals that occur. But now, let us explore the damping mechanism based on the underlying physics by considering a slightly viscous fluid. Following Ffowcs Williams (Chap. 2, pp. 78 in Crighton et al. [15]), sound waves in such a medium are described by the operator

$$O = \frac{\partial^2}{\partial t^2} - 2\beta\nabla^2\frac{\partial}{\partial t} - c^2\nabla^2. \tag{2.118}$$

Here, the small constant β is proportional to the kinematic viscosity of the fluid, neglecting squares of β. Ffowcs Williams simply states the operator (2.118) without giving any derivation or reference. This gap can be filled by considering the modified wave equation derived by Morse and Ingard (see [2, p. 282, Eq. (6.4.22)]). If thermal conduction is neglected, this wave equation in its homogeneous form reads as follows

$$\nabla^2 p = \frac{1}{c^2}\left(\frac{\partial^2}{\partial t^2} - l'_\nu c\frac{\partial}{\partial t}\nabla^2\right)p \text{ or } \frac{\partial^2\varphi}{\partial t^2} - c^2\nabla^2\varphi - l'_\nu c\frac{\partial}{\partial t}\nabla^2\varphi = 0 \tag{2.119}$$

with $p = -\rho\frac{\partial\varphi}{\partial t}$, where the characteristic length of viscosity is expressed by (see Morse and Ingard [2, Eq. 6.4.20])

$$l'_\nu = \frac{\eta + \frac{4}{3}\mu}{\rho c}. \tag{2.120}$$

Here, η is the bulk or volume viscosity and μ the dynamic or shear viscosity. Comparing expressions (2.118) and (2.119), we find that

$$2\beta = l'_\nu c = \frac{\eta + \frac{4}{3}\mu}{\rho} \tag{2.121}$$

and is proportional to the kinematic viscosity μ/ρ, as claimed by Ffowcs Williams when η is neglected.

After this introduction, we now consider a radiating source moving in a viscous fluid. For simplicity, let the source move with constant positive velocity $u > 0$ along the x-axis, exactly as described in Sect. 2.2.1. The inhomogeneous wave equation corresponding to expressions (2.118) or (2.119)—now written for the acoustic potential—is then

$$\Delta\varphi - \frac{1}{c^2}\frac{\partial^2\varphi}{\partial t^2} + \frac{2\beta}{c^2}\Delta\frac{\partial\varphi}{\partial t} = \frac{Q_1}{\rho}e^{-i\Omega t}\delta(x - ut)\delta(y)\delta(z),\ t \in (-\infty, \infty). \tag{2.122}$$

The temporal Fourier transform of Eq. (2.122) gives for the pressure $\hat{p}(\omega) = i\omega\rho\hat{\varphi}$

$$\left(1 - 2i\omega\frac{\beta}{c^2}\right)\Delta\hat{p} + \left(\frac{\omega}{c}\right)^2\hat{p} = -\frac{Q}{u}e^{i(\omega-\Omega)\frac{x}{u}}\delta(y)\delta(z)$$

with $Q = -i\omega Q_1$ or

$$\Delta\hat{p} + \frac{1}{(1 - i\delta_v)}\left(\frac{\omega}{c}\right)^2\hat{p} = -\frac{Q}{u(1 - i\delta_v)}e^{i(\omega-\Omega)\frac{x}{u}}\delta(y)\delta(z), \tag{2.123}$$

where

$$\delta_v = 2\omega\frac{\beta}{c^2} \tag{2.124}$$

is the correction term for the viscosity.

By defining the modified wave number

$$k_{\text{mod}} = \frac{\omega/c}{\sqrt{1 - i\delta_v}} \tag{2.125}$$

and the modified amplitude

$$Q_{\text{mod}} = \frac{Q}{1 - i\delta_v}, \tag{2.126}$$

we now obtain the solution using the Green's function method as described in Sect. 2.2.1.1

$$\hat{p} = \frac{Q_{\text{mod}}}{4\pi u}\int_{-\infty}^{\infty}\frac{e^{ik_{\text{mod}}\sqrt{(x-\xi)^2+y^2+z^2}}}{\sqrt{(x-\xi)^2+y^2+z^2}}e^{i(\omega-\Omega)\frac{\xi}{u}}\,d\xi \tag{2.127}$$

or

$$\hat{p}(\omega) = \frac{iQ_{\text{mod}}}{4u}e^{i(\frac{\omega-\Omega}{u})x}H_0^{(1)}\left(r\sqrt{k_{\text{mod}}^2 - \left(\frac{\omega-\Omega}{u}\right)^2}\right). \tag{2.128}$$

The sign of the square root in Eq. (2.125) should be chosen in such a way that k_{mod} acquires a positive imaginary part.

The only difference between the undamped solution (2.19) and the damped solution (2.128) is that the wave number and the amplitude are replaced by their modified counterparts. This is the well-known effect that frequencies and amplitudes are slighted changed by the viscosity of the fluid.

The strength of the viscosity depends on the medium being considered. In air, for example, the situation is more difficult since attenuation is not only caused by

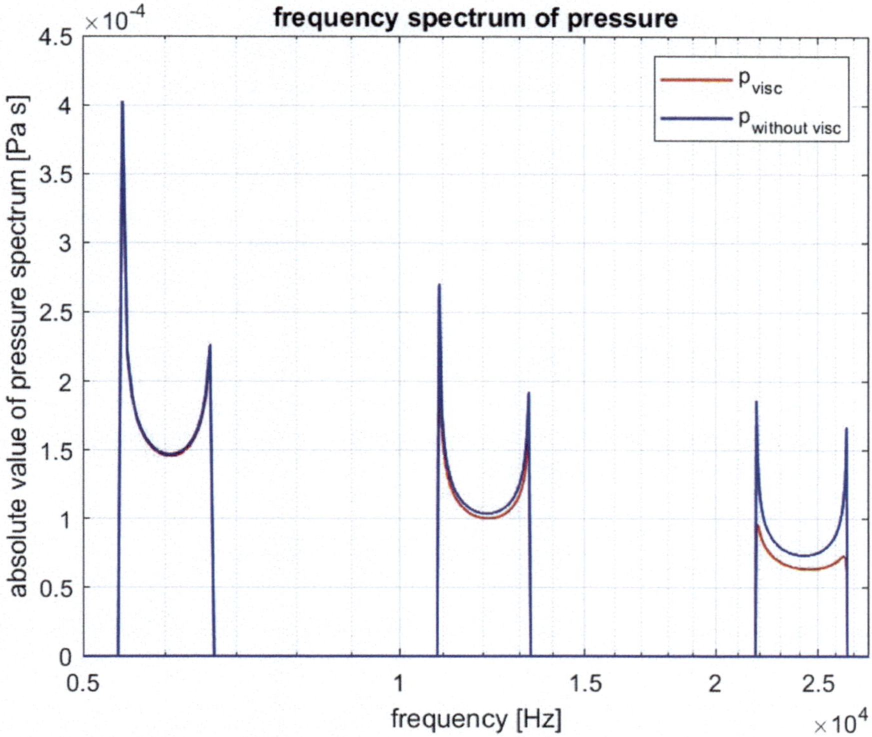

Fig. 2.20 Absolute value of the 3D spectral pressure for the three radiating frequencies $\Omega/(2\pi) =$ 6, 12, 24 kHz; $p_{\text{without\_visc}}$ = Eq. (2.19) without viscosity, p_{visc} = Eq. (2.128) with viscosity; parameters: $(x, y, z) = (10\text{ m}, 10\text{ m}, 10\text{ m})$, $M = 0.1$, $Q = 1\text{ kg/s}^2$

classical absorption, but also by "*the molecular absorption caused by rotational relaxation [...] and by vibrational relaxation of oxygen and nitrogen, respectively*" (quoted from ISO 9613-1 [16, p. 15]). The bulk viscosity and the dynamic viscosity in air at 20° C are both approximately (see Cremer [17, p. 218] and Shang et al. [18, Fig. 9])

$$\eta \approx \mu \approx 1.8 \cdot 10^{-5}\text{ kg/(ms)}. \tag{2.129}$$

The correction term for the viscosity is thus according to Eq. (2.124) $\delta_v = \frac{2\beta}{c^2} 2\pi f \approx 4 \cdot 10^{-8}$ for $f = 20$ Hz and $\delta_v = \frac{2\beta}{c^2} 2\pi f \approx 4 \cdot 10^{-5}$ for $f = 20$ kHz. The „classical absorption“ in air is, therefore, relatively low in almost the entire audio frequency range and only becomes more important for ultrasound. This is shown in Fig. 2.20 for the three frequencies 6, 12 and 24 kHz when the trajectory of the source (i.e., the x-axis) is not far from the receiver, which is located at $(x, y, z) = (10\text{ m}, 10\text{ m}, 10\text{ m})$.

However, the absorption effect becomes stronger with increasing distance between the source and the receiver. This is particularly the case when the noise from overflights is measured on the ground. In Fig. 2.21, the receiver is located at 100 m altitude. All other parameters remain as in Fig. 2.20. In Fig. 2.22, we see the influence on the damping for an overflight altitude of 1000 m at 4, 6, and 8 kHz.

These damping effects are not sufficiently large enough to regularize the integral (2.24) for an infinitely long trajectory in the audio range when using the integral (2.127) instead. On the one hand, the damping effect increases with increasing distance, but on the other hand, the integration interval must be enlarged to achieve sufficient accuracy. In addition, the oscillation of the exponential term $\exp(i(\omega - \Omega)\xi/u)$ is not prevented by the absorption.

However, it should be noted once again that infinite trajectories are of primarily theoretical interest, as only a finite measurement duration is possible in practice. Therefore, the CCI must be extended only over a finite interval, for which a regularization is not necessary in most cases (cf. Sect. 2.3.2).

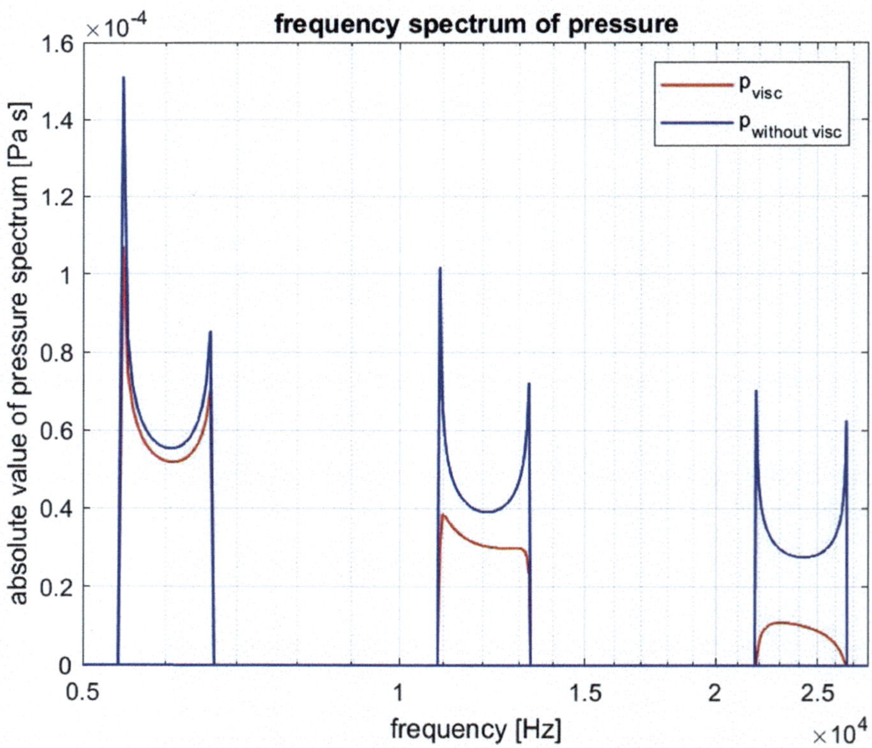

Fig. 2.21 Absolute value of the 3D spectral pressure for the three radiating frequencies $\Omega/(2\pi) =$ 6, 12, 24 kHz; $p_{\text{without\_visc}}$ = Eq. (2.19) without viscosity, p_{visc} = Eq. (2.128) with viscosity; parameters: $(x, y, z) = (0\text{ m}, 0\text{ m}, 100\text{ m})$, $M = 0.1$, $Q = 1\text{ kg/s}^2$

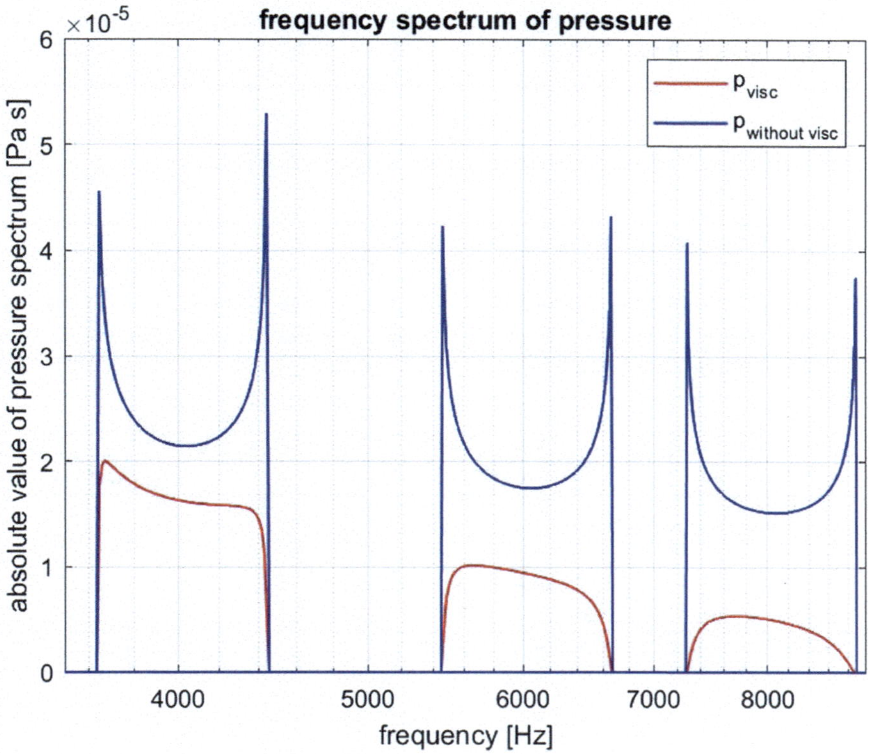

Fig. 2.22 Absolute value of the 3D spectral pressure for the three radiating frequencies $\Omega/(2\pi) = 4, 6, 8$ kHz; $p_{\text{without\_visc}}$ = Eq. (2.19) without viscosity, p_{visc} = Eq. (2.128) with viscosity; parameters: $(x, y, z) = (0$ m, 0 m, 1000 m$)$, $M = 0.1$, $Q = 1$ kg/s^2

It is a complex and broad area to explore additional, potentially more significant attenuation effects, such as molecular absorption, and incorporate them into the wave equation. This would lead us too far from the main focus of the book and hence will not be further addressed.

2.4 Radiating Monopole Moving in Polar Coordinates (2D Case)

In the previous sections, we analyzed the frequency spectrum of moving sources expressed in Cartesian coordinates. The methodology outlined therein can be effectively extended to other separable coordinate systems. We now shift our focus to polar coordinates and will investigate the motion of the source along a circle and an Archimedean spiral. Thus, in this present section, we explore the 2D scenario, i.e., a moving line source. Subsequently, in Sect. 2.5, we will study a moving point source in cylindrical 3D coordinates.

2.4.1 A Line Source Moving Along a Circular Path

2.4.1.1 Derivation of the Spectral Pressure

We consider a radiating line source that moves on a circle of radius r_0 with angular velocity w around the origin. The sound field of the acoustic velocity potential φ_2 is most appropriately described by the 2D wave equation

$$\Delta_2\varphi_2 - \frac{1}{c^2}\frac{\partial^2\varphi_2}{\partial t^2} = \frac{Q_1}{\rho}e^{-i\Omega t}\frac{1}{r}\delta(r-r_0)\delta_c(\phi - wt) \tag{2.130}$$

in the polar coordinates r, ϕ with

$$x = r\cos\phi,\ y = r\sin\phi. \tag{2.131}$$

The polar angle is 2π periodic and thus $\delta_c(\phi - wt)$ is the Dirac comb (see [19]) defined by

$$\delta_c(\phi - wt) = \sum_{n=-\infty}^{\infty}\delta(\phi - wt - 2\pi n) = \frac{1}{2\pi}\sum_{n=-\infty}^{\infty}e^{in(\phi-wt)}. \tag{2.132}$$

It is very important to emphasize explicitly that the periodicity of the polar angle must be taken into account. This is often ignored in the literature, and on the right-hand side of Eq. (2.130), simply the Dirac delta function $\delta(\phi - wt)$ is written. If one overlooks this and works with the conventional Dirac delta function instead of the Dirac comb, erroneous results will be obtained.

The Fourier transform with respect to time for the pressure $\hat{p}_2(\omega) = i\omega\rho\hat{\varphi}_2$ leads to

$$\begin{aligned}\Delta_2\hat{p}_2 + \left(\frac{\omega}{c}\right)^2\hat{p}_2 &= -Q\frac{\delta(r-r_0)}{2\pi r}\int_{-\infty}^{\infty}e^{i(\omega-\Omega)t}\sum_{n=-\infty}^{\infty}e^{in(\phi-wt)}\mathrm{d}t\\ &= -Q\frac{\delta(r-r_0)}{2\pi r}\int_{-\infty}^{\infty}\sum_{n=-\infty}^{\infty}e^{i(\omega-\Omega-nw)t}e^{in\phi}\mathrm{d}t.\end{aligned} \tag{2.133}$$

with $Q = -i\omega Q_1$.

Using the relationship

$$\int_{-\infty}^{\infty} e^{i(\omega-\Omega-nw)t}\mathrm{d}t = 2\pi\,\delta(\omega-\Omega-nw) \tag{2.134}$$

and interchanging the order of integration and summation, we get

$$\Delta_2\hat{p}_2 + \left(\frac{\omega}{c}\right)^2\hat{p}_2 = -Q\frac{\delta(r-r_0)}{r}\sum_{n=-\infty}^{\infty}\delta(\omega-\Omega-nw)e^{in\phi}. \tag{2.135}$$

Once again, we employ the Green's function method (GFM) to solve this inhomogeneous Helmholtz equation. The Green's function in polar coordinates is a solution of

$$\Delta_2 g_2 + \left(\frac{\omega}{c}\right)^2 g_2 = -\frac{\delta(r-r_0)}{r}\delta_c(\phi-\phi_0) \tag{2.136}$$

and is well-known (see, e.g., Duffy [4, p. 279, Eq. (5.1.30)])

$$g_2(r,\phi;r_0,\phi_0) = \frac{i}{4}\sum_{n=-\infty}^{\infty} H_n^{(1)}(kr_>)J_n(kr_<)\cos[n(\phi-\phi_0)]. \tag{2.137}$$

Here, J_n represents the Bessel function of order n, $H_n^{(1)}$ is the Hankel function of first kind and order n, $k = \frac{\omega}{c}$, $r_< = \min\{r, r_0\}$ and $r_> = \max\{r, r_0\}$.

For the further analysis, it is more convenient to work with an expression for $g_2(r,\phi;r_0,\phi_0)$ that involves exponentials instead of trigonometric functions, i.e., with

$$g_2(r,\phi;r_0,\phi_0) = \frac{i}{4}\sum_{n=-\infty}^{\infty} H_n^{(1)}(kr_>)J_n(kr_<)e^{in(\phi-\phi_0)}. \tag{2.138}$$

The equivalence of Eqs. (2.137) and (2.138) is shown in Appendix C.

The GFM requires the convolution of the Green's function (2.138) with the negative right-hand side of the inhomogeneous Helmholtz Eq. (2.135)

$$R(r,\phi) = Q\frac{\delta(r-r_0)}{r}\sum_{n=-\infty}^{\infty}\delta(\omega-\Omega-nw)e^{in\phi}. \tag{2.139}$$

Explicitly written, we need to evaluate the double integral

$$\hat{p}_2 = \int_0^{\infty}\int_0^{2\pi} g_2\big(r,\phi;\hat{r},\hat{\phi}\big)R\big(\hat{r},\hat{\phi}\big)\hat{r}\mathrm{d}\hat{\phi}\mathrm{d}\hat{r}. \tag{2.140}$$

If we compare the right-hand sides of Eqs. (2.135) and (2.136), it becomes clear that only the integration over the polar angle must be performed, since the radial dependence in Eq. (2.135) is only given by the Dirac delta function. We thus obtain

$$\hat{p}_2 = \frac{iQ}{4} \sum_{n=-\infty}^{\infty} H_n^{(1)}(kr_>)J_n(kr_<) \int_0^{2\pi} e^{in(\phi-\hat{\phi})} \sum_{m=-\infty}^{\infty} \delta(\omega - \Omega - mw)e^{im\hat{\phi}} \mathrm{d}\hat{\phi}. \tag{2.141}$$

Assuming that we can interchange the order of integration and summation and taking into account that

$$\int_0^{2\pi} e^{i(m-n)\hat{\phi}} d\hat{\phi} = 2\pi\delta_{nm} \text{ with the Kronecker delta } \delta_{nm} = \begin{cases} 0, \text{ if } n \neq m \\ 1, \text{ if } n = m \end{cases}, \tag{2.142}$$

we obtain

$$\begin{aligned} \hat{p}_2 &= \frac{iQ}{4} \sum_{n=-\infty}^{\infty} H_n^{(1)}(kr_>)J_n(kr_<)e^{in\phi} \sum_{m=-\infty}^{\infty} \delta(\omega - \Omega - mw) \int_0^{2\pi} e^{i(m-n)\hat{\phi}} \mathrm{d}\hat{\phi} \\ &= \frac{i\pi Q}{2} \sum_{n=-\infty}^{\infty} H_n^{(1)}(kr_>)J_n(kr_<)e^{in\phi} \sum_{m=-\infty}^{\infty} \delta(\omega - \Omega - mw)\delta_{nm}. \end{aligned} \tag{2.143}$$

Therefore, the double infinite sum simplifies to a single sum

$$\hat{p}_2 = \frac{i\pi Q}{2} \sum_{n=-\infty}^{\infty} H_n^{(1)}(kr_>)J_n(kr_<)\delta(\omega - \Omega - nw)e^{in\phi}. \tag{2.144}$$

This beautiful formula is the frequency spectrum of a circularly moving line source, which emits a harmonic signal of frequency Ω.

2.4.1.2 Discussion of the Formula for the Spectral Pressure

Obviously, the spectral pressure $\hat{p}_2$ is a pure line spectrum with amplitudes $A_n(\phi)$

$$A_n(\varphi) = \frac{i\pi Q}{2} H_n^{(1)}(k_n r_>)J_n(k_n r_<)e^{in\phi} \tag{2.145}$$

depending on the polar angle ϕ at the discrete frequencies $\omega_n = \Omega + nw$ and $k_n = \omega_n/c$.

When the angular velocity w becomes zero, only the original frequency $\omega_n = \Omega$ is contributing.

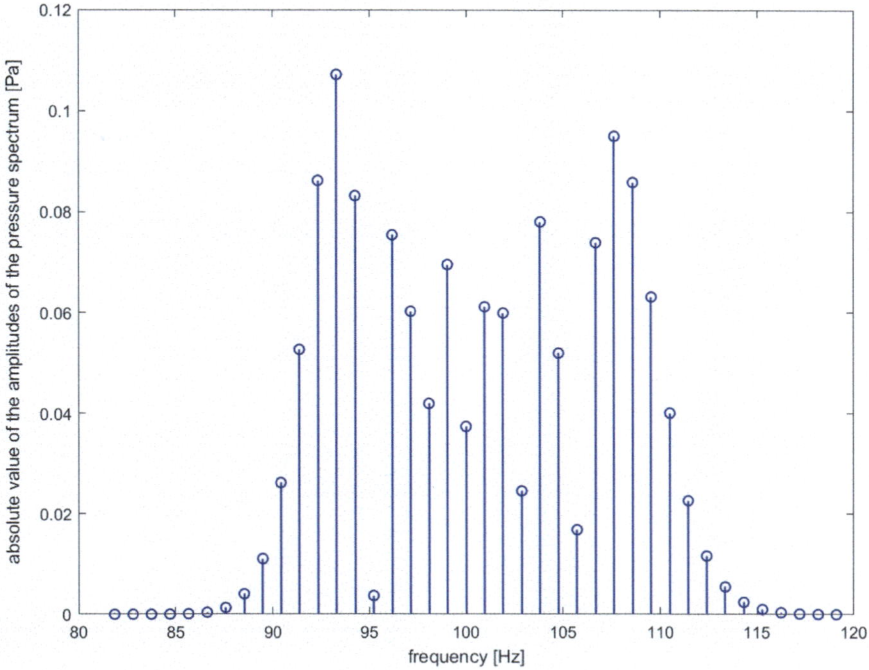

Fig. 2.23 Absolute value $|A_n(\phi)|$ of the spectral pressure; parameters: $r_0 = 5$ m, $r = 10$ m, $\Omega = 2\pi 100\ \text{s}^{-1}$, $w = 6\ \text{s}^{-1}$, $(u = w \cdot r_0 = 30\ \text{m/s})$, $Q = 1\ \text{kg/(ms}^2)$

In Fig. 2.23 we consider the absolute value

$$|A_n(\varphi)| = \frac{\pi Q}{2} \left| H_n^{(1)}\left(\frac{\omega_n}{c} r_>\right) J_n\left(\frac{\omega_n}{c} r_<\right) \right| \tag{2.146}$$

at frequencies $\omega = \omega_n = \Omega + nw$.

The figure illustrates that outside a specific frequency range centered around the radiation frequency of 100 Hz, the spectrum rapidly approaches zero—similar to rectilinear motion, where the spectrum is zero outside the Doppler range (cp. Fig. 2.3). This occurs because when $|n|$ is large enough, the absolute value of the order of the Bessel function becomes greater than the argument. Thus, the Bessel function becomes significantly small, as can be observed from the asymptotic expansion for large orders. First, let us consider the case $n > 0$. Due to Abramowitz and Stegun [7, p. 365, Eq. 9.3.1]

$$J_n(z) \sim \frac{1}{\sqrt{2\pi n}} \left(\frac{ez}{2n}\right)^n \underset{n\to\infty}{\longrightarrow} 0.$$

A small calculation shows that

$$n \gg z = \left(\frac{\omega_n}{c}\right) r_< = \left(\frac{\Omega + nw}{c}\right) r_< \text{ if } n \gg n_{\text{pos}} = \frac{\Omega r_<}{c - w r_<}. \tag{2.147}$$

For the parameters used in Fig. 2.23, we obtain $n_{\text{pos}} \sim 10$. This results in the rule of thumb that we only have to consider $n_{\max} \sim 2n_{\text{pos}}$ lines of the spectrum according to Fig. 2.23, where the lines of the spectrum above 115 Hz have very small numerical values.

For $n < 0$, we take into account (see Abramowitz and Stegun [7, Eq. 9.3.1 and Eq. 9.1.5]) that

$$J_{-n}(z) = (-1)^n J_n(z)$$

and therefore

$$J_n(z) = J_{-|n|}(z) \sim (-1)^n \frac{1}{\sqrt{2\pi |n|}} \left(\frac{ez}{2|n|}\right)^{|n|} \underset{|n|\to\infty}{\longrightarrow} 0.$$

Now we get

$$|n| = -n \gg z = \left(\frac{\Omega + nw}{c}\right) r_< \text{ if } n \ll n_{\text{neg}} = -\frac{\Omega r_<}{c + w r_<}. \tag{2.148}$$

For the parameters used in Fig. 2.23, we obtain $n_{\text{neg}} \sim -8$. This leads us to the rule of thumb that we only have to consider the frequency interval with $n \in \left[-2n_{\text{neg}}, 2n_{\text{pos}}\right]$. A similar observation can be found in the paper of Poletti [20, p. 3, Eqs. (16) and (17)], but without a detailed derivation.

The pressure spectrum (2.144) observed at the origin $r = 0$ is given by

$$\hat{p}_2(r = 0, \phi) = \frac{i\pi Q}{2} \sum_{n=-\infty}^{\infty} H_n^{(1)}(kr_0) J_n(0) \delta(\omega - \Omega - nw) e^{in\phi}. \tag{2.149}$$

Since $J_n(0) = \delta_{n0}$, we immediately get (note that $Q = -i\omega Q_1$)

$$\begin{aligned} \hat{p}_2(r = 0, \phi) &= 2\pi\delta(\omega - \Omega) \frac{iQ}{4} H_0^{(1)}\left(\frac{\omega}{c} r_0\right) \\ &= 2\pi\delta(\omega - \Omega) \frac{i(-i\omega Q_1)}{4} H_0^{(1)}\left(\frac{\omega}{c} r_0\right). \end{aligned} \tag{2.150}$$

This expression describes the sound field of the corresponding *stationary* line source located at distance r_0 from the origin, which is also evident from the inverse Fourier transform of Eq. (2.150), yielding the time-depending pressure field of a stationary harmonic line source

$$p_2(r = 0, t) = \frac{iQ_1}{4} \frac{\partial}{\partial t} \left\{ H_0^{(1)}\left(\frac{\Omega}{c} r_0\right) e^{-i\Omega t} \right\} = \frac{\Omega Q_1}{4} H_0^{(1)}\left(\frac{\Omega}{c} r_0\right) e^{-i\Omega t}. \tag{2.151}$$

An observer at the origin, therefore, perceives no frequency shift from the original emission frequency Ω when the source orbits the observer on a circle with the radius r_0 and with angular speed w. By interchanging the radius of emission r_0 and the radius of observation r and setting r_0 equal to zero, we obtain the same formula

$$p_2(r,t)|_{r_0=0} = \frac{\Omega Q_1}{4} H_0^{(1)}\left(\frac{\Omega}{c} r\right) e^{-i\Omega t} \tag{2.152}$$

due to the reciprocity of the Green's function. The result is immediately obvious since the source is now at the origin and the circular motion is lost.

Let us now examine the pressure in the time domain in more detail.

2.4.1.3 The Pressure in the Time Domain and Comparison with Results from the Literature

The inverse Fourier transform of the spectral pressure (2.144) is easy to perform due to the appearance of the Dirac delta function and yields the time-dependent sound pressure field (note that $Q = -i\omega Q_1$)

$$\begin{aligned} p_2(t) &= \frac{1}{2\pi} \int_{-\infty}^{\infty} \frac{\pi \omega Q_1}{2} \sum_{n=-\infty}^{\infty} H_n^{(1)}\left(\frac{\omega}{c} r_>\right) J_n\left(\frac{\omega}{c} r_<\right) \delta(\omega - \Omega - nw) e^{in\phi} e^{-i\omega t} \mathrm{d}\omega \\ &= \frac{Q_1}{4} \sum_{n=-\infty}^{\infty} H_n^{(1)}\left(\frac{1}{c}(\Omega + nw) r_>\right) J_n\left(\frac{1}{c}(\Omega + nw) r_<\right) (\Omega + nw) e^{in\phi} e^{-i(\Omega+nw)t}. \end{aligned} \tag{2.153}$$

For the acoustic potential φ_2 with $p_2 = -\rho \frac{\partial \varphi_2}{\partial t}$ we obtain the formula

$$\varphi_2(t) = \frac{-iQ_1}{4\rho} \sum_{n=-\infty}^{\infty} H_n^{(1)}\left(\frac{1}{c}(\Omega + nw) r_>\right) J_n\left(\frac{1}{c}(\Omega + nw) r_<\right) e^{in\phi} e^{-i(\Omega+nw)t}. \tag{2.154}$$

This solution can also be written as a Fourier series multiplied by the frequency factor $e^{-i\Omega t}$

$$\varphi_2(t) = e^{-i\Omega t} \sum_{n=-\infty}^{\infty} c_n e^{-inwt} \tag{2.155}$$

with coefficients

$$c_n = \frac{-iQ_1}{4\rho} H_n^{(1)}\left(\frac{1}{c}(\Omega + nw) r_>\right) J_n\left(\frac{1}{c}(\Omega + nw) r_<\right) e^{in\phi}, \tag{2.156}$$

which leads us to a simplified derivation of the solution which we will demonstrate here:

By employing the approach (2.155) with unknown coefficients c_n and substituting it into the wave wave equation (2.130), we obtain the equation

$$e^{-i\Omega t}\sum_{n=-\infty}^{\infty}\left\{\Delta_2 c_n(r,\phi)+\left(\frac{\Omega+nw}{c}\right)^2 c_n(r,\phi)\right\}e^{-inwt}$$
$$=\frac{Q_1}{\rho}e^{-i\Omega t}\frac{1}{2\pi r}\delta(r-r_0)\sum_{n=-\infty}^{\infty}e^{in(\phi-wt)}, \qquad (2.157)$$

where we have used the Fourier series expansion of the Dirac comb (2.132). By equating the coefficients, we get

$$\Delta_2 c_n(r,\phi)+\left(\frac{\Omega+nw}{c}\right)^2 c_n(r,\phi)=\frac{Q_1}{\rho}\frac{1}{2\pi r}\delta(r-r_0)e^{in\phi} \text{ for } n\in\mathbb{Z}. \qquad (2.158)$$

The solution of (2.158) is obtained by convoluting its negative right-hand side with the Green's function (2.138) with $k\to(\Omega+nw)/c$. This leads to

$$c_n=\frac{-Q_1}{2\pi\rho}\sum_{m=-\infty}^{\infty}\frac{i}{4}H_n^{(1)}\left(\frac{1}{c}(\Omega+nw)r_>\right)J_n\left(\frac{1}{c}(\Omega+nw)r_<\right)\int_0^{2\pi}e^{im(\phi-\hat{\phi})}e^{in\hat{\phi}}\mathrm{d}\hat{\phi}$$
$$=\frac{-iQ_1}{4\rho}H_n^{(1)}\left(\frac{1}{c}(\Omega+nw)r_>\right)J_n\left(\frac{1}{c}(\Omega+nw)r_<\right)e^{in\phi} \qquad (2.159)$$

in accordance with Eq. (2.156).

With the relatively simple formula (2.154), the sound field of such a rotating line source can be easily auralized, e.g., with the Matlab function "*sound*". Figure 2.24 shows the real part of the time signal for the same parameters as in Fig. 2.23 and for a time interval of 10 s.

Poletti used a "simple source approach" to derive "series expansions of rotating two- and three-dimensional sound fields" [21]. In [20], he studied the 2D case in detail, and his formulas Eqs. (14) and (15) are nearly identical to our formula (2.153). In our notation, they read

$$p_{\text{polleti}}(t)=\sum_{n=-\infty}^{\infty}H_n^{(\delta_n)}\left(\left|\frac{1}{c}(\Omega+nw)\right|r_>\right)J_n\left(\left|\frac{1}{c}(\Omega+nw)\right|r_<\right)e^{in\phi}e^{-i(\Omega+nw)t} \qquad (2.160)$$

with $\delta_n=\begin{cases}1 \text{ if } k_n=\frac{1}{c}(\Omega+nw)>0\\ 2 \text{ if } k_n=\frac{1}{c}(\Omega+nw)<0\end{cases}.$

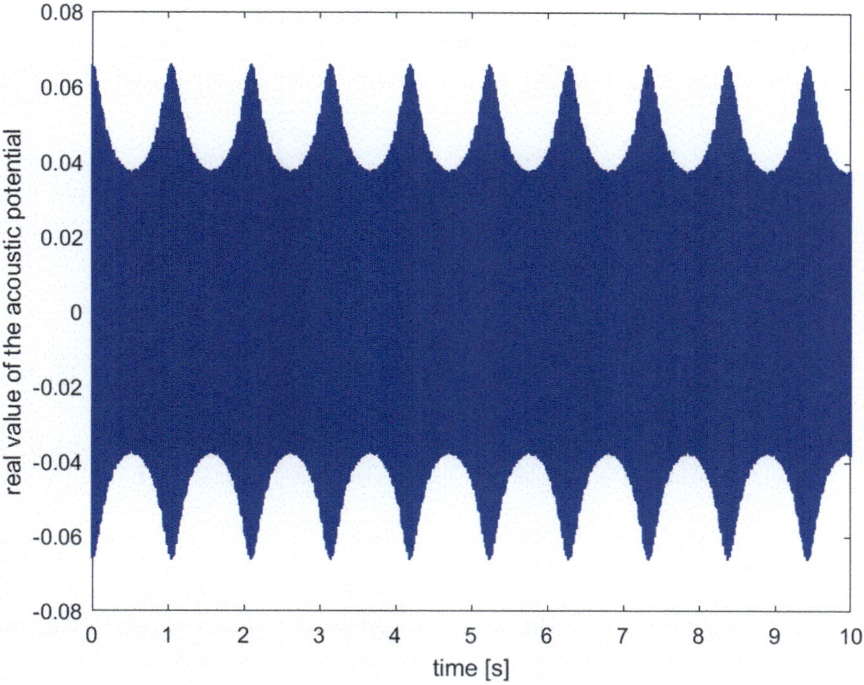

Fig. 2.24 Real part of the acoustic potential (2.154) in the time domain; parameters: $r_0 = 5$ m, $r = 10$ m, $\Omega = 2\pi 100\ \text{s}^{-1}$, $w = 6\ \text{s}^{-1}$, $(u = w \cdot r_0 = 30$ m/s), $\phi = 0$

Thus, if $\delta_n > 1$, the sound field $p_{\text{polleti}}(t)$ is identical to our expression for the acoustic potential (2.154) except for the differently defined amplitudes. However, what happens for $\delta_n < 1$, i.e., for negative wave numbers k_n? To study this case, we use the relations (see Abramowitz and Stegun [7, Eqs. 9.1.35 and 9.1.39])

$$J_n(-z) = (-1)^n J_n(z) \text{ and } H_n^{(1)}(-z) = -(-1)^n H_n^{(2)}(z),\, n \in \mathbb{Z}$$

and thus

$$\begin{aligned} J_n(z)H_n^{(1)}(z) &= \begin{cases} J_n(z)H_n^{(1)}(z) \text{ if } z > 0 \\ -J_n(|z|)H_n^{(2)}(|z|)\; z < 0 \end{cases} \\ &= (-1)^{(\tilde{\delta}+1)} J_n(|z|)H_n^{(\tilde{\delta})}(|z|) \end{aligned}$$

with $\tilde{\delta} = \begin{cases} 1 \text{ if } z > 0 \\ 2 \text{ if } z < 0 \end{cases}$.

It follows that our solution (2.154) can be also written as

$$\varphi_2(t) = \frac{-iQ_1}{4\rho} \sum_{n=-\infty}^{\infty} (-1)^{\delta_n+1} H_n^{(\delta_n)}(|k_n|r_>) J_n(|k_n|r_<) e^{in\phi} e^{-i(\Omega+nw)t}. \tag{2.161}$$

Now, the question arises: why is the factor $(-1)^{\delta_n+1}$ missing in Poletti's Eq. (2.160)? The resolution, as it seems to us, is as follows: Poletti's line source (Eq. (4) of [20])—for "negative frequencies", as formulated by Poletti—solves the wave equation

$$\Delta_2 p_{\text{line}} - \frac{1}{c^2}\frac{\partial^2 p_{\text{line}}}{\partial t^2} = \frac{4}{i}\big[-e^{-i\Omega t}\delta(x - x_s)\delta(y - y_s)\big]. \tag{2.162}$$

However, the expression of the line source for "positive frequencies" (Eq. (5) of Poletti [20]) solves the wave equation (i.e., the complex conjugate of Eq. (2.162))

$$\Delta_2 p_{\text{line}} - \frac{1}{c^2}\frac{\partial^2 p_{\text{line}}}{\partial t^2} = -\frac{4}{i}\big[-e^{+i\Omega t}\delta(x - x_s)\delta(y - y_s)\big]. \tag{2.163}$$

Since one has to start from the same wave equation, a minus sign is missing in Poletti's Eq. (5), and must be inserted into his formula for negative frequencies as shown in Eq. (2.161) by adding $(-1)^{\delta_n+1}$.

Fortunately, the missing factor $(-1)^{\delta_n+1}$ does not affect the results of Polettis's paper [20], as he also derived the same limits (2.147) and (2.148) for the expansion. From this, he concluded that "*negative frequencies do not contribute significantly to the sound field in the subsonic case*" (quoted from [20, p. 3, below Eq. (16)]). Poletti also considers the case of a supersonic source, which is not addressed in this book.

2.4.2 A Line Source Moving Along an Archimedean Spiral

2.4.2.1 Derivation of the Spectral Pressure

Our second example in polar coordinates is a radiating line source that moves along an Archimedean spiral, represented by a radius $r_0(t) = r_s + v_r t$. This radius increases over time with the velocity v_r. Again, we assume that the polar angle is $\phi_0 = wt$ with angular velocity w. Thus, we obtain the standard equation of the Archimedean spiral in polar coordinates (see [22])

$$r_0(\phi_0) = r_s + b\phi_0 \text{ with } b = v_r/w. \tag{2.164}$$

In Cartesian coordinates, the spiral is written as

$$\vec{x}_0(t) = \begin{pmatrix} x_0(t) \\ y_0(t) \end{pmatrix} = \begin{pmatrix} (r_s + v_r t)\cos wt \\ (r_s + v_r t)\sin wt \end{pmatrix}, \; t \geq 0. \tag{2.165}$$

It is important to note that the domain of the spiral is $t \geq 0$ and not the whole range $-\infty < t < \infty$. Therefore, we are forced to introduce a characteristic function on the right-hand side of the wave equation, as shown below.

The absolute value of the velocity of the source can be easily calculated as:

$$v(t) = \left| \frac{\mathrm{d}}{\mathrm{d}t}\vec{x}_0(t) \right| = \sqrt{v_r^2 + w^2(r_s + v_r t)^2}. \tag{2.166}$$

This indicates that the velocity of the source becomes greater with time, resulting in a continuous increase in the Mach number $M = v/c$. For sufficiently large values of t, the source becomes supersonic with $M > 1$. This is a first example of how the spectrum can include both subsonic and supersonic components.

Figure 2.25 illustrates the trajectory of the source starting at $r_s = 10$ m and at time $t = 0$ s.

Similar to the circular motion, the sound field φ_2 is described by the 2D wave equation

$$\Delta_2\varphi_2 - \frac{1}{c^2}\frac{\partial^2\varphi_2}{\partial t^2} = \frac{Q_1}{\rho}e^{-i\Omega t}\frac{1}{r}\delta(r - (r_s + v_r t))\delta_c(\phi - wt)\chi_{[0,t_{\max}]}(t) \tag{2.167}$$

in polar coordinates, and $\delta_c(\phi - wt)$ is again the Dirac comb. Since both arguments of the Dirac delta functions are linear functions of time, the zeros of these functions can be determined explicitly (cf. Eq. (2.8)). As described in Sect. 2.3.2, we need to multiply the right-hand side of the wave equation by the characteristic function $\chi_{[0,t_{\max}]}(t)$ in order to start with $t = 0$. This approach also allows us to consider the signal only up to the maximum observation time $t_{\max}$ and to exclude, if necessary, an excessive increase in speed over time.

As usual, we start with the temporal Fourier transform of the wave equation, yielding for the spectral pressure $\hat{p}_2(\omega) = i\omega\rho\hat{\varphi}_2$

$$\Delta_2\hat{p}_2 + \left(\frac{\omega}{c}\right)^2\hat{p}_2 = -Q\int_{-\infty}^{\infty} e^{i(\omega-\Omega)t}\frac{\delta(r - (r_s + v_r t))}{r}\frac{1}{2\pi}\sum_{n=-\infty}^{\infty} e^{in(\phi - wt)}\chi_{[0,t_{\max}]}(t)\mathrm{d}t. \tag{2.168}$$

with $Q = -i\omega Q_1$.

The substitution $\xi = v_r t$, $\xi_{\max} = v_r t_{\max}$ leads to

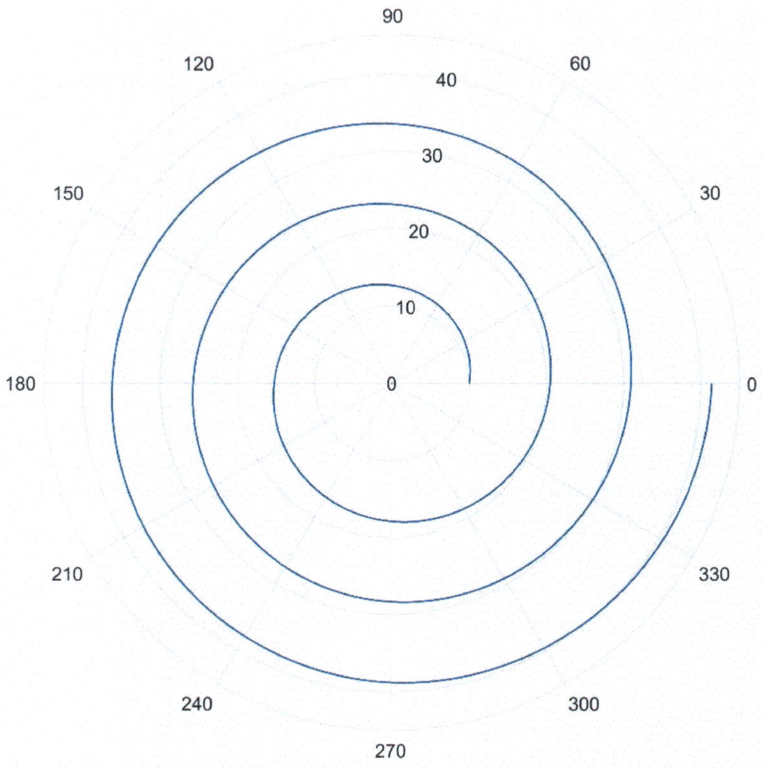

Fig. 2.25 The Archimedean spiral with parameters: $r_s = 10$ m, $v_r = 5$ m/s, $w = 3\ \mathrm{s}^{-1}, 0 \leq \varphi_0 \leq 3 \cdot 2\pi$

$$\Delta_2 \hat{p}_2 + \left(\frac{\omega}{c}\right)^2 \hat{p}_2 = -\frac{Q}{2\pi v_r r} \int_{-\infty}^{\infty} e^{i(\omega-\Omega)\frac{\xi}{v_r}} \delta(r - r_s - \xi) \sum_{n=-\infty}^{\infty} e^{in\left(\phi - \frac{w}{v_r}\xi\right)} \chi_{[0,\xi_{\max}]}(\xi) \mathrm{d}\xi. \tag{2.169}$$

Here, we have used the definition of the characteristic function by means of the Heaviside unit step function

$$\begin{aligned}\chi_{[0,t_{\max}]}(t) &= H(t) - H(t - t_{\max}) = H\left(\frac{\xi}{v_r}\right) - H\left(\frac{\xi - \xi_{\max}}{v_r}\right) \\ &= H(\xi) - H(\xi - \xi_{\max}) = \chi_{[0,\xi_{\max}]}(\xi).\end{aligned}$$

Due to the occurrence of the Dirac delta function, the integral can be solved immediately

$$\Delta_2\hat{p}_2 + \left(\frac{\omega}{c}\right)^2 \hat{p}_2 = -\frac{Q}{2\pi v_r r} e^{i(\omega-\Omega)\frac{r-r_s}{v_r}} \sum_{n=-\infty}^{\infty} e^{in\left(\phi - \frac{w}{v_r}(r-r_s)\right)} \chi_{[r_s, r_{\max}]}(r), \quad (2.170)$$

since $\chi_{[0,\xi_{\max}]}(\xi)\big|_{\xi=r-r_s} = H(r-r_s) - H(r-r_s-v_r t_{\max}) = \chi_{[r_s,r_{\max}]}(r)$ with $r_{\max} = r_s + t_{\max} v_r$.

Thus, we obtain the inhomogeneous Helmholtz equation

$$\Delta_2\hat{p}_2 + \left(\frac{\omega}{c}\right)^2 \hat{p}_2 = -f_{\text{RHS}}(r,\phi). \quad (2.171)$$

with the negative right-hand side

$$f_{\text{RHS}}(r,\phi) = \frac{A}{r}\chi_{[r_s,r_{\max}]}(r) \sum_{n=-\infty}^{\infty} e^{i(\omega-\Omega-nw)\frac{r-r_s}{v_r}} e^{in\phi}, A = \frac{Q}{2\pi v_r}. \quad (2.172)$$

Following the same procedure as in the preceding Sect. 2.4.1, we have to convolute the Green's function (2.138) with $f_{\text{RHS}}(r,\phi)$. At first sight, this yields a complicated expression:

$$\begin{aligned}
\hat{p}_2 &= \int_0^\infty \int_0^{2\pi} g_2(r,\phi;\hat{r},\hat{\phi}) f_{\text{RHS}}(\hat{r},\hat{\phi})\, \hat{r}\, \mathrm{d}\hat{\phi}\, \mathrm{d}\hat{r} \\
&= A \int_0^\infty \chi_{[r_s,r_{\max}]}(\hat{r}) \int_0^{2\pi} \sum_{n=-\infty}^{\infty} e^{i(\omega-\Omega-nw)\frac{\hat{r}-r_s}{v_r}} e^{in\hat{\phi}} g_2(r,\phi;\hat{r},\hat{\phi})\, \mathrm{d}\hat{\phi}\, \mathrm{d}\hat{r} \\
&= A \int_0^\infty \chi_{[r_s,r_{\max}]}(\hat{r}) \int_0^{2\pi} \left[\sum_{n=-\infty}^{\infty} e^{i(\omega-\Omega-nw)\frac{\hat{r}-r_s}{v_r}} e^{in\hat{\phi}} \right] \\
&\quad \times \left\{ \frac{i}{4} \sum_{m=-\infty}^{\infty} H_m^{(1)}(k\hat{r}_>) J_m(k\hat{r}_<) e^{im(\phi-\hat{\phi})} \right\} \mathrm{d}\hat{\phi}\, \mathrm{d}\hat{r} \\
&= \frac{i}{4} A \sum_{n=-\infty}^{\infty} \sum_{m=-\infty}^{\infty} \int_0^\infty \chi_{[r_s,r_{\max}]}(\hat{r}) e^{i(\omega-\Omega-nw)\frac{\hat{r}-r_s}{v_r}} H_m^{(1)}(k\hat{r}_>) J_m(k\hat{r}_<) e^{im\phi} \int_0^{2\pi} e^{i(n-m)\hat{\phi}}\, \mathrm{d}\hat{\phi}\, \mathrm{d}\hat{r}
\end{aligned} \quad (2.173)$$

with $\hat{r}_< = \min\{r,\hat{r}\}$ and $\hat{r}_> = \max\{r,\hat{r}\}$.

Using the relation (2.174)

$$\int_0^{2\pi} e^{i(m-n)\hat{\phi}} \mathrm{d}\hat{\phi} = 2\pi\delta_{nm} \text{ with the Kronecker delta } \delta_{nm} = \begin{cases} 0, & \text{if } n \neq m \\ 1, & \text{if } n = m \end{cases}, \quad (2.174)$$

expression (2.173) simplifies significantly as it allows us to eliminate one integral and one infinite sum

$$\hat{p}_2 = \frac{\pi i}{2} A \sum_{n=-\infty}^{\infty} e^{in\phi} \int_0^{\infty} \chi_{[r_s, r_{\max}]}(\hat{r}) e^{i(\omega - \Omega - nw)\frac{\hat{r}-r_s}{v_r}} H_n^{(1)}(k\hat{r}_>) J_n(k\hat{r}_<) \mathrm{d}\hat{r}. \tag{2.175}$$

Thus, we get the nice result that the pressure spectrum of the Archimedean spiral is given by

$$\hat{p}_2 = \frac{i}{4} \frac{Q}{v_r} \sum_{n=-\infty}^{\infty} e^{in\phi} \int_{r_s}^{r_{\max}} e^{i(\omega - \Omega - nw)\frac{\hat{r}-r_s}{v_r}} H_n^{(1)}(k\hat{r}_>) J_n(k\hat{r}_<) \mathrm{d}\hat{r}, \; r_{\max} = t_{\max} v_r + r_s. \tag{2.176}$$

Considering the definitions of $\hat{r}_<$ and $\hat{r}_>$, we obtain explicitly

$$\hat{p}_2 = \frac{i}{4} \frac{Q}{v_r} \sum_{n=-\infty}^{\infty} e^{in\phi} J_n\left(\frac{\omega}{c} r\right) \int_{r_s}^{r_{\max}} e^{i(\omega - \Omega - nw)\frac{\hat{r}-r_s}{v_r}} H_n^{(1)}\left(\frac{\omega}{c}\hat{r}\right) \mathrm{d}\hat{r} \text{ for } 0 \le r < r_s, \tag{2.177}$$

$$\begin{aligned} \hat{p}_2 = &\frac{i}{4} \frac{Q}{v_r} \sum_{n=-\infty}^{\infty} e^{in\phi} H_n^{(1)}\left(\frac{\omega}{c} r\right) \int_{r_s}^{r} e^{i(\omega - \Omega - nw)\frac{\hat{r}-r_s}{v_r}} J_n\left(\frac{\omega}{c}\hat{r}\right) \mathrm{d}\hat{r} \\ &+ \frac{i}{4} \frac{Q}{v_r} \sum_{n=-\infty}^{\infty} e^{in\phi} J_n\left(\frac{\omega}{c} r\right) \int_{r}^{r_{\max}} e^{i(\omega - \Omega - nw)\frac{\hat{r}-r_s}{v_r}} H_n^{(1)}\left(\frac{\omega}{c}\hat{r}\right) d\hat{r} \text{ for } r_s < r < r_{\max}, \end{aligned} \tag{2.178}$$

$$\hat{p}_2 = \frac{i}{4} \frac{Q}{v_r} \sum_{n=-\infty}^{\infty} e^{in\phi} H_n^{(1)}\left(\frac{\omega}{c} r\right) \int_{r_s}^{r_{\max}} e^{i(\omega - \Omega - nw)\frac{\hat{r}-r_s}{v_r}} J_n\left(\frac{\omega}{c}\hat{r}\right) \mathrm{d}\hat{r} \text{ for } r_{\max} < r. \tag{2.179}$$

2.4.2.2 Discussion of the Formula for the Spectral Pressure

While the spectral pressure (2.176) must be evaluated numerically as the integral cannot be solved explicitly, analytical results can be obtained if the observer is positioned at the origin.

Spectral pressure in the origin

To obtain a simpler expression, let us assume that the receiver is at the origin, i.e., $r = 0$. Since $J_n(0) = \delta_{n0}$, the spectral pressure (2.177) takes the form

$$\hat{p}_2(r=0) = \frac{i}{4}\frac{Q}{v_r} e^{-i(\omega-\Omega)\frac{r_s}{v_r}} \int_{r_s}^{r_{\max}} e^{i(\omega-\Omega)\frac{\hat{r}}{v_r}} H_0^{(1)}\left(\frac{\omega}{c}\hat{r}\right) \mathrm{d}\hat{r}. \tag{2.180}$$

We compare this result with formula (2.117) for a line source that moves in a straight line along the x-axis from x_0 to x_1. For an observer at the origin, we obtain from Eq. (2.117) with $\Phi(\xi/u) = 0$

$$\hat{p}_2(x=0, y=0) = \frac{iQ}{4u} \int_{x_0}^{x_1} e^{i(\omega-\Omega)\frac{\xi}{u}} H_0^{(1)}\left(\frac{\omega}{c}\xi\right) \mathrm{d}\xi. \tag{2.181}$$

This yields the following interesting result: At the origin, the spectrum of a line source moving along an Archimedean spiral from $r_s = 0$ to $r_{\max}$ with radial velocity v_r (blue curve) is identical to the spectrum of a source moving in a straight line from $x_0 = 0$ to $x_1 = r_{\max}$ (red line) with velocity u. This is sketched in Fig. 2.26.

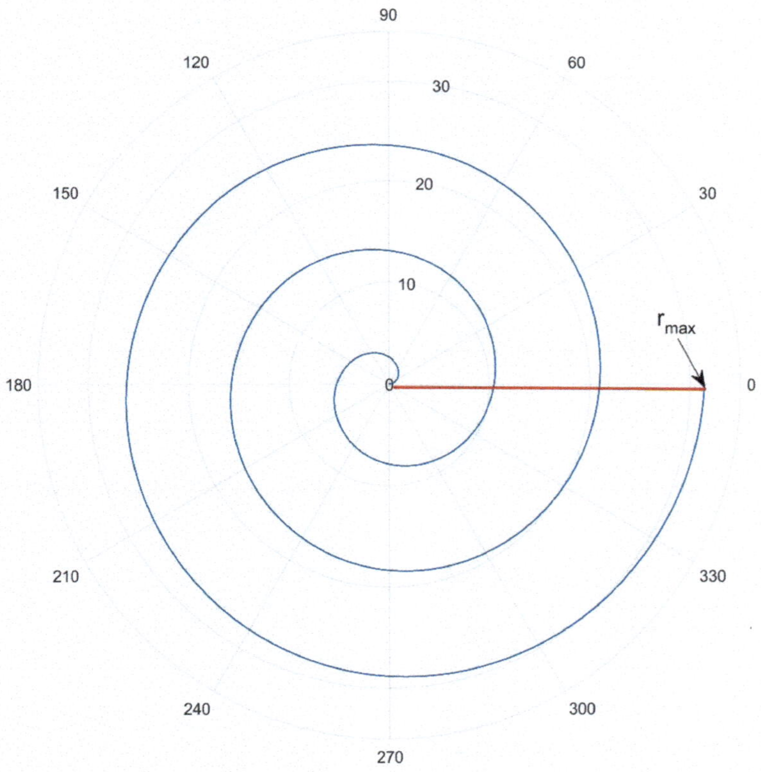

Fig. 2.26 The Archimedean spiral and the corresponding straight line

The integral (2.180) can be solved explicitly if we consider the interval $[r_s, r_{\max}) = [0, \infty)$, i.e.,

$$\hat{p}_2(r=0) = \frac{i}{4}\frac{Q}{v_r} e^{-i(\omega-\Omega)\frac{r_s}{v_r}} \int_0^\infty e^{i(\omega-\Omega)\frac{\hat{r}}{v_r}} H_0^{(1)}\left(\frac{\omega}{c}\hat{r}\right) \mathrm{d}\hat{r}. \tag{2.182}$$

For this, we use the formulas 10.22.59 and 10.22.60 from the NIST *Digital Library of Mathematical Functions* [23], which are given for the order $\mu = 0$ by

$$\int_0^\infty e^{ibt} J_0(at)\mathrm{d}t = \begin{cases} \frac{1}{\sqrt{a^2-b^2}}, 0 \le b < a \\ \frac{i}{\sqrt{b^2-a^2}}, 0 < a < b \end{cases}. \tag{2.183}$$

and by

$$\int_0^\infty e^{ibt} Y_0(at)\mathrm{d}t = \begin{cases} \frac{2i}{\pi}\frac{1}{\sqrt{a^2-b^2}} \arcsin\left(\frac{b}{a}\right), 0 \le b < a \\ \frac{1}{\sqrt{b^2-a^2}}\left[-1 + \frac{2i}{\pi}\ln\left(\frac{a}{b+\sqrt{b^2-a^2}}\right)\right], 0 < a < b \end{cases}, \tag{2.184}$$

where Y_0 is the Neumann function of order zero.

Since $H_0^{(1)}(at) = J_0(at) + iY_0(at)$, we obtain

$$\begin{aligned} I(a,b) :&= \int_0^\infty e^{ibt} H_0^{(1)}(at)\mathrm{d}t \\ &= \begin{cases} \frac{1}{\sqrt{a^2-b^2}}\left\{1 - \frac{2}{\pi}\arcsin\left(\frac{b}{a}\right)\right\},\ 0 \le b < a,\ (\alpha)/(\alpha-) \\ \frac{-2}{\pi}\frac{1}{\sqrt{b^2-a^2}}\ln\left(\frac{a}{b+\sqrt{b^2-a^2}}\right),\ 0 < a < b,\ (\beta)/(\beta-) \end{cases}. \end{aligned} \tag{2.185}$$

The meaning of the symbols $(\alpha)/(\alpha-)$ and $(\beta)/(\beta-)$ are explained in the definitions (2.188) and (2.189) and in Fig. 2.27.

By choosing $a := \frac{\omega}{c}$ and $b := \frac{\omega-\Omega}{v_r}$, one obtains the analytical solution of the integral. (2.182):

$$\hat{p}_2(r=0) = \frac{i}{4}\frac{Q}{v_r} e^{-i(\omega-\Omega)\frac{r_s}{v_r}} I\left(\frac{\omega}{c}, \frac{\omega-\Omega}{v_r}\right). \tag{2.186}$$

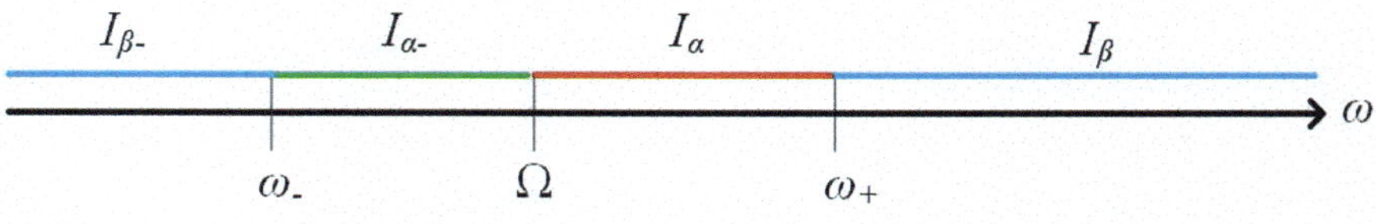

Fig. 2.27 Frequency intervals considered

The zeros of the square root are the Doppler frequencies $\omega_- = \Omega/(1+M_r)$ and $\omega_+ = \Omega/(1-M_r)$ with $M_r = v_r/c$ since (cf. Eq. (2.74))

$$-\kappa^2(\omega) := b^2 - a^2 = \left(\frac{\omega-\Omega}{v_r}\right)^2 - \left(\frac{\omega}{c}\right)^2 = \frac{1-M_r^2}{v_r^2}(\omega-\omega_-)(\omega-\omega_+). \tag{2.187}$$

The inequalities in formula (2.185) determine the valid frequency intervals $I_{\alpha,\beta}$ for the subcases (α) and (β) which are given by (see Fig. 2.27)

$$I_{\alpha,\beta} = \begin{cases} 0 \le b < a \Leftrightarrow \Omega \le \omega < \omega_+ & (\alpha) \\ 0 < a < b \Leftrightarrow \omega > \omega_+ & (\beta) \end{cases}. \tag{2.188}$$

If $0 < \omega < \Omega$ and thus $b := \frac{\omega-\Omega}{v_r} < 0$, the formulas given in Eqs. (2.183)–(2.185) also apply. This can be seen by considering $b_- := -b > 0$ and taking the complex conjugate of the left-hand sides of Eqs. (2.183) and (2.184). Alternatively, this can be shown with the help of formulas 7,8 and 11,12 on page 731 of Gradshteyn and Ryzhik [24]. For $b < 0$, we obtain the additional subcases $(\alpha-)$ and $(\beta-)$ which are given by

$$I_{\alpha-,\beta-} = \begin{cases} b < 0 \text{ and } 0 \le |b| < a \Leftrightarrow \omega_- \le \omega < \Omega & (\alpha-) \\ b < 0 \text{ and } 0 \le a < |b| \Leftrightarrow 0 \le \omega < \omega_- & (\beta-) \end{cases}. \tag{2.189}$$

All four intervals cover the entire frequency interval, as shown in Fig. 2.27. In summary, we obtain from Eqs. (2.186)

$$\hat{p}_2(r=0) = \frac{i}{4}\frac{Q}{v_r} e^{-i(\omega-\Omega)\frac{r_s}{v_r}} I\left(\frac{\omega}{c}, \frac{\omega-\Omega}{v_r}\right) \tag{2.190}$$

with

$$I\left(\frac{\omega}{c}, \frac{\omega-\Omega}{v_r}\right) = \begin{cases} \frac{1}{\sqrt{\left(\frac{\omega}{c}\right)^2 - \left(\frac{\omega-\Omega}{v_r}\right)^2}}\left\{1 - \frac{2}{\pi}\arcsin\left(\frac{\omega-\Omega}{\omega M_r}\right)\right\}, & \omega \in I_\alpha \cup I_{\alpha-} \\ \frac{-2/\pi}{\sqrt{\left(\frac{\omega-\Omega}{v_r}\right)^2 - \left(\frac{\omega}{c}\right)^2}} \ln\left[\frac{\omega}{c}\Big/\left[\left(\frac{\omega-\Omega}{v_r}\right) + \sqrt{\left(\frac{\omega-\Omega}{v_r}\right)^2 - \left(\frac{\omega}{c}\right)^2}\right]\right], & \omega \in I_\beta \cup I_{\beta-} \end{cases}. \tag{2.191}$$

For the special case that $\omega = \Omega$, we have to use the first line of Eq. (2.191) and get the simple formula

$$\hat{p}_2(r = 0, \omega = \Omega) = \frac{i}{4}\frac{Q}{v_r}\frac{c}{\Omega} = \frac{i}{4}Q\frac{1}{\Omega M_r}. \tag{2.192}$$

We now compare the numerical solution of the integral representation (2.182) with the analytical solution (2.190) and (2.191) using Matlab. Such a comparison can be important because the integral in (2.182) becomes highly oscillating for $r_{\max} \to \infty$, and we, therefore, need to determine a suitable finite value for the upper limit $r_{\max} = t_{\max} v_r + r_s$ instead of infinity.

In the following ***numerical study***, we have chosen $r_{\max} = 100\ m$ as the upper limit of the integral. If we increase $r_{\max}$ up to approximately 5000 m, no differences can be observed between the curves. From $r_{\max} = 8000$ m onwards, however, the integral solution (2.182) delivers deviating results due to the oscillating nature of the integrand.

In Figs. 2.28, 2.29, and 2.30, the absolute value, the real part and the imaginary part of the frequency curves are calculated by means of the integral (2.182) and compared with the corresponding analytical solutions (2.190) and (2.191). Note that the Doppler frequencies for the selected parameters are $\omega_-/(2\pi) \approx 330$ Hz and $\omega_+/(2\pi) \approx 396$ Hz.

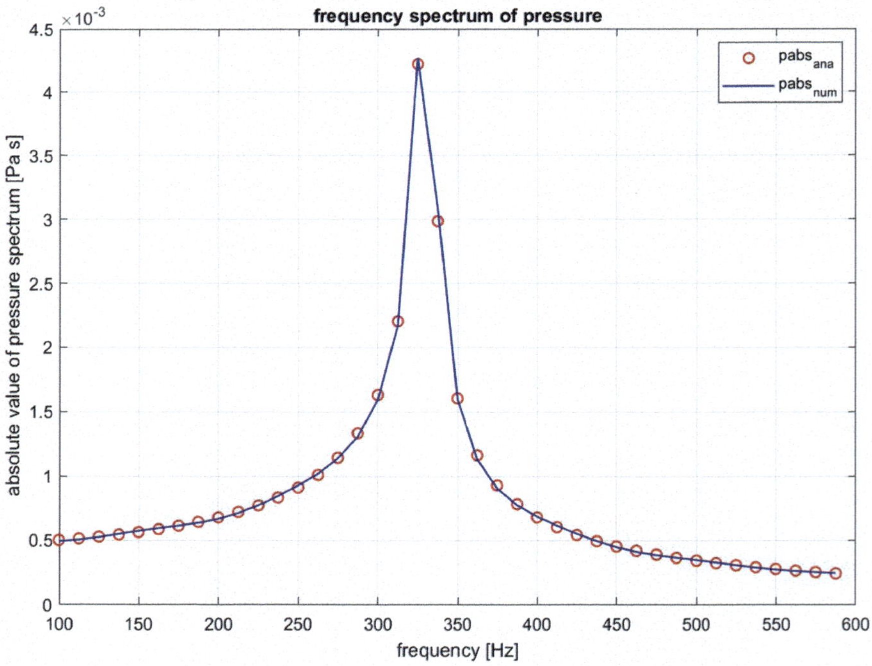

Fig. 2.28 Absolute value of the spectral pressure $\hat{p}_2(r = 0)$; blue curve = Eq. (2.182), red circles = Eqs. (2.190)/(2.191); parameters: $r_s = 0$ m, $v_r = M_r * 343$ m/s, $M_r = 0.09$, $\Omega = 2\pi 360\ \text{s}^{-1}$, $r_{\max} = 100$ m, $Q = 1\ \text{kg}/(\text{ms}^2)$

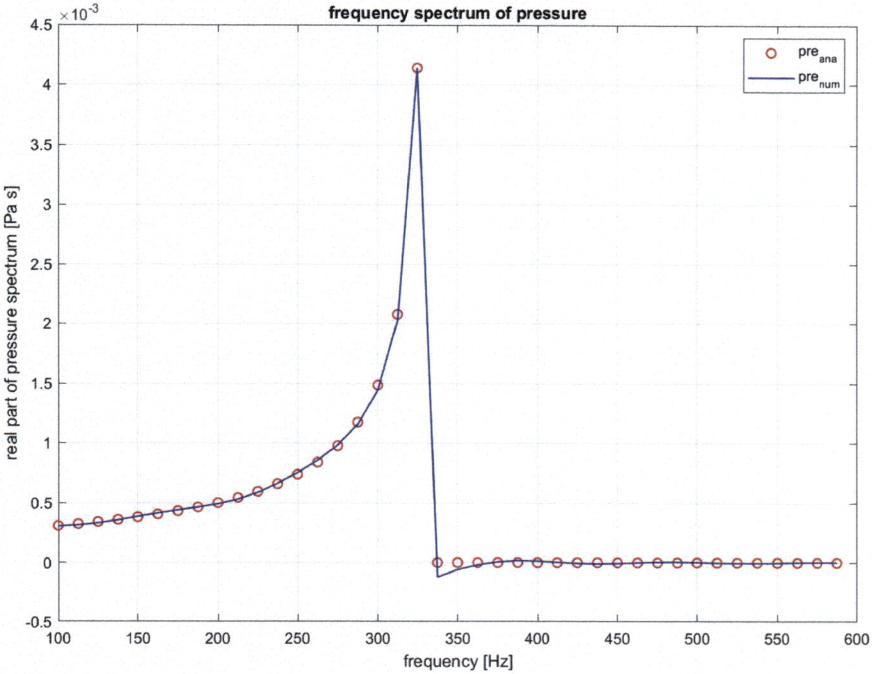

Fig. 2.29 Real part of the spectral pressure $\hat{p}_2(r = 0)$; blue curve = Eq. (2.182), red circles = Eqs. (2.190)/(2.191); parameters: $r_s = 0$ m, $v_r = M_r * 343$ m/s, $M_r = 0.09$, $\Omega = 2\pi 360$ s^{-1},$r_{\max} = 100$ m, $Q = 1$ kg/(ms^2)

The figures demonstrate that the numerical and the analytical solutions agree very well across the considered frequency range. It has no influence on the results whether we calculate the spectrum with the first or the second line of Eq. (2.191). The results remain the same for the entire frequency range under consideration. The reason lies in the following fact: The arcsine function can be converted into its logarithmic form for complex arguments (see [25]) by

$$\arcsin(z) = -i \ln\left(\sqrt{1 - z^2} + iz\right).$$

Matlab automatically performs this conversion, ensuring that both formulas produce identical results.

The figures show only one peak at the lower Doppler frequency $\omega_-/(2\pi)$ near 330 Hz. But what happens at the higher Doppler frequency $\omega_+/(2\pi)$? The answer is that it is a removable singularity. The limit can be calculated from the second line of Eq. (2.191) and results in the finite value

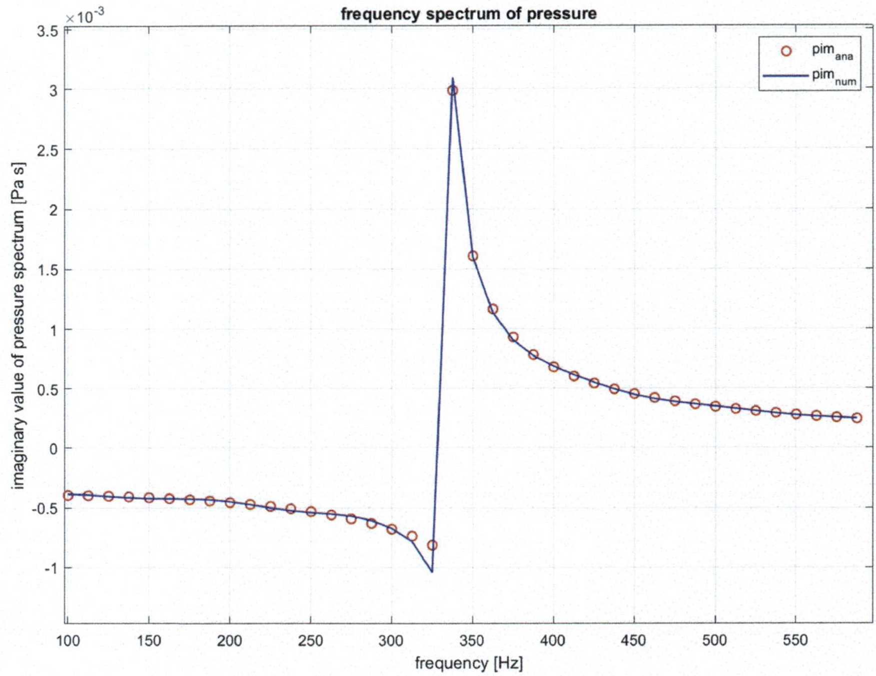

Fig. 2.30 Imaginary part of the spectral pressure $\hat{p}_2(r = 0)$; blue curve = Eq. (2.182), red circles = Eqs. (2.190)/(2.191); parameters: $r_s = 0$ m, $v_r = M_r * 343$ m/s, $M_r = 0.09$, $\Omega = 2\pi 360$ s^{-1}, $r_{\max} = 100$ m, $Q = 1$ kg/(ms^2)

$$I = \frac{-2/\pi}{\sqrt{\left(\frac{\omega-\Omega}{v_r}\right)^2 - \left(\frac{\omega}{c}\right)^2}} \ln\left[\frac{\omega}{c} \Big/ \left[\left(\frac{\omega-\Omega}{v_r}\right) + \sqrt{\left(\frac{\omega-\Omega}{v_r}\right)^2 - \left(\frac{\omega}{c}\right)^2}\right]\right] \xrightarrow[\omega\to\omega_+]{} \frac{2}{\pi}(1-M_r)\frac{c}{\Omega}, \tag{2.193}$$

whereby the detailed calculation is presented in Appendix D. Using Eq. (2.190), we finally obtain

$$\hat{p}_2(r = 0, \omega \to \omega_+) = \frac{i}{4}\frac{Q}{v_r} e^{-i\frac{\omega_+}{c}v_r} I(\omega \to \omega_+) = \frac{i}{2\pi} Q \frac{1-M_r}{M_r}\frac{1}{\Omega} e^{-i\frac{\omega_+}{c}v_r}. \tag{2.194}$$

In Table 2.3 we compare this limit value at the higher Doppler frequency $\omega_+/(2\pi) \approx 396$ Hz with the numerical result from Eq. (2.182) and with the analytical expression (2.191), 2nd line, using the same parameters as in the last three figures.

Table 2.3 Limit for $\omega \to \omega_+$

Integral (2.182)	Expression (2.191), 2nd line	Limit (2.194)
−3.6492e−07 + 7.1728e−04i	−2.8349e−13 + 7.1144e−04i	0.0000e + 00 + 7.1144e−04i

We can observe that the analytical expression and the exact limit are almost identical. The result of the integral differs slightly because we have used the finite upper bound $r_{\max} = 100$ instead of $r_{\max} = \infty$.

The absence of a peak at the higher Doppler frequency is easily explained. Since the observer is at the origin and the radius of the trajectory increases with time (see Eq. (2.165) and Fig. 2.25), the source turns further and further away from the receiver so that only the lower Doppler frequency can be observed.

2.5 Radiating Monopole Moving in Cylindrical Coordinates (3D Case)

In the previous Sect. 2.4, we explored the frequency spectrum of moving line sources described in two-dimensional polar coordinates. In this section, we transition from two to three dimensions to address a moving point source in three-dimensional cylindrical coordinates. Firstly, we derive the frequency spectrum of a source moving along a circular trajectory and afterward that of a source moving along a circular helix.

2.5.1 A Point Source Moving Along a Circular Path

2.5.1.1 Derivation of the Spectral Pressure

We consider a radiating point source in three-dimensional space that moves on a circle of radius r_0 with angular velocity w around the origin – exactly as in the two-dimensional case of the last section. However, the sound field φ_3 must now satisfy the 3D wave equation

$$\Delta\varphi_3 - \frac{1}{c^2}\frac{\partial^2\varphi_3}{\partial t^2} = \frac{Q_1}{\rho} e^{-i\Omega t} \frac{1}{r} \delta(r - r_0)\delta_c(\phi - wt)\delta(z - z_0) \tag{2.195}$$

in the cylindrical coordinates r, ϕ, z with

$$x = r\cos\phi, \; y = r\sin\phi, \; z = z. \tag{2.196}$$

Again, the polar angle is 2π periodic, and thus, $\delta_c(\phi - wt)$ is the Dirac comb. Without loss of generality, the source is located in the xy-plane, i.e., $z_0 = 0$. The further derivation is similar to that in polar coordinates but is presented briefly for the sake of clarity.

The Fourier transform with respect to time for the pressure $\hat{p}_3(\omega) = i\omega\rho\hat{\varphi}_3$ leads to

$$\Delta\hat{p}_3 + \left(\frac{\omega}{c}\right)^2 \hat{p}_3 = -Q\frac{\delta(r-r_0)}{2\pi r}\delta(z)\int_{-\infty}^{\infty} e^{i(\omega-\Omega)t}\sum_{n=-\infty}^{\infty} e^{in(\phi-wt)}\mathrm{d}t$$
$$= -Q\frac{\delta(r-r_0)}{2\pi r}\delta(z)\int_{-\infty}^{\infty}\sum_{n=-\infty}^{\infty} e^{i(\omega-\Omega-nw)t}e^{in\phi}\mathrm{d}t. \tag{2.197}$$

with $Q = -i\omega Q_1$.

Using the known relationship (2.134)

$$\int_{-\infty}^{\infty} e^{i(\omega-\Omega-nw)t}\mathrm{d}t = 2\pi\delta(\omega-\Omega-nw)$$

and exchanging the order of integration and summation, we get

$$\Delta_3\hat{p}_3 + \left(\frac{\omega}{c}\right)^2 \hat{p}_3 = -Q\frac{\delta(r-r_0)}{r}\delta(z)\sum_{n=-\infty}^{\infty}\delta(\omega-\Omega-nw)e^{in\phi}. \tag{2.198}$$

Now, we have two options to apply the Green's function method (GFM) for solving this inhomogeneous Helmholtz equation. We can work directly with the 3D Green's function in cylindrical coordinates, or we can first perform a Fourier transform that allows the use of the 2D Green's function in polar coordinates, as in Sect. 2.4. We choose the second approach.

The Fourier transform in z of the spectral pressure is defined by

$$\tilde{p}(k_z) := F_z[\hat{p}_3(z)] = \int_{-\infty}^{\infty}\hat{p}_3(z)e^{-ik_z z}\mathrm{d}z. \tag{2.199}$$

The Fourier transform of the Helmholtz Eq. (2.198) is, therefore, given by

$$\Delta_{r,\varphi}\tilde{p} + \left[\left(\frac{\omega}{c}\right)^2 - k_z^2\right]\tilde{p} = -Q\frac{\delta(r-r_0)}{r}\sum_{n=-\infty}^{\infty}\delta(\omega-\Omega-nw)e^{in\phi}, \tag{2.200}$$

where $\Delta_{r,\phi}$ is the 2D Laplace operator in polar coordinates. If we compare Eq. (2.200) with Eq. (2.135), we immediately see that both equations are identical, if we replace the wave number $k = \omega/c$ in (2.135) by the new wave number $\kappa = \sqrt{\left(\frac{\omega}{c}\right)^2 - k_z^2}$. That is very convenient, since we can now apply the Green's function method together with the Green's function (2.138) in polar coordinates as in Sect. 2.4.1. The solution is given immediately by Eq. (2.144)

$$\tilde{p} = \frac{i\pi Q}{2} \sum_{n=-\infty}^{\infty} H_n^{(1)}(\kappa r_>) J_n(\kappa r_<) \delta(\omega - \Omega - nw) e^{in\phi}, \tag{2.201}$$

where we have only replaced $\hat{p}_2$ by $\tilde{p}$ and k by κ. Finally, the inverse Fourier transform in z gives the frequency spectrum of a circularly moving source in three-dimensional space, which emits a harmonic signal of frequency Ω:

$$\begin{aligned}\hat{p}_3(r, \varphi, z) &= \frac{iQ}{4} \sum_{n=-\infty}^{\infty} \delta(\omega - \Omega - nw) e^{in\phi} \\ &\times \int_{-\infty}^{\infty} H_n^{(1)}\left(r_> \sqrt{k^2 - k_z^2}\right) J_n\left(r_< \sqrt{k^2 - k_z^2}\right) e^{ik_z z} \, \mathrm{d}k_z \end{aligned} \tag{2.202}$$

with $r_< = \min\{r, r_0\}$ and $r_> = \max\{r, r_0\}$.

2.5.1.2 Discussion of the Formula for the Spectral Pressure

In contrast to the spectrum of the rotating source (2.144) in 2D, the formula for the spectral pressure $\hat{p}_3$ in 3D contains not only an infinite series but also an infinite Fourier integral, which complicates the evaluation. In Sect. 2.6, we will discover that by using spherical coordinates, we can derive a formula that avoids an infinite integral and instead involves an infinite double series.

Pressure spectrum at the axis of the cylindrical coordinate system

We verify formula (2.202) by placing the receiver on the axis of the cylindrical coordinate system, i.e., $r = 0$. It follows that $r_< = \min\{r, r_0\} = 0$ and $r_> = \max\{r, r_0\} = r_0$. Since $J_n(0) = \delta_{n0}$, we immediately get

$$\hat{p}_3(r = 0, z) = \frac{iQ}{4} \delta(\omega - \Omega) \int_{-\infty}^{\infty} H_0^{(1)}\left(r_0 \sqrt{k^2 - k_z^2}\right) e^{ik_z z} \mathrm{d}k_z. \tag{2.203}$$

Considering that $H_0^{(1)}\left(r_0 \sqrt{k^2 - k_z^2}\right)$ is an even function of the variable k_z, we obtain

$$\int_{-\infty}^{\infty} H_0^{(1)}\left(r_0 \sqrt{k^2 - k_z^2}\right) e^{ik_z z} \mathrm{d}k_z = 2 \int_{0}^{\infty} H_0^{(1)}\left(r_0 \sqrt{k^2 - k_z^2}\right) \cos(k_z z) \mathrm{d}k_z. \tag{2.204}$$

From Erdélyi [26, p. 59], formula 59 with $\alpha = r_0,\ \beta = k,\ x = k_z,\ y = z$, we find that

$$\int_0^\infty H_0^{(1)}\left(r_0\sqrt{k^2-k_z^2}\right)\cos(k_z z)\mathrm{d}k_z = -i\left(r_0^2+z^2\right)^{-1/2} e^{ik\sqrt{r_0^2+z^2}}, \tag{2.205}$$

if $r_0 > 0,\ z > 0$, and $\pi > \arg\sqrt{k^2-k_z^2} \geq 0$. A similar result is Eq. (2.57). By introducing (2.204) and (2.205) into (2.203), we finally get

$$\hat{p}_3(r=0,z) = 2\pi\,\delta(\omega-\Omega)\frac{Q}{4\pi}\frac{e^{ik\sqrt{r_0^2+z^2}}}{\sqrt{r_0^2+z^2}} = 2\pi\,\delta(\omega-\Omega)\frac{-i\omega Q_1}{4\pi}\frac{e^{ik\sqrt{r_0^2+z^2}}}{\sqrt{r_0^2+z^2}}. \tag{2.206}$$

Similar to Sect. 2.4.1, Eq. (2.150), this expression describes the sound field of the corresponding *stationary* point source located at the distance r_0 from the origin at height $z_0 = 0$, which can be observed through the inverse Fourier transform of Eq. (2.206). As expected, this leads to the time-dependent pressure field of a stationary harmonic point source

$$p_3(r=0,t) = \frac{Q_1}{4\pi}\frac{\partial}{\partial t}\left\{\frac{e^{i\frac{\Omega}{c}\sqrt{r_0^2+z^2}}}{\sqrt{r_0^2+z^2}}e^{-i\Omega t}\right\} = \frac{-i\Omega Q_1}{4\pi}\frac{e^{i\frac{\Omega}{c}\sqrt{r_0^2+z^2}}}{\sqrt{r_0^2+z^2}}e^{-i\Omega t}. \tag{2.207}$$

Like in the 2D case, an observer located on the axis will not perceive any frequency shift from the original emission frequency Ω, since the source surrounds the observer on a circle of radius r_0 and height $z_0 = 0$, so that the distance between source and receiver $\sqrt{r_0^2+z^2}$ always remains the same.

2.5.2 A Point Source Moving Along a Circular Helix

It is possible to extend the analysis of a source moving along an Archimedean spiral from 2D (see Sect. 2.4.2) to 3D. However, we find it more interesting to consider, as a second example, a source winding along a circular helix of radius r_0 and pitch $2\pi v_z/w$

$$\vec{x}_0(t) = \begin{pmatrix} x_0(t) \\ y_0(t) \\ z_0(t) \end{pmatrix} = \begin{pmatrix} r_0\cos(wt) \\ r_0\sin(wt) \\ v_z t \end{pmatrix} \text{ for } t \in (-\infty,\infty). \tag{2.208}$$

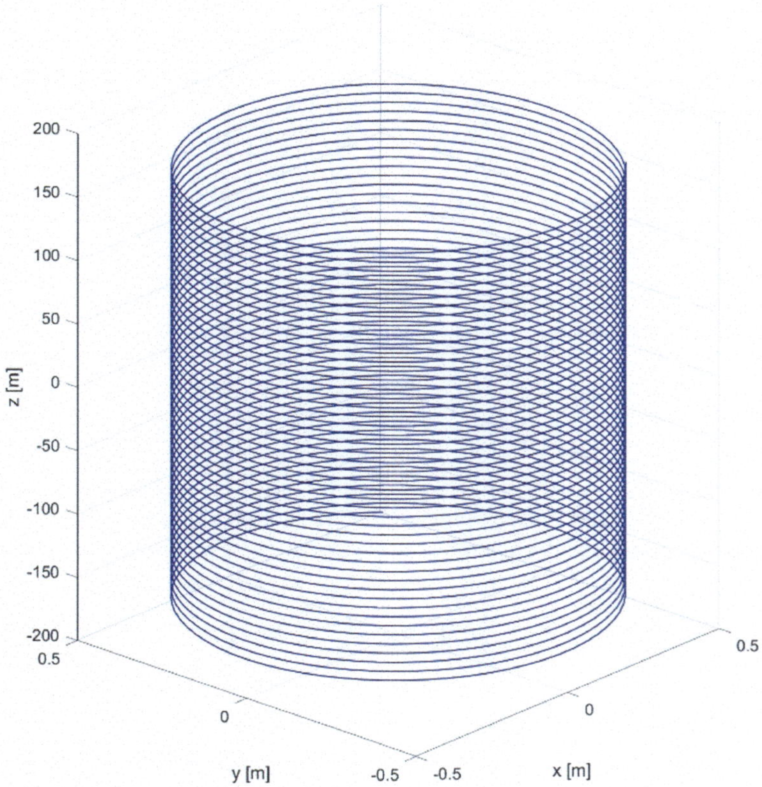

Fig. 2.31 A circular helix with parameters: $r_0 = 0.5$ m, $w = 30$ s^{-1}, $v_z = 34.3$ m/s, $t \in (-5, 5)$s

Here, the polar angle is $\phi_0 = wt$ with angular velocitay w, and the velocity in z-direction is v_z.

The modulus of the velocity of the source has the constant value

$$\left|\frac{d\vec{x}_0(t)}{dt}\right| = \sqrt{(wr_0)^2 + v_z^2}. \tag{2.209}$$

A sketch of such a helix is shown in Fig. 2.31.

2.5.2.1 Derivation of the Spectral Pressure

Similar to circular motion, the sound field φ_3 is described by the 3D wave equation

$$\Delta\varphi_3 - \frac{1}{c^2}\frac{\partial^2\varphi_3}{\partial t^2} = \frac{Q_1}{\rho}e^{-i\Omega t}\frac{1}{r}\delta(r - r_0)\delta_c(\phi - wt)\delta(z - v_z t), t \in (-\infty, \infty) \tag{2.210}$$

in cylindrical coordinates, where again $\delta_c(\phi - wt)$ is the Dirac comb. The difference to a purely circular path is, therefore, that the height $z_0(t)$ of the source now also changes with time. As usual, we start with the temporal Fourier transform of the wave equation which gives

$$\Delta\hat{p}_3 + \left(\frac{\omega}{c}\right)^2 \hat{p}_3 = -Q\frac{\delta(r - r_0)}{2\pi r} \int_{-\infty}^{\infty} \sum_{n=-\infty}^{\infty} e^{i(\omega-\Omega-nw)t} e^{in\phi} \delta(z - v_z t) \mathrm{d}t \tag{2.211}$$

for the pressure $\hat{p}_3(\omega) = i\omega\rho\hat{\varphi}_3$ with $Q = -i\omega Q_1$.

We use the substitution $\tau = v_z t$ and obtain for the integral in (2.211)

$$I := \frac{1}{v_z} \sum_{n=-\infty}^{\infty} e^{in\phi} \int_{-\infty}^{\infty} e^{i(\omega-\Omega-nw)\frac{\tau}{v_z}} \delta(z - \tau) \mathrm{d}\tau,$$

where we assume that the order of integration and summation can be interchanged. The integral is immediately solved by the occurrence of the Dirac delta function

$$I := \frac{1}{v_z} \sum_{n=-\infty}^{\infty} e^{i(\omega-\Omega-nw)\frac{z}{v_z}} e^{in\phi}, \tag{2.212}$$

so that we obtain the inhomogeneous Helmholtz equation

$$\Delta\hat{p}_3 + \left(\frac{\omega}{c}\right)^2 \hat{p}_3 = -Q\frac{\delta(r - r_0)}{2\pi r v_z} \sum_{n=-\infty}^{\infty} e^{i(\omega-\Omega-nw)\frac{z}{v_z}} e^{in\phi}. \tag{2.213}$$

Following the same procedure as in the preceding Sect. 2.5.1, we firstly perform a Fourier transform in z (see Eq. 2.199) of the Helmholtz Eq. (2.213), which leads to

$$\begin{aligned} \Delta_{r,\varphi}\tilde{p} + \left[\left(\frac{\omega}{c}\right)^2 - k_z^2\right]\tilde{p} &= -Q\frac{\delta(r - r_0)}{v_z r} \sum_{n=-\infty}^{\infty} \delta\left(k_z - \frac{\omega - \Omega - nw}{v_z}\right) e^{in\phi} \\ &=: -R(r, \phi), \end{aligned} \tag{2.214}$$

where we have taken into account that

$$\int_{-\infty}^{\infty} e^{i(\omega-\Omega-nw)\frac{z}{v_z}} e^{-ik_z z} \mathrm{d}z = 2\pi\delta\left(k_z - \frac{\omega - \Omega - nw}{v_z}\right). \tag{2.215}$$

Secondly, we apply the Green's function method and convolute the right-hand side $R(r, \phi)$ of the transformed Helmholtz equation with the Green's function (2.138) in polar coordinates as in the preceding section with the following result

$$\begin{aligned}\tilde{p} &= \int_0^\infty \int_0^{2\pi} g_2(r,\phi;\hat{r},\hat{\phi})R(\hat{r},\hat{\phi})\hat{r}\mathrm{d}\hat{\phi}\mathrm{d}\hat{r} \\ &= \frac{iQ}{4v_z}\sum_{n=-\infty}^{\infty} H_n^{(1)}(\kappa r_>)J_n(\kappa r_<)\int_0^{2\pi}\sum_{m=-\infty}^{\infty}\delta\left(k_z - \frac{\omega-\Omega-mw}{v_z}\right)e^{im\hat{\phi}}e^{in(\varphi-\hat{\phi})}\mathrm{d}\hat{\phi} \\ &= \frac{iQ}{4v_z}2\pi\sum_{n=-\infty}^{\infty} H_n^{(1)}(\kappa r_>)J_n(\kappa r_<)\delta\left(k_z - \frac{\omega-\Omega-nw}{v_z}\right)e^{in\phi},\end{aligned} \tag{2.216}$$

where we have used Eq. (2.217)

$$\int_0^{2\pi} e^{i(m-n)\hat{\phi}}\mathrm{d}\hat{\phi} = 2\pi\delta_{nm} \text{ with the Kronecker delta } \delta_{nm} = \begin{cases} 0, & \text{if } n \neq m \\ 1, & \text{if } n = m \end{cases}, \tag{2.217}$$

and $\kappa = \sqrt{\left(\frac{\omega}{c}\right)^2 - k_z^2}$, $r_< = \min\{r, r_0\}$ and $r_> = \max\{r, r_0\}$. Finally, the inverse Fourier transform in z yields the frequency spectrum of a helically moving source in three-dimensional space that radiates a harmonic signal of frequency Ω:

$$\begin{aligned}\hat{p}_3 = \frac{Q}{v_z}\frac{i}{4}\sum_{n=-\infty}^{\infty} H_n^{(1)}&\left(r_>\sqrt{\left(\frac{\omega}{c}\right)^2 - \left(\frac{\omega-\Omega-nw}{v_z}\right)^2}\right) \\ &\times J_n\left(r_<\sqrt{\left(\frac{\omega}{c}\right)^2 - \left(\frac{\omega-\Omega-nw}{v_z}\right)^2}\right)e^{iz\left(\frac{\omega-\Omega-nw}{v_z}\right)}e^{in\phi}.\end{aligned} \tag{2.218}$$

Surprisingly and fortunately, the spectrum of the helix is even simpler than that of the circular orbit (2.202) since it no longer contains an infinite integral! This result was also recently derived in reference [27].

2.5.2.2 A Simplified Derivation of the 3D-Source Moving in a Straight Line at Constant Speed

In Sect. 2.2.1, we derived the spectral pressure of a radiating monopole moving with constant velocity on a straight line using Cartesian coordinates. However, even when we use cylindrical coordinates, we obtain a very convenient and elegant solution by means of the expansion of the Hankel function $H_0^{(1)}$. If the angular velocity w is allowed to approach zero, the trajectory is simplified to a straight line parallel to the z-axis

$$\vec{x}_0(t) = \begin{pmatrix} r_0 \cos(wt) \\ r_0 \sin(wt) \\ v_z t \end{pmatrix} \underset{w \to 0}{\longrightarrow} \begin{pmatrix} r_0 \\ 0 \\ v_z t \end{pmatrix} \text{for } t \in (-\infty, \infty) \tag{2.219}$$

and we obtain directly from (2.218)

$$\hat{p}_3 = \frac{Q}{v_z}\frac{i}{4} e^{iz\left(\frac{\omega-\Omega}{v_z}\right)} \sum_{n=-\infty}^{\infty} H_n^{(1)}\left(r_> \sqrt{\left(\frac{\omega}{c}\right)^2 - \left(\frac{\omega-\Omega}{v_z}\right)^2}\right) \times J_n\left(r_< \sqrt{\left(\frac{\omega}{c}\right)^2 - \left(\frac{\omega-\Omega}{v_z}\right)^2}\right) e^{in\phi}. \tag{2.220}$$

On the other hand, the expansion for the Hankel function of order zero in 2D polar coordinates is given by (see Morse and Feshbach [28, p. 827, Eq. (7.2.51)])

$$H_0^{(1)}(\kappa|\vec{r} - \vec{r}_0|) = \sum_{n=-\infty}^{\infty} H_n^{(1)}(\kappa r_>) J_n(\kappa r_<) e^{in(\phi-\phi_0)} \tag{2.221}$$

with $r = |\vec{r}|,\ r_0 = |\vec{r}_0|,\ r_< = \min\{r, r_0\},\ r_> = \max\{r, r_0\}$.

By choosing $\kappa = \sqrt{\left(\frac{\omega}{c}\right)^2 - \left(\frac{\omega-\Omega}{v_z}\right)^2}$, $\vec{r} = (x, y)^T$, $\vec{r}_0 = (x_0, 0)^T$, $\phi_0 = 0$, and introducing (2.221) into (2.220), we obtain

$$\hat{p}_3 = \frac{Q}{v_z}\frac{i}{4} e^{iz\left(\frac{\omega-\Omega}{v_z}\right)} H_0^{(1)}\left(\sqrt{(x-x_0)^2 + y^2} \sqrt{\left(\frac{\omega}{c}\right)^2 - \left(\frac{\omega-\Omega}{v_z}\right)^2}\right). \tag{2.222}$$

We compare this result with the solution (2.19) from Sect. 2.2.1, where the source moves along the x-axis. A complete agreement can be observed if we rotate the coordinate system so that the x-axis in Sect. 2.2.1 becomes the z-axis in the present chapter (with $x_0 = 0$). If we moreover replace $\phi - wt$ in (2.210) by $\phi - \phi_0 - wt$, where (r_0, ϕ_0) corresponds to (x_0, y_0), we obtain the general case

$$\hat{p}_3 = \frac{Q}{v_z}\frac{i}{4} e^{iz\left(\frac{\omega-\Omega}{v_z}\right)} H_0^{(1)}\left(\sqrt{(x-x_0)^2 + (y-y_0)^2} \sqrt{\left(\frac{\omega}{c}\right)^2 - \left(\frac{\omega-\Omega}{v_z}\right)^2}\right). \tag{2.223}$$

Here, (x_0, y_0) is the point at which the straight trajectory penetrates orthogonally into the (x, y)-plane.

2.5.2.3 Pressure Spectrum at the Axis of the Cylindrical Coordinate System ($r = 0$)

Let us look at what happens on the axis of the cylindrical coordinate system:

We check formula (2.218) by placing the receiver on the axis of the cylindrical coordinate system, i.e., $r = 0$. It follows that $r_< = \min\{r, r_0\} = 0$ and $r_> = \max\{r, r_0\} = r_0$.

Since $J_n(0) = \delta_{n0}$, we immediately get

$$\hat{p}_3(r = 0, z) = \frac{Q}{v_z}\frac{i}{4}H_0^{(1)}\left(r_0\sqrt{\left(\frac{\omega}{c}\right)^2 - \left(\frac{\omega - \Omega}{v_z}\right)^2}\right)e^{iz\left(\frac{\omega-\Omega}{v_z}\right)} \tag{2.224}$$

If this formula is compared with Eq. (2.19), the following relationship results: The spectrum of a helix with radius r_0 at the axis is the same as the spectrum of a source moving in a straight line along the z-axis with the constant velocity v_z at field points (r_0, z). In contrast, the spectrum of a circular orbit at the origin is simply that of a stationary source (cf. the previous Sect. 2.5.1).

2.5.2.4 Convergence of the Series and Numerical Evaluation

The convergence behavior of the infinite series (2.218) is now to be examined. The radicand of the square root in Eq. 2.218 can be written in the following form

$$\begin{aligned}\mathrm{rad}(\omega, \omega_{n-}, \omega_{n+}) &:= \left(\frac{\omega}{c}\right)^2 - \left(\frac{\omega - \Omega - nw}{v_z}\right)^2 \\ &= \frac{M_z^2 - 1}{v_z^2}(\omega - \omega_{n-})(\omega - \omega_{n+}),\ M_z = v_z/c\end{aligned} \tag{2.225}$$

containing the $2n$ Doppler frequencies in z-direction

$$\omega_{n-} = (\Omega + nw)/(1 + M_z),\ \omega_{n+} = (\Omega + nw)/(1 - M_z). \tag{2.226}$$

So we have

$$\begin{aligned}\hat{p}_3 = \frac{Q}{v_z}\frac{i}{4}\sum_{n=-\infty}^{\infty} & H_n^{(1)}\left(r_>\sqrt{\mathrm{rad}(\omega, \omega_{n-}, \omega_{n+})}\right) \\ & \times J_n\left(r_<\sqrt{\mathrm{rad}(\omega, \omega_{n-}, \omega_{n+})}\right)e^{iz\left(\frac{\omega-\Omega-nw}{v_z}\right)}e^{in\phi}.\end{aligned} \tag{2.227}$$

If we consider the case with $M_z = v_z/c < 1$, we find that

$$\text{rad}(\omega, \omega_{n-}, \omega_{n+}) = \begin{cases} > 0 \text{ if } \exists n : \omega \in I_n := (\omega_{n-}, \omega_{n+}) \\ < 0 \text{ if } \forall n \in \mathbb{Z} : \omega \notin I_n := (\omega_{n-}, \omega_{n+}) \end{cases}. \tag{2.228}$$

Therefore, the argument of the Hankel function is only real if ω is located in one of the $2n$ "Doppler intervals" $I_n := (\omega_{n-}, \omega_{n+})$. If this is not the case, the square root becomes imaginary, and the Hankel function approaches zero exponentially, as explained in Sect. 2.2.1.3 below Eq. (2.42).

This also applies to negative n with $\Omega + nw < 0$, where we only have to note that $\omega_{n-} > \omega_{n+}$ and thus the Doppler intervals become $I_n := (\omega_{n+}, \omega_{n-})$.

On the other hand, the Bessel function $J_n(z)$ becomes very small when the absolute value $|n|$ of the order of the Bessel function becomes larger than the argument, as shown in Sect. 2.4.1 below Fig. 2.23. We make a rough estimate for a frequency point within each Doppler interval by selecting the harmonic mean $\omega_{n,har} = 2(1/\omega_{n+} + 1/\omega_{n-})^{-1} = \Omega + nw$:

$$\begin{aligned} \text{rad}\big(\omega_{n,har}, \omega_{n-}, \omega_{n+}\big) &= \frac{M_z^2 - 1}{v_z^2}\left(\Omega + nw - \frac{\Omega + nw}{1 + M_z}\right)\left(\Omega + nw - \frac{\Omega + nw}{1 - M_z}\right) \\ &= \frac{M_z^2 - 1}{v_z^2}(\Omega + nw)^2\left(1 - \frac{1}{1 + M_z}\right)\left(1 - \frac{1}{1 - M_z}\right) \\ &= \frac{M_z^2}{v_z^2}(\Omega + nw)^2 = \left(\frac{\Omega + nw}{c}\right)^2. \end{aligned} \tag{2.229}$$

Consequently, the requirement

$$|n| \gg r_< \sqrt{\text{rad}\big(\omega_{n,\text{har}}, \omega_{n-}, \omega_{n+}\big)} = r_<(\Omega + nw)/c$$

leads to

$$n \gg n_{\max} = \frac{r_< \Omega / c}{1 - r_< w/c} \text{ for positive } n \tag{2.230}$$

and

$$n \ll n_{\min} = -\frac{r_< \Omega / c}{1 + r_< w/c} \text{ for negative } n. \tag{2.231}$$

This result is identical to the estimates (2.147) and (2.148) for circular motion in 2D. The Bessel functions thus ensure that the spectral contributions within the Doppler intervals converge to zero for a sufficiently large number $|n|$. A more detailed mathematical investigation of the convergence behavior of the infinite series (2.218) is not performed here. Instead, we investigate the behavior of the series numerically.

Figures 2.32, 2.33, 2.34, and 2.35 display the spectral pressure for a source radiating at a frequency of 200 Hz and traveling along a helix with radius $r_0 = 0.5$ m, vertical Mach number $M_z = v_z/c = 0.1$, and azimuthal Mach number $M_\varphi =$

$wr_0/c = 2\pi \cdot 30r_0/c$. The source describes 30 turns over a vertical z-distance of 34.3 m in one second. Three pressure curves are shown for comparison: The green curve is calculated using the series (2.227), performing the summation from $n = -10$ to $n = 10$. In Sect. 2.3.2, we derived the *Cartesian convolution integral* (CCI) for finite observation times and claimed that the CCI perfectly agrees with the results of the infinite series. To show this, the curve with the blue circles is created according to the CCI (2.116) with an observation time of ten seconds ($t_0 = -5$ s, $t_1 = 5$ s):

$$p_{\text{num}}\text{helix} = \frac{Q}{4\pi v_z} \int\limits_{z_0=v_z t_0}^{z_1=v_z t_1} \frac{e^{i\frac{\omega}{c}\sqrt{(x-r_0\cos(w\zeta/v_z))^2+(y-r_0\sin(w\zeta/v_z))^2+(z-\zeta)^2}} \cdot e^{i(\omega-\Omega)\frac{\zeta}{v_z}}}{\sqrt{(x-r_0\cos(w\zeta/v_z))^2+(y-r_0\sin(w\zeta/v_z))^2+(z-\zeta)^2}}\, d\zeta \tag{2.232}$$

For comparison, the curve generated by a source moving at velocity v_z in a straight line along the z-axis is also shown (red curve, see Eq. (2.223)).

As stated, both methods—the CCI of Sect. 2.3.2 and the series solution—match perfectly and thus confirm each other. Since the receiver is close to the z-axis at

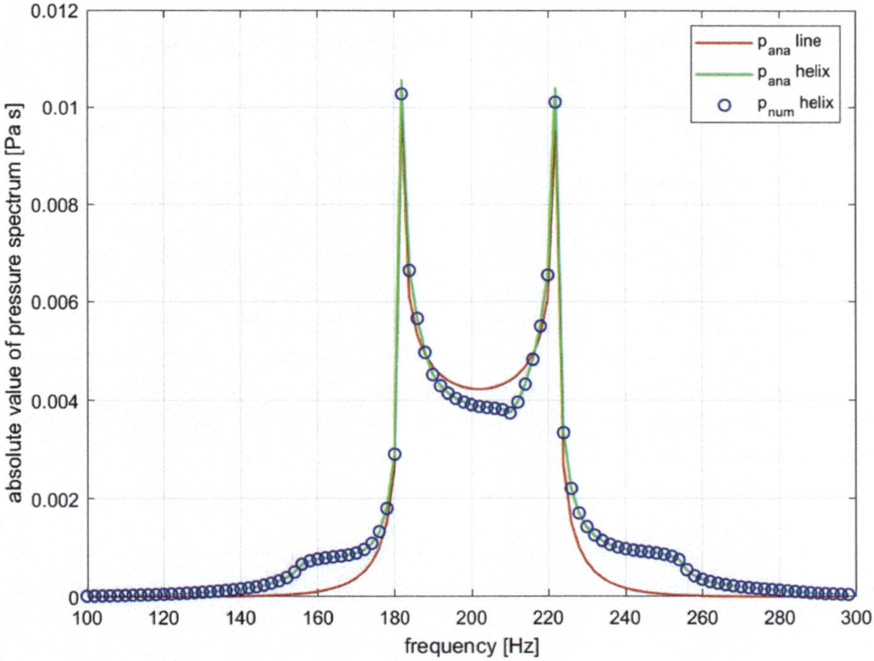

Fig. 2.32 Receiver inside the helix and near the z-axis: $(x, y, z) = (0.1\text{ m}, 0\text{ m}, 0.1\text{ m})$. Absolute value of the spectral pressure of a helix: $p_{\text{ana}}\text{line}$ = analytical expression (2.223), $p_{\text{ana}}\text{helix}$ = series (2.227), $p_{\text{num}}\text{helix}$ = integral (2.232). Parameters: $r_0 = 0.5$ m, $w = 2\pi \cdot 30\text{ s}^{-1}$, $v_z = 34.3$ m/s, $\Omega = 2\pi 200$ 1/s, $Q = 1$ kg/s^2

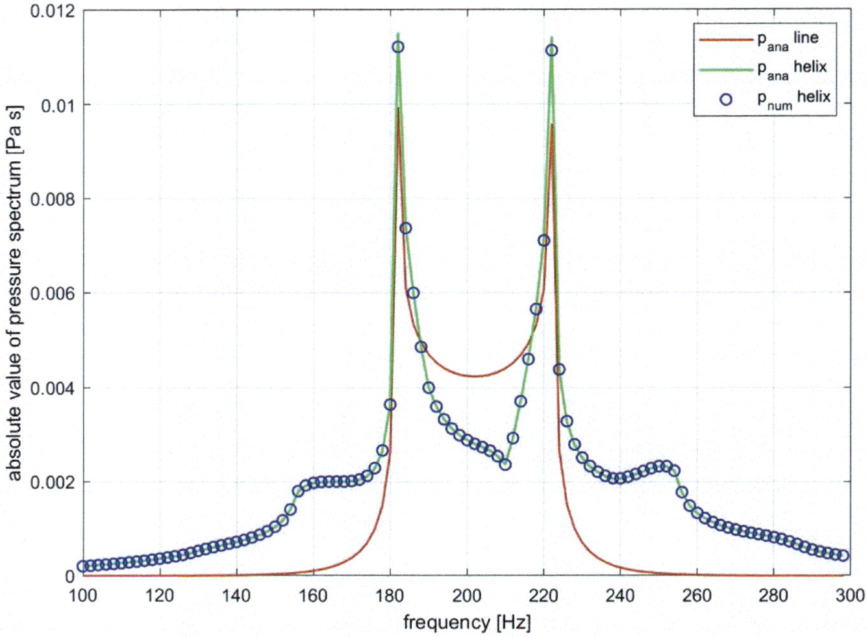

Fig. 2.33 Receiver still inside the helix: $(x, y, z) = (0.25 \text{ m}, 0 \text{ m}, 0.1 \text{ m})$. Absolute value of the spectral pressure of a helix: p_{ana}line = analytical expression (2.223), p_{ana}helix = series (2.227), p_{num}helix = integral (2.232). Parameters: $r_0 = 0.5$ m, $w = 2\pi \cdot 30 \text{ s}^{-1}$, $v_z = 34.3$ m/s, $\Omega = 2\pi 200$ 1/s, $Q = 1 \text{ kg/s}^2$.

$(x, y, z) = (0.1 \text{ m}, 0 \text{ m}, 0.1 \text{ m})$, the red curve produced by a rectilinearly moving source gives similar results, which are consistent with Sect. 2.5.2.3.

In the other three pictures, the receiver moves further outwards:

In Fig. 2.33, the receiving point $(x, y, z) = (0.25 \text{ m}, 0 \text{ m}, 0.1 \text{ m})$ is still inside the helix in the middle between the z-axis and the radius $r_0 = 0.5\ m$ of the helix.

In Fig. 2.34, the receiving point $(x, y, z) = (0.7 \text{ m}, 0.7 \text{ m}, 0.1 \text{ m})$ is outside but close to the helix:

In Fig. 2.35, the receiving point $(x, y, z) = (5 \text{ m}, 0 \text{ m}, 0.1 \text{ m})$ is outside but further away from the helix.

In the last figure, we have omitted the spectrum p_{ana}line of a straight line, as it now exhibits stronger deviations from the other curves due to its greater distance from the axis.

All figures demonstrate excellent agreement between the two distinct methods—the series solution and the CCI—which holds true for other parameters as well.

The convergence is good enough because the value of n_{max} of Eq. (2.230) for the selected parameters is always less than three so that the summation from $n = -10$ to $n = 10$ is sufficient to ensure the convergence of the series. The observation time of ten seconds was clearly adequate for evaluating the CCI.

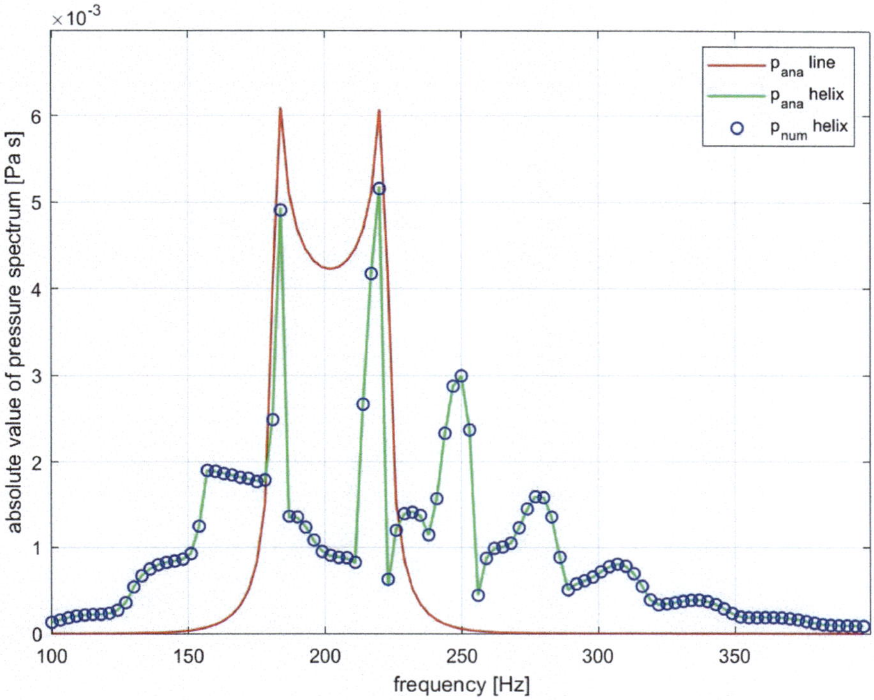

Fig. 2.34 Receiver outside and near the helix: $(x, y, z) = (0.7\text{ m}, 0.7\text{ m}, 0.1\text{ m})$. Absolute value of the spectral pressure of a helix: p_{ana}line = analytical expression (2.223), p_{ana}helix = series (2.227), p_{num}helix = integral (2.232). Parameters: $r_0 = 0.5$ m, $w = 2\pi \cdot 30\text{ s}^{-1}$, $v_z = 34.3$ m/s, $\Omega = 2\pi 200$ 1/s, $Q = 1$ kg/s^2.

The main peaks can easily be identified as the Doppler frequencies $f_{0-} = \Omega/(1 + M_z)/(2\pi) \approx 182$ Hz and $f_{0+} = \Omega/(1 - M_z)/(2\pi) \approx 222$ Hz and can be seen in all four figures. The side lobes are caused by the other Doppler frequencies $f_{n\pm} = (\Omega + nw)/(1 \mp M_z)/(2\pi)$ for $n \neq 0$. However, since they overlap, it becomes difficult to distinguish them easily.

If the theory is further extended in cylindrical coordinates, the radial coordinate could also vary linearly with time, which would lead to the wave equation

$$\Delta_3\varphi_3 - \frac{1}{c^2}\frac{\partial^2\varphi_3}{\partial t^2} = -Qe^{-i\Omega t}\frac{1}{r}\delta(r - r_s - v_s t)\delta_c(\phi - wt)\delta(z - v_z t). \qquad (2.233)$$

Similar to Sect. 2.4.2, where an Archimedean spiral was studied, a conical helix is described here in 3D. However, when the Green's function method is employed to Eq. (2.233), we obtain a complicated solution for the frequency spectrum that involves an infinite series and two infinite integrals. In the upcoming Sect. 2.6.2, we demonstrate that a more manageable solution can be obtained by working in spherical coordinates.

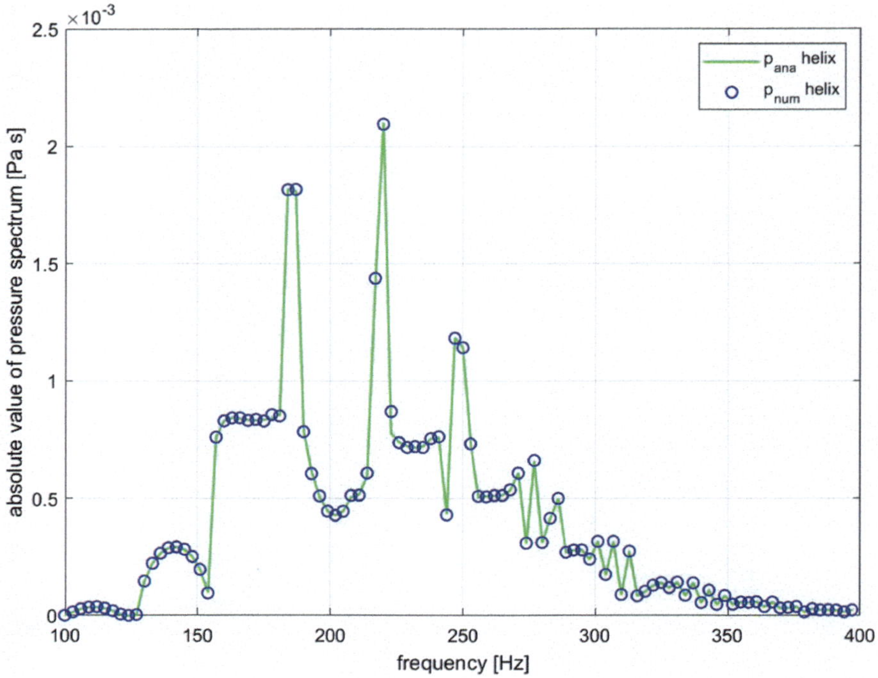

Fig. 2.35 Receiver outside but further away from the helix: $(x, y, z) = (5\text{ m}, 0\text{ m}, 0.1\text{ m})$. Absolute value of the spectral pressure of a helix: p_{ana}helix = series (2.227), p_{num}helix = integral (2.232). Parameters: $r_0 = 0.5$ m, $w = 2\pi \cdot 30\text{ s}^{-1}$, $v_z = 34.3$ m/s, $\Omega = 2\pi 200\ 1/\text{s}$, $Q = 1\text{ kg/s}^2$

2.6 Radiating Monopole Moving in Spherical Coordinates (3D Case)

In this section, we treat a moving point source in three-dimensional spherical coordinates. First, we again derive the frequency spectrum of a source traveling along a circular trajectory, and then we analyze a source moving along a conical helix.

2.6.1 A Point Source Moving Along a Circular Path

2.6.1.1 Derivation of the Spectral Pressure

We consider a radiating point source in three-dimensional space that moves on a circle of radius r_0 with angular velocity w around the origin in a plane parallel to the xy-plane—just as in cylindrical coordinates discussed in Sect. 2.5.1. In this case, however, the sound field needs to satisfy the 3D wave equation

$$\Delta\varphi_3 - \frac{1}{c^2}\frac{\partial^2\varphi_3}{\partial t^2} = \frac{Q_1}{\rho}e^{-i\Omega t}\frac{1}{r^2\sin\theta}\delta(r-r_0)\delta_c(\phi-wt)\delta(\theta-\theta_0) \tag{2.234}$$

in spherical coordinates r, θ, ϕ with

$$x = r\sin\theta\cos\phi,\ y = r\sin\theta\sin\phi,\ z = r\cos\theta. \tag{2.235}$$

Obviously, $r\sin\theta$ is the projection of the radius onto the xy-plane, i.e., the radius in cylindrical coordinates, and $r\cos\theta$ is the projection of the radius onto the z-axis.

The trajectory of the source is described by the curve

$$\begin{pmatrix} x_0(t) \\ y_0(t) \\ z_0(t) \end{pmatrix} = \begin{pmatrix} r_0\sin\theta_0\cos(wt) \\ r_0\sin\theta_0\sin(wt) \\ r_0\cos\theta_0 \end{pmatrix} \text{ for } t \in (-\infty, \infty). \tag{2.236}$$

The azimuthal angle of the source $\phi_0 = wt$ is 2π periodic. Thus, we once again work with the Dirac comb

$$\delta_c(\phi - wt) = \sum_{n=-\infty}^{\infty}\delta(\phi - wt - n2\pi) = \frac{1}{2\pi}\sum_{n=-\infty}^{\infty}e^{in(\varphi-wt)},$$

since the subsequent temporal Fourier transform ranges from minus infinity to plus infinity.

The Fourier transform with respect to time for the pressure $\hat{p}_3(\omega) = i\omega\rho\hat{\varphi}_3$ leads to

$$\begin{aligned}\Delta\hat{p}_3 + \left(\frac{\omega}{c}\right)^2\hat{p}_3 &= -Q\frac{\delta(r-r_0)}{2\pi r^2\sin\theta}\delta(\theta-\theta_0)\int_{-\infty}^{\infty}e^{i(\omega-\Omega)t}\sum_{n=-\infty}^{\infty}e^{in(\phi-wt)}\mathrm{d}t \\ &= -Q\frac{\delta(r-r_0)}{2\pi r^2\sin\theta}\delta(\theta-\theta_0)\int_{-\infty}^{\infty}\sum_{n=-\infty}^{\infty}e^{i(\omega-\Omega-nw)t}e^{in\phi}\mathrm{d}t\end{aligned} \tag{2.237}$$

with $Q = -i\omega Q_1$.

Using the known relationship

$$\int_{-\infty}^{\infty}e^{i(\omega-\Omega-nw)t}\mathrm{d}t = 2\pi\delta(\omega-\Omega-nw)$$

and exchanging the order of integration and summation, we get

$$\Delta\hat{p}_3 + \left(\frac{\omega}{c}\right)^2\hat{p}_3 = -Q\frac{\delta(r-r_0)}{r^2\sin\theta}\delta(\theta-\theta_0)\sum_{n=-\infty}^{\infty}\delta(\omega-\Omega-nw)e^{in\phi} =: -R. \tag{2.238}$$

The negative right-hand side R must be convoluted with the Green's function for the Helmholtz equation in spherical coordinates, which is given by

$$g_3(r,\theta,\phi;\hat{r},\hat{\theta},\hat{\phi}) = ik\sum_{n=0}^{\infty}\sum_{m=-n}^{n} h_n^{(1)}(k\hat{r}_>)j_n(k\hat{r}_<)Y_n^m(\theta,\phi)Y_n^{m*}(\hat{\theta},\hat{\phi}), \quad (2.239)$$

where the asterisk denotes the conjugate complex. Here, j_n and $h_n^{(1)}$ are the spherical Bessel and Hankel function of order n, $k = \frac{\omega}{c}$, $\hat{r}_< = \min\{r, \hat{r}\}$ and $\hat{r}_> = \max\{r, \hat{r}\}$. $Y_n^m(\theta, \phi)$ are the spherical harmonics defined by

$$Y_n^m(\theta,\phi) = A(n,m)P_n^{|m|}(\cos\theta)e^{im\phi}, \quad (2.240)$$

where

$$A(n,m) = \sqrt{\frac{(2n+1)}{4\pi}\frac{(n-|m|)!}{(n+|m|)!}},$$

and the $P_n^{|m|}(\cos\theta)$ are the associated Legendre polynomials (see Abramowitz and Stegun [7, Chapt. 8 and 10]). The representation of the Green's function (2.239) containing $|m|$ used here can be found in Colton and Kress [29], Theorem 2.7 and 2.10. However, it can easily be converted into the usual representation without the absolute value of m (see for example Williams [30, p. 259]).

The convolution leads to

$$\hat{p}_3 = \int_0^{\infty}\int_0^{2\pi}\int_0^{\pi} g_3(r,\theta,\phi;\hat{r},\hat{\theta},\hat{\phi})R(\hat{r},\hat{\theta},\hat{\phi})\hat{r}^2\sin\hat{\theta}\mathrm{d}\hat{\theta}\mathrm{d}\hat{\phi}\mathrm{d}\hat{r}. \quad (2.241)$$

Using the relation

$$\int_0^{2\pi} e^{i(m-n)\hat{\phi}}d\hat{\phi} = 2\pi\,\delta_{nm} \text{ with the Kronecker delta } \delta_{nm} = \begin{cases} 0, & \text{if } n \neq m \\ 1, & \text{if } n = m \end{cases}, \quad (2.242)$$

a straightforward calculation yields

$$\hat{p}_3 = 2\pi ikQ\sum_{n=0}^{\infty}\sum_{m=-n}^{n} A^2(n,m)j_n(kr_<)h_n^{(1)}(kr_>)P_n^{|m|}(\cos\theta)P_n^{|m|}(\cos\theta_0)$$
$$\times\,\delta(\omega-\Omega-mw)e^{im\phi} \quad (2.243)$$

with $r_< = \min\{r, r_0\}$,$r_> = \max\{r, r_0\}$.

This representation of the spectral pressure does not include any integral, as found in the expression (2.202) in cylindrical coordinates

$$\hat{p}_3(r,\phi,z) = \frac{iQ}{4}\sum_{n=-\infty}^{\infty}\delta(\omega-\Omega-nw)e^{in\phi}$$
$$\times\int_{-\infty}^{\infty}H_n^{(1)}\left(\sqrt{k^2-k_z^2}\,r_>\right)J_n\left(k\sqrt{k^2-k_z^2}\,r_<\right)e^{ik_z z}\,\mathrm{d}k_z.$$

Therefore, the version in spherical coordinates seems to be preferable for further evaluation.

2.6.1.2 Discussion of the Formula for the Spectral Pressure

Spectral pressure in the origin

As a first check, we consider the case where the receiver is at the origin, i.e., $r \to 0$. Since $j_n(0) = \delta_{n0}$, the spectral pressure takes the form

$$\begin{aligned}\hat{p}_3(r=0) &= 2\pi\delta(\omega-\Omega)ikQA^2(0,0)h_0^{(1)}(kr_0)P_0^0(\cos\theta)P_0^0(\cos\theta_0)\\ &= 2\pi\delta(\omega-\Omega)Q\frac{e^{i\frac{\omega}{c}r_0}}{4\pi r_0},\end{aligned} \tag{2.244}$$

since

$$A^2(0,0) = 1/(4\pi),\ h_0^{(1)}(kr_0) = -ie^{ikr_0}/(kr_0),\ \text{and}P_0^0(\cos\theta) = P_0^0(\cos\theta_0) = 1.$$

At the origin, we, therefore, observe the sound from a stationary point source located at a distance r_0 from the origin without any Doppler shift since the distance to the observer remains constant during the movement.

Pressure in the time domain

Due to the occurrence of the Dirac delta functions $\delta(\omega - \Omega - mw)$, the inverse Fourier transform of Eq. (2.243) can be performed immediately and gives the time-dependent pressure (note that $Q = -i\omega Q_1$)

$$\begin{aligned}p_3 &= \int_{-\infty}^{\infty}\frac{\omega^2}{c}Q_1\sum_{n=0}^{\infty}\sum_{m=-n}^{n}A^2(n,m)j_n(kr_<)h_n^{(1)}(kr_>)\\ &\quad\times P_n^{|m|}(\cos\theta)P_n^{|m|}(\cos\theta_0)\delta(\omega-\Omega-mw)e^{im\phi}e^{-i\omega t}\,\mathrm{d}\omega\\ &= Q_1 c\sum_{n=0}^{\infty}\sum_{m=-n}^{n}\tilde{k}_m^2A^2(n,m)j_n(\tilde{k}_m r_<)h_n^{(1)}(\tilde{k}_m r_>)\\ &\quad\times P_n^{|m|}(\cos\theta)P_n^{|m|}(\cos\theta_0)e^{-i(\Omega+mw)t}e^{im\phi}\end{aligned} \tag{2.245}$$

with $\tilde{k}_m = (\Omega + mw)/c$.

For the acoustic potential φ_3 we obtain the formula

$$\varphi_3 = -\frac{i}{\rho} Q_1 \sum_{n=0}^{\infty} \sum_{m=-n}^{n} \tilde{k}_m A^2(n, m) j_n(\tilde{k}_m r_<) h_n^{(1)}(\tilde{k}_m r_>) \\ \times P_n^{|m|}(\cos\theta) P_n^{|m|}(\cos\theta_0) e^{-i(\Omega+mw)t} e^{im\phi}, \qquad (2.246)$$

since $p_3 = -\rho \frac{\partial \varphi_3}{\partial t}$.

Comparison with results found in the literature

These results for the circular orbit in spherical coordinates have already been found in the literature by Poletti [21] and Mao et al. [31]. Similar to Sect. 2.4.1.3 for the circular orbit in 2D, it can be observed that the solution (2.246) is identical to that presented by Poletti in [21] in his Eqs. (40) and (42), which are reproduced in compact form in Eq. (A8) of Mao et al. [31]. This can be easily recognized by choosing $\phi_0 = 0$ and considering the following properties of the spherical Bessel functions (see Abramowitz and Stegun [7, Eqs. 10.1.34 and 10.1.36])

$$j_n(-z) = (-1)^n j_n(z) \text{ and } h_n^{(1)}(-z) = (-1)^n h_n^{(2)}(z),\, n \in \mathbb{Z}$$

and thus[3]

$$j_n(z) h_n^{(1)}(z) = \begin{cases} j_n(z) h_n^{(1)}(z) & \text{if } z > 0 \\ j_n(|z|) h_n^{(2)}(|z|) & \text{if } z < 0 \end{cases} = j_n(|z|) h_n^{(\tilde{\delta}_n)}(|z|) \qquad (2.247)$$

with $\tilde{\delta}_n = \begin{cases} 1 \; if \; z > 0 \\ 2 \; if \; z < 0 \end{cases}$. The differently defined amplitudes are caused by different right-hand sides of the wave equation. The acoustic pressure given by Poletti [21] and Mao et al. [31] corresponds to our acoustic velocity potential.

Poletti determined his solution with a simple source approach. In addition, the frequency spectrum of the sound field is given in Eqs. (9.13) of [31] agrees with our result (2.243), written in terms of the acoustic potential (cf. Appendix A in [31]).

In these and many other papers, such as in [32–41], etc., the solution for the pressure radiated from a point source on a circular orbit is evaluated and discussed in detail. We will, therefore, not dwell on it. We will instead focus on the conical helix, which is also analyzed in reference [27].

[3] Note the additional factor $(-1)^{(\tilde{\delta}+1)}$ in the formulas for the "cylindrical" Bessel functions below Eq. (2.160).

2.6.2 A Point Source Moving Along a Conical Helix

Now we consider a point source winding along the conical helix

$$\begin{pmatrix} x_0(t) \\ y_0(t) \\ z_0(t) \end{pmatrix} = \begin{pmatrix} (r_s + v_r t)\sin\theta_0\cos(wt) \\ (r_s + v_r t)\sin\theta_0\sin(wt) \\ (r_s + v_r t)\cos\theta_0 \end{pmatrix} \text{ for } t \in [t_0, t_{\max}], \tag{2.248}$$

where the radius $r_0(t) = r_s + v_r t$ increases over time with the velocity v_r. Again, we assume that the azimuthal angle is $\phi_0 = wt$ with angular velocity w. We note that the helix is only defined for $t \geq 0$ and not for the whole range $-\infty < t < \infty$. Therefore, we are forced to insert a characteristic function into the right-hand side of the wave equation as shown below (cf. the 2D case of an Archimedean spiral in Sect. 2.4.2).

The absolute value of the speed of the source can be easily calculated

$$\left|\frac{\mathrm{d}\vec{x}_0(t)}{\mathrm{d}t}\right| = \sqrt{v_r^2 + w^2[\sin\theta_0(r_s + v_r t)]^2} \tag{2.249}$$

and increases as time progresses.

A sketch of such a conical helix is shown in Fig. 2.36.

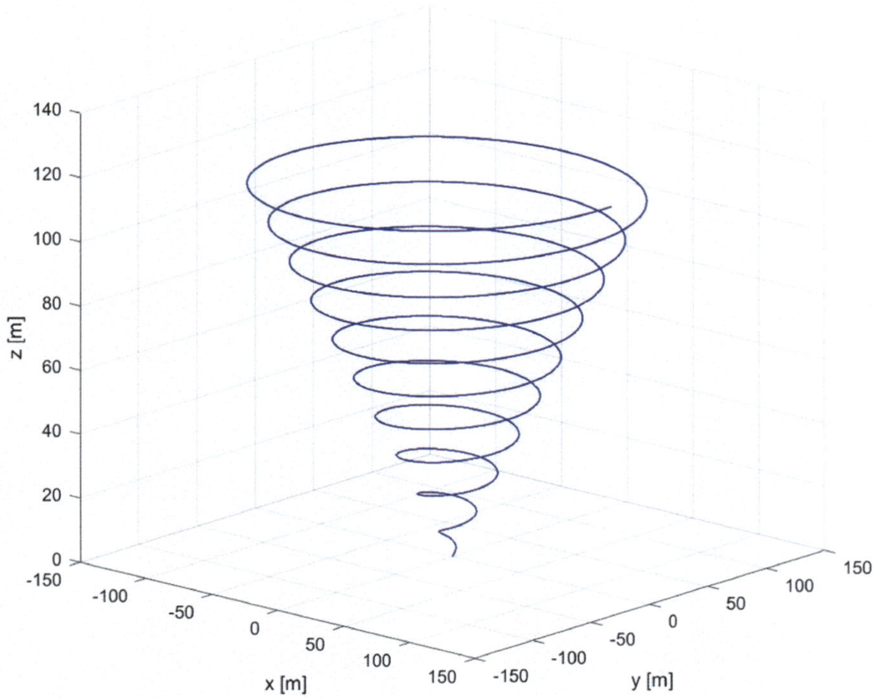

Fig. 2.36 A conical helix with parameters: $r_s = 0$, $v_r = 0.05 \cdot c$, $w = 2\pi\ \mathrm{s}^{-1}$, $\theta_0 = \pi/4$, $t \in (0, 10)$ s

2.6.2.1 Derivation of the Spectral Pressure

The sound field φ_3 is described by the 3D wave equation

$$\Delta\varphi_3 - \frac{1}{c^2}\frac{\partial^2\varphi_3}{\partial t^2} = \frac{Q_1}{\rho}e^{-i\Omega t}\frac{1}{r^2\sin\theta}\delta(r-(r_s+v_rt))\delta_c(\phi-wt) \times \delta(\theta-\theta_0)\chi_{[0,t_{\max}]}(t) \tag{2.250}$$

in spherical coordinates. For simplification, we have chosen $t_0 = 0$.

The Fourier transform with respect to time leads to the Helmholtz equation for the spectral pressure $\hat{p}_3(\omega) = i\omega\rho\hat{\varphi}_3$

$$\begin{aligned}\Delta\hat{p}_3 + \left(\frac{\omega}{c}\right)^2\hat{p}_3 &= -Q\frac{\delta(\theta-\theta_0)}{2\pi r^2\sin\theta}\int\limits_{-\infty}^{\infty} e^{i(\omega-\Omega)t}\delta(r-(r_s+v_rt)) \\ &\quad\times\sum_{n=-\infty}^{\infty} e^{in(\phi-\omega t)}\chi_{[0,t_{\max}]}(t)\,\mathrm{d}t \\ &= -Q\frac{\delta(\theta-\theta_0)}{2\pi r^2\sin\theta}\sum_{n=-\infty}^{\infty} e^{in\phi}\int\limits_{-\infty}^{\infty} e^{i(\omega-\Omega-n\omega)t} \\ &\quad\times\frac{1}{v_r}\delta\left(\frac{r-r_s}{v_r}-t\right)\chi_{[0,t_{\max}]}(t)\,\mathrm{d}t,\end{aligned} \tag{2.251}$$

with $Q = -i\omega Q_1$. Thanks to the Dirac delta function, the last integral can be evaluated so that we obtain

$$\begin{aligned}\Delta\hat{p}_3 + \left(\frac{\omega}{c}\right)^2\hat{p}_3 &= -Q\frac{\delta(\theta-\theta_0)}{2\pi v_r r^2\sin\theta}\sum_{n=-\infty}^{\infty} e^{in\phi}e^{i(\omega-\Omega-nw)\frac{r-r_s}{v_r}}\chi_{[r_s,r_{\max}]}(r) \\ &=: -R(r,\theta,\phi)\end{aligned} \tag{2.252}$$

with $r_{\max} = r_s + v_rt_{\max}$.

The right-hand side $R(r,\theta,\phi)$ must be convoluted with the Green's function for the Helmholtz equation in spherical coordinates $g_3\big(r,\theta,\phi;\hat{r},\hat{\theta},\hat{\phi}\big)$, which is given by Eq. (2.239):

$$\hat{p}_3 = \int\limits_0^{\infty}\int\limits_0^{2\pi}\int\limits_0^{\pi} g_3\big(r,\theta,\phi;\hat{r},\hat{\theta},\hat{\phi}\big)R\big(\hat{r},\hat{\theta},\hat{\phi}\big)\hat{r}^2\sin\hat{\theta}\,\mathrm{d}\hat{\theta}\,\mathrm{d}\hat{\phi}\,\mathrm{d}\hat{r}.$$

Using the relationship

$$\int\limits_0^{2\pi} e^{i(m-n)\hat{\phi}}\mathrm{d}\hat{\varphi} = 2\pi\delta_{nm} \text{ with the Kronecker delta } \delta_{nm} = \begin{cases}0, & \text{if } n\neq m\\ 1, & \text{if } n = m\end{cases}, \tag{2.253}$$

a straight forward calculation leads to the following representation of the acoustical pressure

$$\hat{p}_3 = \\ \frac{ikQ}{v_r}\sum_{n=0}^{\infty}\sum_{m=-n}^{n} A^2(n,m)P_n^{|m|}(\cos\theta)P_n^{|m|}(\cos\theta_0)e^{im\phi}\int_{r_s}^{r_{\max}} j_n(kr_<)h_n^{(1)}(kr_>)e^{i(\omega-\Omega-mw)\frac{\hat{r}-r_s}{v_r}}\,\mathrm{d}\hat{r} \tag{2.254}$$

with the known abbreviation $A^2(n,m) = \frac{(2n+1)}{4\pi}\frac{(n-|m|)!}{(n+|m|)!}$ and $r_< = \min\{r,\hat{r}\}$,$r_> = \max\{r,\hat{r}\}$. Formula (2.254) is the three-dimensional counterpart to the spectrum of the two-dimensional Archimedean spiral (2.176).

2.6.2.2 Discussion of the Formula for the Spectral Pressure

Obviously, the spectral pressure (2.254) requires numerical evaluation as the integral cannot be solved explicitly. However, we obtain some interesting results when the observer is positioned at the origin.

Spectral pressure in the origin

To obtain a simpler expression, we assume that the receiver is located at the origin, i.e., $r \to 0$. Since $j_n(0) = \delta_{n0}$, $A^2(0,0) = 1/(4\pi)$, and $P_0^0(\cos\theta) = 1$, the spectral pressure takes the form

$$\hat{p}_3(r=0) = \frac{ikQ}{4\pi v_r}\int_{r_s}^{r_{\max}} h_0^{(1)}(k\hat{r})e^{i(\omega-\Omega)\frac{\hat{r}-r_s}{v_r}}\,\mathrm{d}\hat{r} = \frac{Q}{v_r}\int_{r_s}^{r_{\max}} \frac{e^{ik\hat{r}}}{4\pi\hat{r}}e^{i(\omega-\Omega)\frac{\hat{r}-r_s}{v_r}}\,\mathrm{d}\hat{r}, \tag{2.255}$$

where we have taken into account that $h_0^{(1)}(k\hat{r}) = e^{ik\hat{r}}/(ik\hat{r})$.

We compare this result with the analogous solution for the Archimedean spiral (see Eq. 2.180) in 2D:

$$2\text{D}: \hat{p}_2(r=0) = \frac{Q}{v_r}\int_{r_s}^{r_{\max}} \frac{i}{4}H_0^{(1)}\left(\frac{\omega}{c}\hat{r}\right)e^{i(\omega-\Omega)\frac{\hat{r}-r_s}{v_r}}\,\mathrm{d}\hat{r}, \tag{2.256}$$

$$3\text{D}: \hat{p}_3(r=0) = \frac{Q}{v_r}\int_{r_s}^{r_{\max}} \frac{e^{i(\frac{\omega}{c})\hat{r}}}{4\pi\hat{r}}e^{i(\omega-\Omega)\frac{\hat{r}-r_s}{v_r}}\,\mathrm{d}\hat{r}. \tag{2.257}$$

We note that these expressions differ only in that the radial dependency is described by either the 2D Green's function $g_2 = \frac{i}{4}H_0^{(1)}\left(\frac{\omega}{c}\hat{r}\right)$ or the 3D Green's function $g_3 = \frac{e^{i(\frac{\omega}{c})\hat{r}}}{4\pi\hat{r}}$.

In addition, we compare Eq. (2.267) with formula (2.111) for a point source that moves in a straight line along the x-axis from x_0 to x_1. For an observer at the origin ($r = 0$), we obtain from Eq. (2.111) with $\Phi(\xi/u) = \psi(\xi/u) = 0$

$$\hat{p}_3(x = 0, y = 0, z = 0) = \frac{Q}{4\pi u} \int\limits_{x_0=ut_0}^{x_1=ut_1} \frac{e^{i\frac{\omega}{c}|\xi|}}{|\xi|} e^{i(\omega-\Omega)\frac{\xi}{u}} \mathrm{d}\xi. \tag{2.258}$$

Just as in 2D for the Archimedean spiral in Sect. 2.4.2.2, this yields the following interesting result in 3D: At the origin, the spectrum of a point source moving along a conical helix from $r_s = 0$ to $r_{\max}$ with radial velocity v_r is identical to the spectrum of a source moving in a straight line from $x_0 = 0$ to $x_1 = r_{\max}$ with velocity u (cf. Fig. 2.26). Except for the prefactor $\exp(-i(\omega - \Omega)r_s/v_r)$, the spectrum also remains the same, even if $r_s = x_0$ is not equal to zero.

If we also assume that $r_{\max} = \infty$, then the pressure at the origin can be written as the exponential-integral function $\mathrm{Ei}(x)$ (see [24, p. 925]). For this purpose, we write Eq. (2.255) in the form

$$\begin{aligned} \hat{p}_3(r = 0) &= \frac{Q}{4\pi v_r} e^{-i(\omega-\Omega)\frac{r_s}{v_r}} \int\limits_{r_s}^{\infty} \frac{e^{i\left(\frac{\omega}{c}+\frac{\omega-\Omega}{v_r}\right)\hat{r}}}{\hat{r}} \mathrm{d}\hat{r} \\ &= \frac{Q}{4\pi v_r} e^{-i(\omega-\Omega)\frac{r_s}{v_r}} \int\limits_{r_s}^{\infty} \frac{e^{i\frac{M_r+1}{v_r}\left(\omega-\frac{\Omega}{M_r+1}\right)\hat{r}}}{\hat{r}} \mathrm{d}\hat{r} \end{aligned} \tag{2.259}$$

with $M_r = v_r/c$.

Due to [24, Eqs. 8.231 and 8.233] we have

$$\begin{aligned} \mathrm{Ei}(ixy) = \mathrm{ci}(xy) + i\,\mathrm{si}(xy) &= -\int\limits_{x}^{\infty} \frac{\cos ty}{t} \mathrm{d}t - i \int\limits_{x}^{\infty} \frac{\sin ty}{t} \mathrm{d}t \\ &= -\int\limits_{x}^{\infty} \frac{e^{ity}}{t} \mathrm{d}t. \end{aligned} \tag{2.260}$$

By substituting $t = \hat{r},\ x = r_s,\ y = \frac{1+M_r}{v_r}\left(\omega - \frac{\Omega}{1+M_r}\right)$ we get

$$\hat{p}_3(r = 0) = \frac{-Q}{4\pi v_r} e^{-i(\omega-\Omega)\frac{r_s}{v_r}} \mathrm{Ei}\left[i\left(\omega - \frac{\Omega}{1+M_r}\right)(1+M_r)\frac{r_s}{v_r}\right]. \tag{2.261}$$

Since the exponential-integral function is not very well known and its solution is restricted to an observer at the origin $r = 0$, we perform the numerical evaluation for arbitrary values of r with the help of the CCI in the next Sect. 2.6.2.3.

Time-dependent potential in the origin

At the origin, we obtain a very simple expression for the time-varying acoustic potential. To perform the inverse Fourier transform with respect to time, we write Eq. (2.255) in the form

$$\hat{p}_3(r=0) = \frac{Q}{4\pi v_r} \int_0^{r_{\max}} \frac{1}{\hat{r}} e^{-i\Omega \frac{\hat{r}}{v_r}} e^{i\omega\left(\frac{\hat{r}}{c}+\frac{\hat{r}}{v_r}\right)} \mathrm{d}\hat{r} \tag{2.262}$$

with $r_s = 0$ for simplicity. The corresponding acoustic potential is given by (cf. Eq. (2.5))

$$\begin{aligned} \hat{\varphi}_3(r=0) &= \frac{\hat{p}_3(r=0)}{i\omega\rho} \\ &= \frac{-Q_1}{4\pi\rho v_r} \int_0^{r_{\max}} \frac{1}{\hat{r}} e^{-i\Omega \frac{\hat{r}}{v_r}} e^{i\omega\left(\frac{\hat{r}}{c}+\frac{\hat{r}}{v_r}\right)} \mathrm{d}\hat{r},\; Q = -i\omega Q_1. \end{aligned} \tag{2.263}$$

We only consider the ω-dependent part

$$F(\omega) = e^{i\frac{\omega}{c}\hat{r}\left(1+\frac{1}{M_r}\right)}. \tag{2.264}$$

The inverse Fourier transform yields

$$\begin{aligned} f(t) &= \frac{1}{2\pi} \int_{-\infty}^{\infty} F(\omega) e^{-i\omega t} \mathrm{d}\omega = \frac{1}{2\pi} \int_{-\infty}^{\infty} e^{i\frac{\omega}{c}\hat{r}\left(1+\frac{1}{M_r}\right)} e^{-i\omega t} \mathrm{d}\omega \\ &= \delta\left[\frac{\hat{r}}{c}\left(1+\frac{1}{M_r}\right) - t\right] = \frac{v_r}{1+M_r}\delta\left[\hat{r} - \frac{v_r}{1+M_r}t\right]. \end{aligned} \tag{2.265}$$

Thus, we obtain

$$\begin{aligned} \varphi_3(r=0,t) &= \frac{1}{2\pi} \int_{-\infty}^{\infty} \hat{\varphi}_3(r=0,\omega) e^{-i\omega t} \mathrm{d}\omega \\ &= \frac{-Q_1}{4\pi\rho} \frac{1}{1+M_r} \int_0^{r_{\max}} \frac{1}{\hat{r}} e^{-i\Omega \frac{\hat{r}}{v_r}} \delta\left[\hat{r} - \frac{v_r}{1+M_r}t\right] \mathrm{d}\hat{r} \end{aligned} \tag{2.266}$$

or

$$\varphi_3(r=0,t) = \frac{-Q_1}{4\pi\rho v_r t} e^{-i\left(\frac{\Omega}{1+M_r}\right)t} \chi_{[0,t_{\max}(1+M_r)]}(t), \tag{2.267}$$

where we have used that

$$\chi_{[0,r_{\max}]}(\hat{r})\Big|_{\hat{r}=\frac{v_r}{1+M_r}t} = \chi_{[0,t_{\max}(1+M_r)]}(t).$$

To summarize, a signal with the lower Doppler frequency $\Omega/(1+M_r)$ can be heard at the origin as the source continuously moves away from the origin on the helical orbit. The signal weakens over time as it gets further from the origin, and it can be detected up to the extended time $t = (1+M_r)t_{\max}$.

2.6.2.3 Comparison with the Cartesian Convolution Integral and Numerical Evaluation

We now derive the Cartesian convolution integral (CCI) for the conical helix as described in Sect. 2.3. For simplicity, we assume in the following that $r_s = 0$ and obtain from Eq. (2.248) the curve

$$\begin{pmatrix} x_0(t) \\ y_0(t) \\ z_0(t) \end{pmatrix} = \begin{pmatrix} v_r t \sin\theta_0 \cos(wt) \\ v_r t \sin\theta_0 \sin(wt) \\ v_r t \cos\theta_0 \end{pmatrix} = \begin{pmatrix} \tilde{v}_r t \cos(wt) \\ \tilde{v}_r t \sin(wt) \\ \tilde{v}_z t \end{pmatrix} \text{ for } t \in (0, t_{\max}), \tag{2.268}$$

where $\tilde{v}_r = v_r \sin\theta_0$ is the projection of the velocity of the source in the r-direction v_r onto the xy-plane and $\tilde{v}_z = v_r \cos\theta_0$ is the projection of v_r onto the z-axis.

The wave equation for the conical helix in Cartesian coordinates reads

$$\begin{aligned} \Delta\varphi_3 - \frac{1}{c^2}\frac{\partial^2\varphi_3}{\partial t^2} = \frac{Q_1}{\rho} e^{-i\Omega t}\delta(x - \tilde{v}_r t \cos wt) \\ \times \delta(y - \tilde{v}_r t \sin wt)\delta(z - \tilde{v}_z t)\chi_{[0,t_{\max}]}(t). \end{aligned} \tag{2.269}$$

Proceeding as usual, the temporal Fourier transform for the pressure $\hat{p}(\omega) = i\omega\rho\hat{\varphi}_3$ yields

$$\begin{aligned} \Delta\hat{p} + \left(\frac{\omega}{c}\right)^2 \hat{p} &= -Q \int_{-\infty}^{\infty} e^{i(\omega-\Omega)t}\delta(x - \tilde{v}_r t \cos wt)\delta(y - \tilde{v}_r t \sin wt)\delta(z - \tilde{v}_z t)\chi_{[0,t_{\max}]}(t)\mathrm{d}t \\ &= -\frac{Q}{\tilde{v}_z} e^{i(\omega-\Omega)\frac{z}{\tilde{v}_z}}\delta\left(x - \frac{\tilde{v}_r}{\tilde{v}_z} z \cos\left(\frac{w}{\tilde{v}_z} z\right)\right)\delta\left(y - \frac{\tilde{v}_r}{\tilde{v}_z} z \sin\left(\frac{w}{\tilde{v}_z} z\right)\right)\chi_{[0,z_{\max}]}(z) \\ &=: R(x, y, z) \end{aligned} \tag{2.270}$$

with $z_{\max} = \tilde{v}_z t_{\max}$ and $Q = -i\omega Q_1$. The convolution with the Green's function

$$\hat{g}(x, y, z) = \frac{1}{4\pi\sqrt{x^2+y^2+z^2}} e^{i\frac{\omega}{c}\sqrt{x^2+y^2+z^2}} \tag{2.271}$$

gives

$$\hat{p} = \int_{-\infty}^{\infty}\int_{-\infty}^{\infty}\int_{-\infty}^{\infty} \hat{g}(x-\xi, y-\eta, z-\zeta)R(\xi,\eta,\zeta)\mathrm{d}\xi\mathrm{d}\eta\mathrm{d}\zeta.$$

This results in

$$\hat{p} = \frac{Q}{4\pi\tilde{v}_z}\int_0^{z_{\max}} \frac{e^{i\frac{\omega}{c}\sqrt{\left(x-\frac{\tilde{v}_r}{\tilde{v}_z}\zeta\cos\left(\frac{w}{\tilde{v}_z}\zeta\right)\right)^2+\left(y-\frac{\tilde{v}_r}{\tilde{v}_z}\zeta\sin\left(\frac{w}{\tilde{v}_z}\zeta\right)\right)^2+(z-\zeta)^2}}\, e^{i(\omega-\Omega)\frac{\zeta}{\tilde{v}_z}}}{\sqrt{\left(x-\frac{\tilde{v}_r}{\tilde{v}_z}\zeta\cos\left(\frac{w}{\tilde{v}_z}\zeta\right)\right)^2+\left(y-\frac{\tilde{v}_r}{\tilde{v}_z}\zeta\sin\left(\frac{w}{\tilde{v}_z}\zeta\right)\right)^2+(z-\zeta)^2}}\mathrm{d}\zeta. \tag{2.272}$$

This formula was also recently derived in [27]. At first glance, this expression appears to be somewhat complicated. Therefore, we want to check whether we get the simple expression (2.255) at the origin. To this end, we set $(x, y, z) = (0, 0, 0)$ and obtain for the square root in the last equation

$$\begin{aligned} \text{root} &= \sqrt{\left(\frac{\tilde{v}_r}{\tilde{v}_z}\zeta\cos\left(\frac{w}{\tilde{v}_z}\zeta\right)\right)^2+\left(\frac{\tilde{v}_r}{\tilde{v}_z}\zeta\sin\left(\frac{w}{\tilde{v}_z}\zeta\right)\right)^2+\zeta^2} \\ &= \sqrt{\left(\frac{\tilde{v}_r}{\tilde{v}_z}\zeta\right)^2+\zeta^2} = \zeta\sqrt{\left(\frac{\tilde{v}_r}{\tilde{v}_z}\right)^2+1} \end{aligned}$$

Since $\tilde{v}_r = v_r\sin\theta_0$ and $\tilde{v}_z = v_r\cos\theta_0$, we immediately obtain

$$\begin{aligned} \text{root} &= \zeta\sqrt{\left(\frac{\tilde{v}_r}{\tilde{v}_z}\right)^2+1} = \frac{\zeta}{\tilde{v}_z}\sqrt{\tilde{v}_r^2+\tilde{v}_z^2} \\ &= \frac{\zeta}{\tilde{v}_z}v_r = \frac{\zeta}{\cos\theta_0}. \end{aligned} \tag{2.273}$$

Thus, the solution (2.272) is simplified to

$$\begin{aligned} \hat{p} &= \frac{Q}{4\pi\tilde{v}_z}\int_0^{z_{\max}} \frac{e^{i\frac{\omega}{c}\frac{\zeta}{\cos\theta_0}}e^{i(\omega-\Omega)\frac{\zeta}{\tilde{v}_z}}}{\zeta/\cos\theta_0}\mathrm{d}\zeta \\ &= \frac{Q}{4\pi v_r\cos\theta_0}\int_0^{r_{\max}\cos\theta_0} \frac{e^{i\frac{\omega}{c}\frac{\zeta}{\cos\theta_0}}e^{i\frac{(\omega-\Omega)}{v_r}\frac{\zeta}{\cos\theta_0}}}{\zeta/\cos\theta_0}\mathrm{d}\zeta, \end{aligned} \tag{2.274}$$

since $z_{\max} = \tilde{v}_z t_{\max} = v_r\cos\theta_0 t_{\max} = r_{\max}\cos\theta_0$. The substitution $\hat{r} = \zeta/\cos\theta_0$ gives

$$\hat{p} = \frac{Q}{4\pi v_r}\int_0^{r_{\max}} \frac{e^{i\frac{\omega}{c}\hat{r}}e^{i\frac{(\omega-\Omega)}{v_r}\hat{r}}}{\hat{r}}\mathrm{d}\hat{r}, \tag{2.275}$$

which is identical to the formula (2.255) for $r_s = 0$ obtained from the solution in spherical coordinates at the origin. Both approaches, therefore, confirm each other in their correctness.

We now compare the numerical solution of the CCI (2.272) with the series solution in spherical coordinates (2.254) using Matlab. The question arises as to how many terms in the expansion (2.254) have to be considered to obtain a sufficiently accurate solution. Similar to Sect. 2.4.1, the approximation of the Bessel function for high order n shows that the spherical Bessel functions also tend to zero when the order n becomes larger than the argument: $n > n_{\text{lim}} = (\omega/c)r \geq (\omega/c)r_<$. Thus, the outer sum in the series (2.254) is restricted to $n_{\max} > n_{\text{lim}}$ terms.

The source always starts at the origin so that $r_s = 0$ and radiates at a frequency of 200 Hz. The polar angle is chosen to $\theta_0 = \pi/4$ and $Q = 1\ \text{kg/s}^2$.

We consider two cases:

Case 1: The source should be always *subsonic.*

For that reason, we choose for the radial and azimuthal velocities

$$v_r = 0.05 \cdot c = 17.15\ \text{m/s and } w = 2\pi\ \text{s}^{-1}.$$

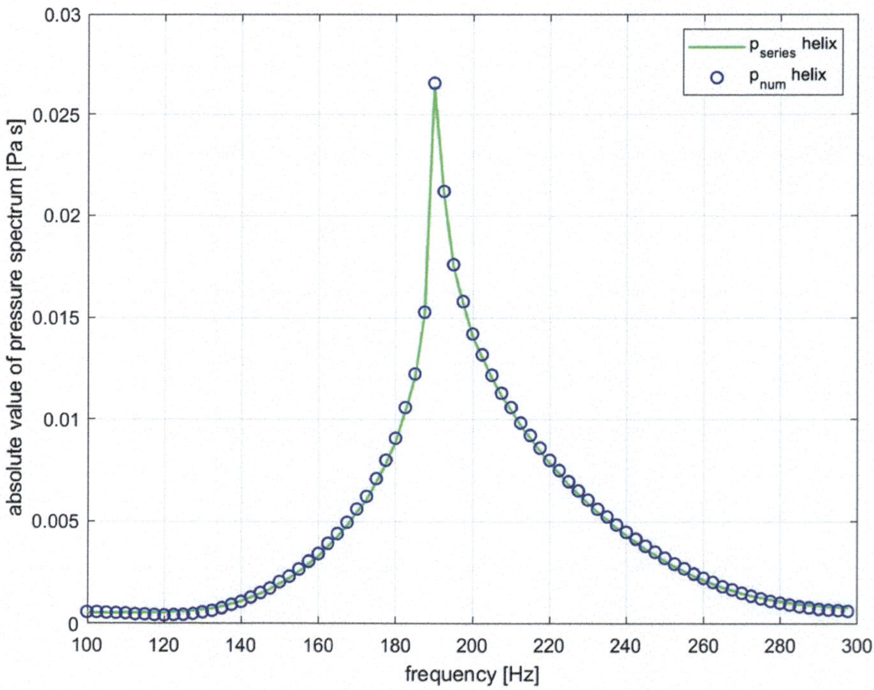

Fig. 2.37 Receiver inside the helix and near the origin: $P_1(x, y, z) = (0.1, 0.1, 0.1)m$. Absolute value of the spectral pressure for the conical helix: p_{ana}helix = series (2.254), p_{num}helix = CCI (2.272). Parameters of case 1 with $n_{\max} = 20$

The observation time is one second. The height of the helix is $\tilde{v}_z \cdot 1\text{ s} = v_r \cos\theta_0 \cdot 1\text{ s} = 12.1$ m and, due to Eq. (2.249), the maximal speed is

$$v_{\max} = 78.1 \text{ m/s at } t = t_{\max} = 1 \text{ s},$$

so that $M_{\max} = v_{\max}/c \approx 0.23 < 1$.

Figures 2.37, 2.38, 2.39, and 2.40 plot the spectral pressure for four different receiver points:

$P_1(x, y, z) = (0.1, 0.1, 0.1)m$, i.e., near the origin,
$P_2(x, y, z) = (1, 1, 10)m$, i.e., still inside and near the top of the helix,
$P_3(x, y, z) = (0, 0, 20)m$, i.e., above the trajectory of the helix,
$P_4(x, y, z) = (5, 0, 2)m$, i.e., located laterally to the helix.

Two pressure curves are shown for comparison: The green curve is calculated using the series (2.254) performing the summation from $n = 0$ to $n = n_{\max}$. The blue circles are created according to the CCI (2.272).

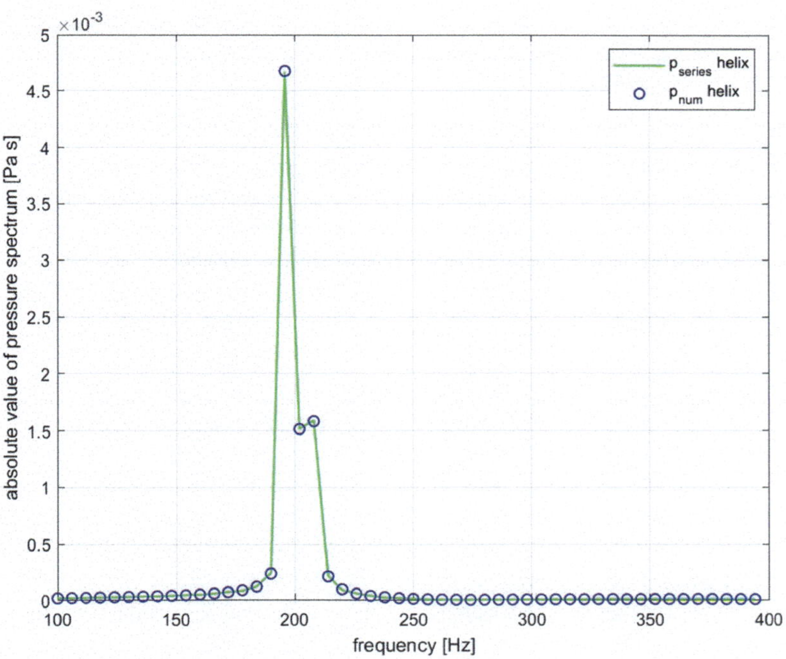

Fig. 2.38 Receiver still inside the helix: $P_2(x, y, z) = (1, 1, 10)m$. Absolute value of the spectral pressure for the conical helix: p_{ana}helix = series (2.254), p_{num}helix = CCI (2.272). Parameters of case 1 with $n_{\max} = 70$

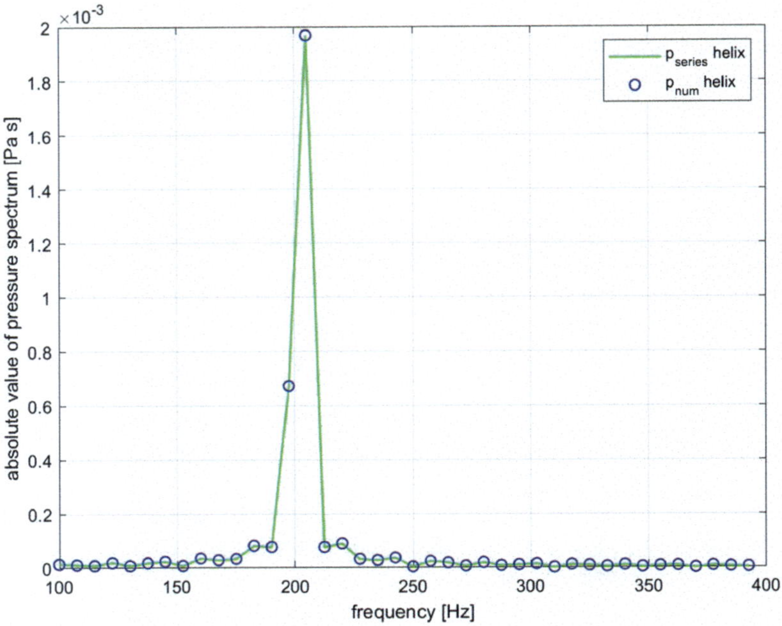

Fig. 2.39 Receiver outside and well above the helix: $P_3(x, y, z) = (0, 0, 20)m$. Absolute value of the spectral pressure for the conical helix: p_{ana}helix = series (2.254), p_{num}helix = CCI (2.272). Parameters of case 1 with $n_{max} = 150$

We observe that both methods—the CCI and the series solution—perfectly match and thus confirm each other. The peak frequency is below the radiating frequency of 200 Hz since the source is primarily moving away from the observer.

In the other three figures, the receiver moves further outwards.

In Fig. 2.38, the receiving point $P_2(x, y, z) = (1, 1, 10)m$ is still inside and near the top of the helix. However, now $n_{lim} = 56$ and, therefore, we have increased the value of n_{max} to $n_{max} = 70$. The main peak frequency is below the radiation frequency, but a minor peak is clearly visible just above 200 Hz. The reason for this is that the source first approaches the observer and then moves away from the observer.

In Fig. 2.39, the receiving point $P_3(x, y, z) = (0, 0, 20)m$ is outside and well above the helix. Now $n_{lim} = 147$ and, therefore, we have increased the value of n_{max} to 150. We see that the peak frequency is slightly above 200 Hz, as the source flies exclusively toward the observer.

In Fig. 2.40 the receiving point $P_4(x, y, z) = (5, 0, 2)m$ is located laterally to the helix. Now $n_{lim} = 39$ and, therefore, we choose $n_{max} = 40$. Two peak frequencies can be clearly recognized.

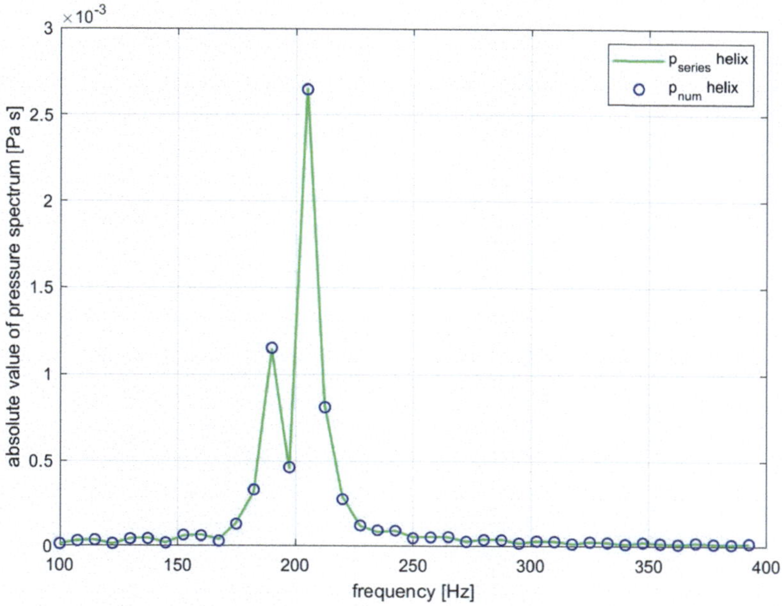

Fig. 2.40 Receiver located laterally to the helix: $P_4(x, y, z) = (5, 0, 2)m$.Absolute value of the spectral pressure for the conical helix: p_{ana}helix = series (2.254), p_{num}helix = CCI (2.272). Parameters of case 1 with $n_{max} = 40$

Case 2: The source is first *subsonic* and then *supersonic*.

Compared to case 1, we only change two parameters: The azimuthal velocity now becomes $w = 6\pi\ s^{-1}$, which corresponds to three revolutions in one second, and the observation time becomes $t_{max} = 3$ s. Due to Eq. (2.249), this leads to a maximal velocity $v_{max} = 686$ m/s at $t = t_{max} = 3$ s corresponding to a maximal Mach number of 2. The source thus starts with subsonic speed and ends with supersonic speed. Figure 2.41 shows the results for an observer at $P_4(x, y, z) = (5, 0, 2)m$. Obviously, $n_{lim} = 39$ and $n_{max} = 40$ are the same as in Fig. 2.40. Compared to the last example in Fig. 2.40, the spectrum in Fig. 2.41 has more pronounced sidelobes.

As with the circular helix discussed in Sect. 2.5.2, all figures show excellent agreement between the two very different methods—the series solution and the CCI. However, the infinite series has a more complicated structure compared to the CCI, as it includes integrals and infinite series. In addition, the series is only applicable to a specific trajectory of the source, whereas the CCI offers greater flexibility and allows the calculation of source spectra for various trajectories. This advantage is demonstrated in the subsequent section through another example in which a source traveling along an elliptical helix is considered.

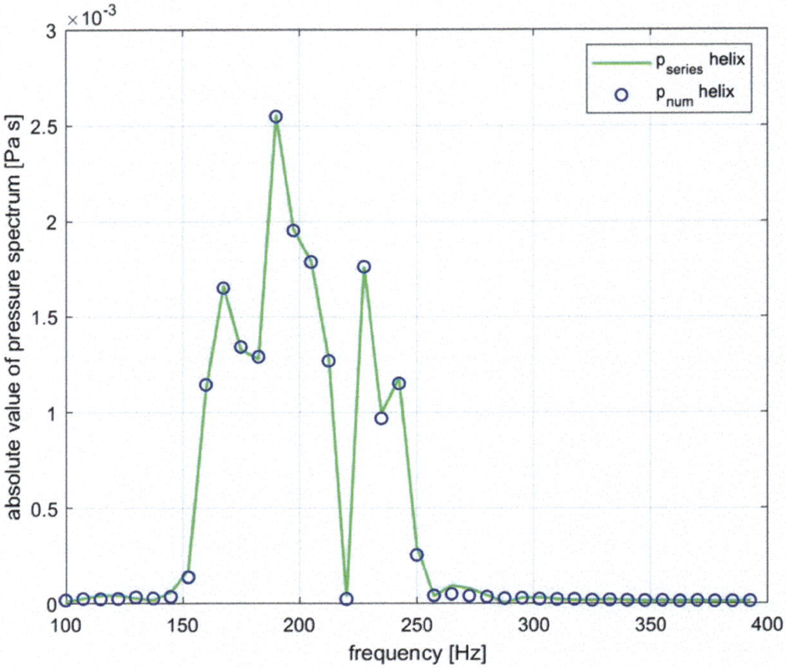

Fig. 2.41 Receiver located laterally to the helix: $P_4(x, y, z) = (5, 0, 2)m$. Absolute value of the spectral pressure for the conical helix: p_{ana}helix = series (2.254), p_{num}helix = CCI (2.272). Parameters of ***case 2*** with $n_{\max} = 40$

2.7 Radiating Monopole Moving Along an Elliptical Helix

In this section, we use the Cartesian convolution integral (CCI) to derive the frequency spectrum of a source traveling along an elliptical helix.

The elliptic cylinder coordinates $\mu,\ \varphi,\ z$ are formulated by (see, e.g., Temme [42, p. 263])

$$\begin{pmatrix} x \\ y \\ z \end{pmatrix} = \begin{pmatrix} a \cosh \mu \cos \phi \\ a \sinh \mu \sin \phi \\ z \end{pmatrix}. \tag{2.276}$$

We now consider a point source moving along the elliptical helix at the time-dependent position

$$\begin{pmatrix} x_0(t) \\ y_0(t) \\ z_0(t) \end{pmatrix} = \begin{pmatrix} a \cosh \mu_0 \cos(wt) \\ a \sinh \mu_0 \sin(wt) \\ v_z t \end{pmatrix} = \begin{pmatrix} r_x \cos(wt) \\ r_y \sin(wt) \\ v_z t \end{pmatrix}. \tag{2.277}$$

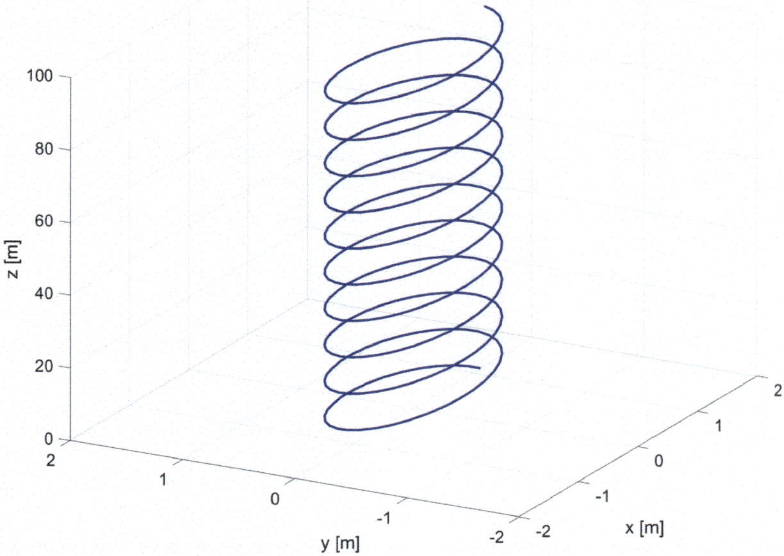

Fig. 2.42 An elliptical helix with parameters: $a = 1$ m, $\mu = 0.5$, $v_z = 10$ m/s, $w = 2\pi$ s^{-1}, $t \in (0, 10)$s

Here, the polar angle is $\phi_0 = wt$ with angular velocity w, and the velocity in z-direction is v_z. The trajectory of the source is located on an elliptical cylinder with semi-axes $r_x = a \cosh \mu_0$ and $r_y = a \sinh \mu_0$, where μ_0 is a non-negative real number (see Temme [42, p. 263]).

A sketch of such an elliptical helix is shown in Fig. 2.42.

2.7.1 Derivation of the Spectral Pressure

The wave equation for the acoustical potential generated by a point source moving on an elliptical helix in *Cartesian coordinates* reads

$$\Delta\varphi - \frac{1}{c^2}\frac{\partial^2\varphi}{\partial t^2} = \frac{Q_1}{\rho} e^{-i\Omega t}\delta(x - r_x \cos wt)\delta\big(y - r_y \sin wt\big)\delta(z - v_z t)\chi_{[t_0,t_1]}(t). \tag{2.278}$$

It is also possible to formulate the wave equation in the *elliptic cylinder coordinates* $\mu,\ \varphi,\ z$ and to derive a solution represented as a series expansion in Mathieu functions by proceeding similarly as in Sects. 2.5 and 2.6, where we found a series expansion in Bessel and Hankel functions for the circular and the conical helix.

However, the use of Mathieu functions is not widespread and not well known. Moreover, the series expansion has a more complicated structure than in cylindrical coordinates (see, e.g., Mechel [43] or Barakat [44]) since the Mathieu functions and the circular functions resulting from the normal temporal Fourier transform are not orthogonal. Therefore, such a derivation is not performed here and is not used for further numerical analysis.

Instead, we use the CCI, which provides very good solutions compared to the solutions from series expansions in special coordinate systems.

If we proceed as in the previous sections, then the time Fourier transform of the wave Eq. (2.278) yields

$$\begin{aligned}\Delta\hat{p}+\left(\frac{\omega}{c}\right)^2\hat{p}&=-Q\int\limits_{-\infty}^{\infty}e^{i(\omega-\Omega)t}\delta(x-r_x\cos wt)\delta\big(y-r_y\sin wt\big)\delta(z-v_zt)\chi_{[t_0,t_1]}(t)\mathrm{d}t\\&=-\frac{Q}{v_z}e^{i(\omega-\Omega)\frac{z}{v_z}}\delta\left(x-r_x\cos\left(\frac{w}{v_z}z\right)\right)\delta\left(y-r_y\sin\left(\frac{w}{v_z}z\right)\right)\chi_{[z_0,z_1]}(z)\\&=:-R(x,y,z)\end{aligned}\tag{2.279}$$

for the pressure $\hat{p}(\omega)=i\omega\rho\hat{\varphi}$ with $Q=-i\omega Q_1$ and $z_{0,1}=v_z t_{0,1}$. The convolution with the 3D Green's function

$$\hat{g}(x,y,z)=\frac{1}{4\pi\sqrt{x^2+y^2+z^2}}e^{i\frac{\omega}{c}\sqrt{x^2+y^2+z^2}}.\tag{2.280}$$

is defined by

$$\hat{p}=\int\limits_{-\infty}^{\infty}\int\limits_{-\infty}^{\infty}\int\limits_{-\infty}^{\infty}\hat{g}(x-\xi,y-\eta,z-\zeta)R(\xi,\eta,\zeta)\mathrm{d}\xi\mathrm{d}\eta\mathrm{d}\zeta$$

and results in

$$\hat{p}=\frac{Q}{4\pi v_z}\int\limits_{z_0}^{z_1}\frac{e^{i\frac{\omega}{c}\sqrt{\left(x-r_x\cos\left(\frac{w}{v_z}\zeta\right)\right)^2+\left(y-r_y\sin\left(\frac{w}{v_z}\zeta\right)\right)^2+(z-\zeta)^2}}e^{i(\omega-\Omega)\frac{\zeta}{v_z}}}{\sqrt{\left(x-r_x\cos\left(\frac{w}{v_z}\zeta\right)\right)^2+\left(y-r_y\sin\left(\frac{w}{v_z}\zeta\right)\right)^2+(z-\zeta)^2}}\mathrm{d}\zeta.\tag{2.281}$$

This formula also follows directly from the general formulation of the CCI (2.115) by selecting $\Phi(t)=r_x\cos(wt)$ and $\psi(t)=r_y\sin(wt)$.

Now, we compare the numerical solution of the CCI for the elliptical helix (2.281) and for the circular helix (see Eq. (2.116) or Eq. (2.232))

$$p_{\mathrm{cyl}} = \frac{Q}{4\pi v_z} \int\limits_{z_0=v_z t_0}^{z_1=v_z t_1} \frac{e^{i\frac{\omega}{c}\sqrt{\left(x-r_0\cos\left(\frac{w\zeta}{v_z}\right)\right)^2+\left(y-r_0\sin\left(\frac{w\zeta}{v_z}\right)\right)^2+(z-\zeta)^2}}\, e^{i(\omega-\Omega)\frac{\zeta}{v_z}}}{\sqrt{\left(x-r_0\cos\left(\frac{w\zeta}{v_z}\right)\right)^2+\left(y-r_0\sin\left(\frac{w\zeta}{v_z}\right)\right)^2+(z-\zeta)^2}}\, \mathrm{d}\zeta \tag{2.282}$$

2.7.2 Numerical Examples

We consider an elliptical helix with parameters $a = 1$ m, $\mu = 0.2$, $v_z = 34.3$ m/s, $w = 2\pi \cdot 3\ \mathrm{s}^{-1}$ and $t \in (0, 3)$s. Thus, we obtain for the radii $r_x = a \cosh \mu \approx 1$ m, $r_y = a \sinh \mu \approx 0.2$ m and for the vertical Mach number $M_z = v_z/c = 0.1$. The source describes nine turns over the entire vertical z-distance of 102.9 m in three seconds of observation time, which corresponds to three revolutions per second

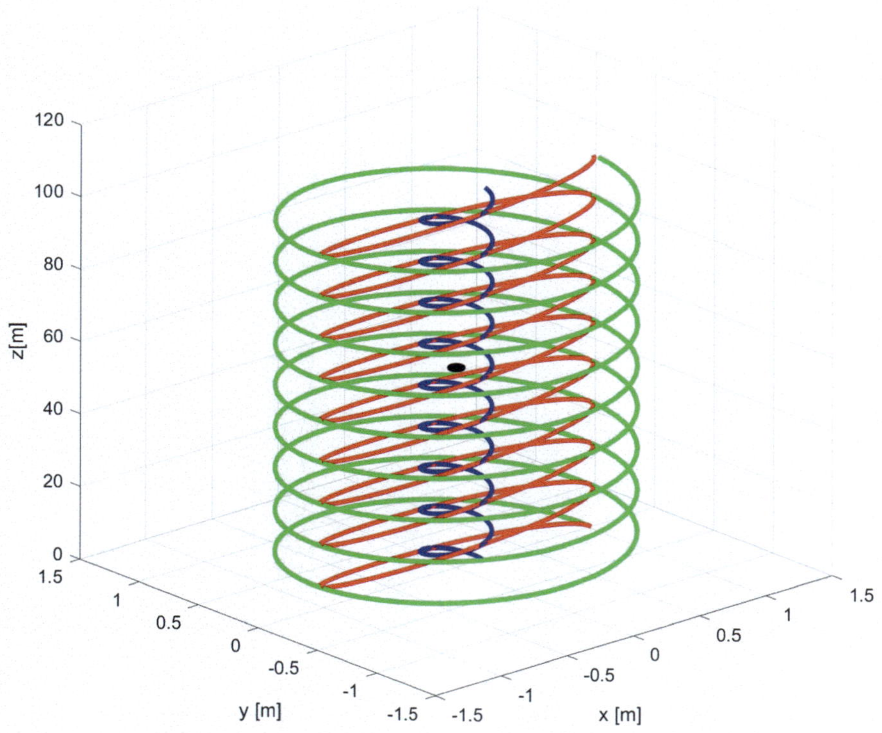

Fig. 2.43 *Red curve*: elliptical helix with parameters $a = 1$ m, $\mu = 0.2$, $r_x = 1$ m, $r_y = 0.2$ m, $v_z = 34.3$ m/s, $w = 2\pi \cdot 3\ \mathrm{s}^{-1}$, observation time $t \in (0, 3)$s; *green curve:* exterior cylindrical helix with radius of 1 m; *blue curve*: interior cylindrical helix with radius of 0.2 m; *black circle*: microphone at $(x, y, z) = (0$ m, 0 m, 55 m)

($w = 2\pi \cdot 3\ \mathrm{s}^{-1}$). Obviously, at $t = 0$, the source starts at the origin and travels in the positive z-direction. We also consider the corresponding exterior (with radius $r_x =$ 1 m) and interior (with radius $r_y = 0.2$ m) circular helices. These three trajectories are shown in Fig. 2.43. The red curve is the elliptical helix, and the green and blue ones are the circular helices. The little black circle symbolizes the position of the receiver or microphone.

Figures 2.44, 2.45, and 2.46 show the spectral pressure for a source radiating at a frequency of 200 Hz and traveling along the three helices for different positions of the receiver. The spectrum depicted in Fig. 2.44 corresponds to the scenario described in Fig. 2.43.

The receiver is placed approximately in the middle of the central axis (see the black circle in Fig. 2.43) and, thus, the two Doppler frequencies are mainly determined by the vertical Mach number $M_z = v_z/c = 0.1$. (cf. Sect. 2.5.2.4) with $f_{0-} = \Omega/(1 + M_z)/(2\pi) \approx 182$ Hz and $f_{0+} = \Omega/(1 - M_z)/(2\pi) \approx 222$ Hz. As expected, the pressure curve of the elliptical helix lies between the curves of the small and large circular helix and varies somewhat more strongly. The blue curve provides higher values than the green curve, as the inner circular helix is closer to the observer than the outer circular helix.

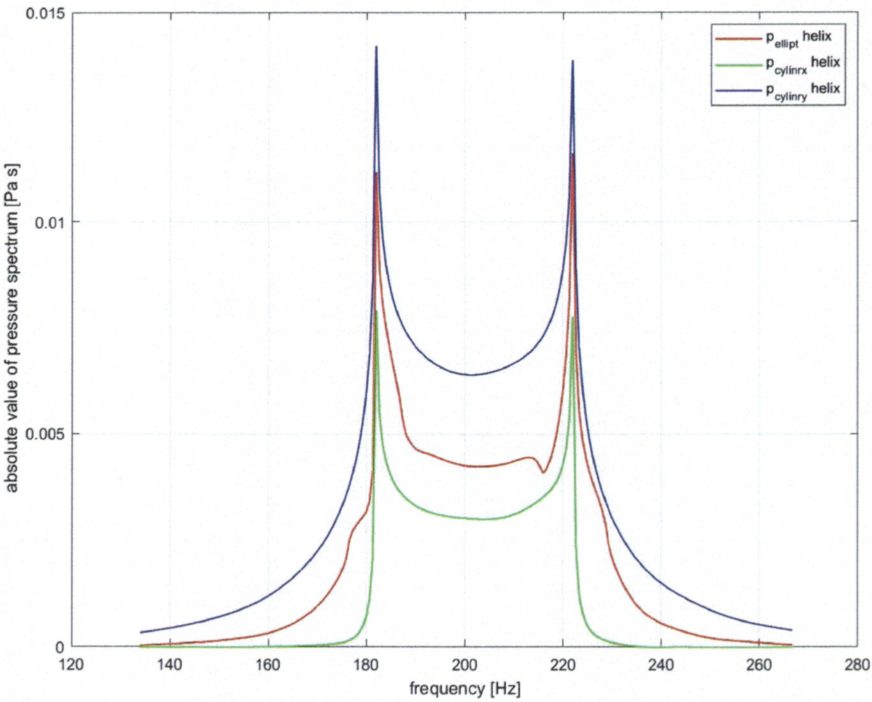

Fig. 2.44 Absolute value of the spectral pressure of the three helices as described in Fig. 2.43 with receiver at $(x, y, z) = (0\ \mathrm{m}, 0\ \mathrm{m}, 55\ \mathrm{m})$

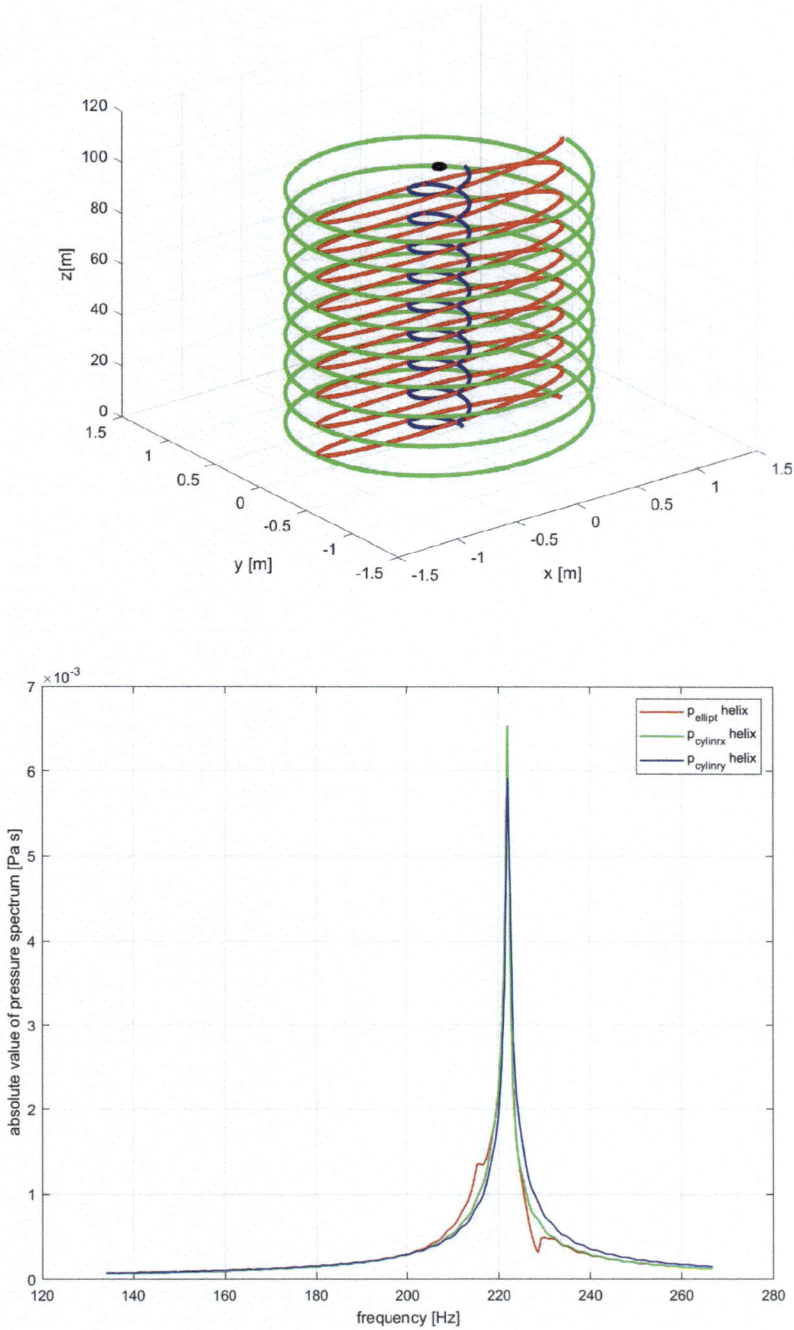

Fig. 2.45 Upper figure: Trajectories and position of receiver, which is now placed at $(x, y, z) = (0\text{ m}, 0\text{ m}, 105\text{ m})$. Lower figure: absolute value of the spectral pressure of the three helices as described in Fig. 2.43

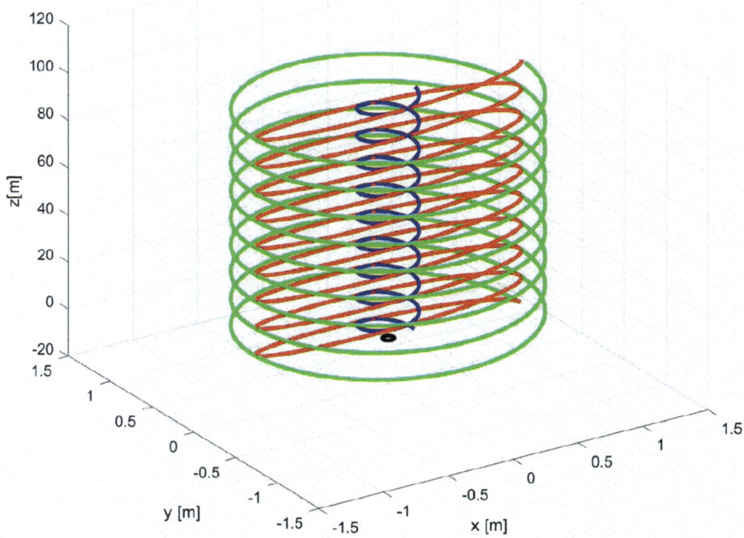

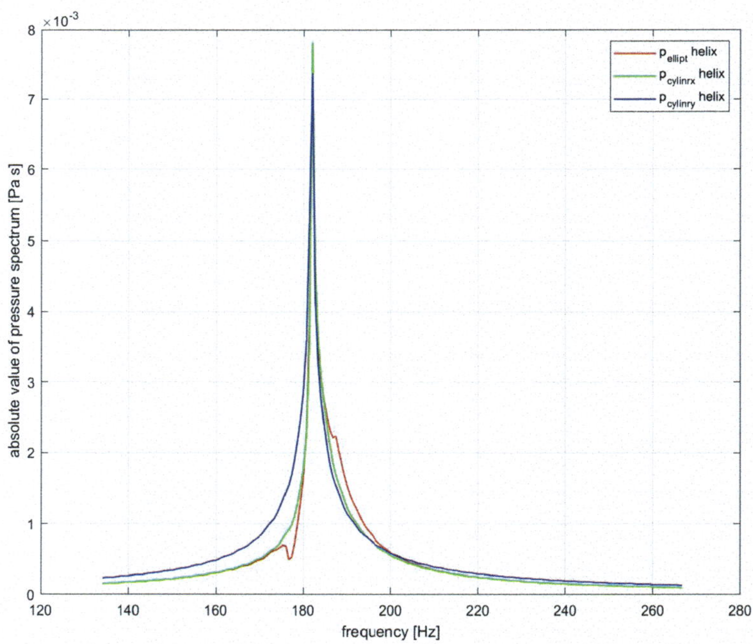

Fig. 2.46 Upper figure: Trajectories and position of receiver, which is now placed at $(x, y, z) = (0 \text{ m}, 0 \text{ m}, -1 \text{ m})$. Lower figure: absolute value of the spectral pressure of the three helices as described in Fig. 2.43

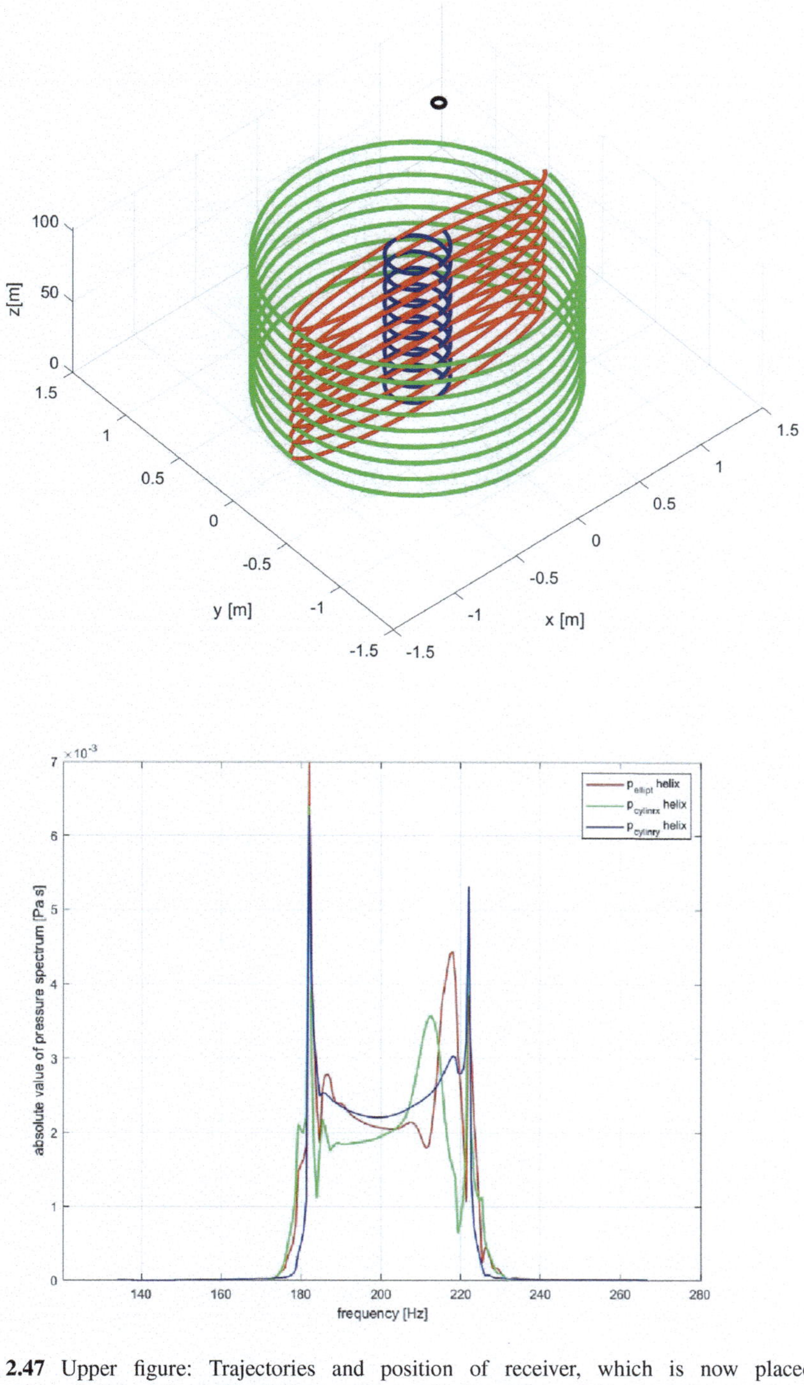

Fig. 2.47 Upper figure: Trajectories and position of receiver, which is now placed at $(x, y, z) = (1.3\text{ m}, 1.3\text{ m}, 55\text{ m})$. Lower figure: absolute value of the spectral pressure of the three helices as described in Fig. 2.43

Now let us move the receiver to a position above all helices at $(x, y, z) = (0\text{ m}, 0\text{ m}, 105\text{ m})$.

Since the source exclusively moves toward the receiver, only the higher Doppler frequency can be detected. Here again, the pressure of the elliptical helix fluctuates slightly more than the pressure on the circular helices.

If we now position the microphone below the three trajectories at $(x, y, z) = (0\text{ m}, 0\text{ m}, -1\text{ m})$, we obtain, as expected, only the lower Doppler frequency, see Fig. 2.46.

However, if we remove the receiver from the inner axis and position it next to the helices at $(x, y, z) = (1.3\text{ m}, 1.3\text{ m}, 55\text{ m})$, all nine turns contribute to the Doppler shift in addition to the vertical movement, and we get a wild and difficult to interpret frequency response, which is shown in Fig. 2.47.

2.8 Moving Sources Above an Impedance Plane

In Sects. 2.1–2.7, we studied the spectra of sound sources moving in free space without boundaries. In this section, we now consider sources traveling in a half-space above an infinite plane in the frequency domain, while Chap. 3 is devoted to half-space solutions in the time domain. Since we are working in the frequency domain, we can describe the boundary condition at the plane located at $z = 0$ by the acoustic impedance

$$Z(\omega) = \hat{p}(\omega)/v_n(\omega), \tag{2.283}$$

where $v_n(\omega)$ is the normal velocity. We assume that the normal unit vector $\vec{n} = (0, 0, -1)^T$ is directed away from the acoustics domain and points into the negative z direction (see Fig. 2.48).

The impedance boundary condition implies that the ground is locally reacting. For a non-locally reacting ground, which is characterized by a semi-infinite fluid medium, other techniques must be applied (see Sect. 3.2). Due to Eq. (2.283), the boundary condition at the plane is described by

$$\frac{\partial \hat{p}}{\partial z} + \Gamma \hat{p} = 0 \text{ with } \Gamma = \frac{i\omega/c}{Z_0} \text{ at } z = 0, \tag{2.284}$$

since $\frac{\partial \hat{p}}{\partial z} = -\frac{\partial \hat{p}}{\partial \vec{n}} = -i\omega\rho v_n$. $Z_0 = Z/(\rho c)$ is the normalized impedance. This situation is shown in Fig. 2.48, where V_+ is the acoustical space above z with $z > 0$ and V_- is the non-physical space below z with $z < 0$.

The solution for stationary sources above impedance planes is given in Ochmann and Piscoya [45, Chaps. 5 and 6] in one, two and three dimensions. By combining the frequency domain technique presented in this book with a modified version of the "*General impedance formula*" given in Theorem 4 of [45] on p. 39, the spectral sound field of sources moving on arbitrary trajectories above an impedance plane can

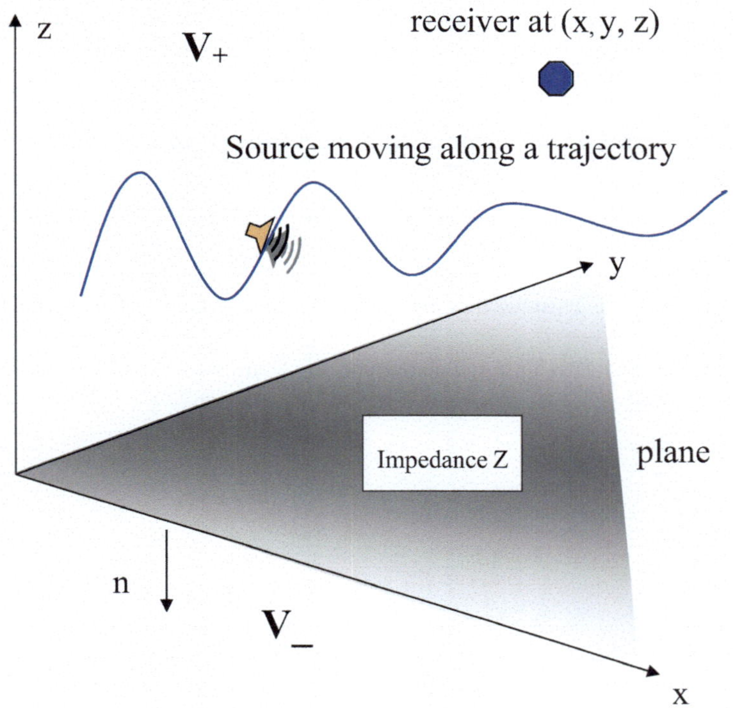

Fig. 2.48 Radiating source moving along a trajectory above an impedance plane

be derived. In this way, we can effectively merge the theory for calculating spectra of moving sources with the theory that applies to half-spaces. We now reformulate Theorem 4 in such a way that it can be applied to the moving sources studied here:

Theorem 2.8.1: Impedance Formula (IF) Let $u_0(z) = u_0(x, y, z)$ be a "*basis solution*" of the inhomogeneous Helmholtz equation in *free space* in one, two or three dimensions (cf., e.g., Eqs. (2.13), (2.46), (2.62))

$$\Delta u_0 + k^2 u_0 = f(x, y, z). \tag{2.285}$$

Then, the composite function (only the z-argument is written)

$$\overline{u}(z) = u_0(z) + u_0(-z) + 2\Gamma \int_0^\infty e^{\Gamma\tau} u_0(-z - \tau)\,\mathrm{d}\tau \tag{IF1}$$

or alternatively

$$\overline{u}(z) = u_0(z) + u_0(-z) + 2i\Gamma \int_0^\infty e^{i\Gamma\tau} u_0(-z - i\tau)\,\mathrm{d}\tau \tag{IF2}$$

satisfies the impedance condition at the plane between both half-spaces at $z = 0$

$$\frac{\partial \overline{u}}{\partial z} + \Gamma\overline{u} = 0 \quad \text{for} \quad z = 0) \tag{IBC}$$

where we have to assume that the integrals appearing in (IF1) and (IF2) are convergent.

If the superposition solution $\overline{u}$ also solves the Helmholtz Eq. (2.285), restricted to the acoustic half-space V_+, then we have found a solution of the half-space problem (HSP)

$$\Delta\overline{u} + k^2\overline{u} = f(x, y, z) \quad \text{in} \quad V_+, \tag{HSPa}$$

$$\frac{\partial \overline{u}}{\partial z} + \Gamma\overline{u} = 0 \quad \text{for} \quad z = 0. \tag{HSPb}$$

The line integral in (IF1) can be considered as a superposition of sources with real source positions in the lower half-space V_- (shortly: real representation). The line integral in (IF2) can be considered as a superposition of sources with complex source positions (shortly: complex representation).

In the original Theorem 4 given in [45], we had the special right-hand side $f(x, y, z) = -\delta(z - h)$ and we were able to show that $\overline{u}$ is a solution of $\Delta\overline{u} + k^2\overline{u} = -\delta(z - h))$ in V_+, since all other singularities of $\overline{u}$—except the singularity at $z = h$—generated by Dirac delta functions are located outside the physical half-space V_+. This is also the case if the right-hand side of the Helmholtz Eq. (2.285) consists of concentrated sources with singularities located only in V_+. Such singularities will be mirrored into the image space V_- by the impedance formulas (IF1) and (IF2), which becomes clear by considering the so-called "driving sources" in [45, Chaps. 4 and 5] and is confirmed in detail in Appendix E for the cases analyzed here.

First, we apply this theorem to a line source moving along a straight line.

2.8.1 *The Spectral Pressure of a Line Source Moving Along a Straight Line*

The spectrum of a line source pulsating with the angular frequency Ω and moving rectilinearly with constant velocity u in the positive x-direction at height $z = h$ in free "space" is given by Eq. (2.51)

$$\hat{p}_{2f}(x,z) = \frac{iQ}{2u} e^{i(\omega-\Omega)\frac{x}{u}} \frac{e^{i\sqrt{\left(\frac{\omega}{c}\right)^2 - \left(\frac{\omega-\Omega}{u}\right)^2}|z-h|}}{\sqrt{\left(\frac{\omega}{c}\right)^2 - \left(\frac{\omega-\Omega}{u}\right)^2}} \tag{2.286}$$

with $Q = -i\omega Q_1$.

To satisfy the impedance boundary condition (2.284), we apply the superposition integral (IF1) or (IF2) and obtain

$$\hat{p}_{2h}(x,z) = \hat{p}_{2f}(x,z) + \hat{p}_{2f}(x,-z) + \sigma_{1,2}, \tag{2.287}$$

where

$$\begin{aligned}\sigma_1 &= 2Q\Gamma \int_0^\infty e^{\Gamma\tau} \hat{p}_{2f}(x,-z-\tau)\mathrm{d}\tau \\ &= \frac{iQ\Gamma}{\kappa u} e^{i(\omega-\Omega)\frac{x}{u}} \int_0^\infty e^{\Gamma\tau} e^{i\kappa(z+h+\tau)}\mathrm{d}\tau \text{ for (IF1)}\end{aligned} \tag{2.288}$$

and

$$\begin{aligned}\sigma_2 &= 2iQ\Gamma \int_0^\infty e^{i\Gamma\tau} \hat{p}_{2f}(x,-z-i\tau)\mathrm{d}\tau \\ &= \frac{iQ\Gamma}{\kappa u} e^{i(\omega-\Omega)\frac{x}{u}} \int_0^\infty i e^{i\Gamma\tau} e^{i\kappa(z+h+i\tau)}\mathrm{d}\tau \text{ for (IF2)}\end{aligned} \tag{2.289}$$

with

$$\kappa = \sqrt{\left(\frac{\omega}{c}\right)^2 - \left(\frac{\omega-\Omega}{u}\right)^2}, \ \Omega > 0. \tag{2.290}$$

Appendix E shows that the representation (2.287) also satisfies the inhomogeneous Helmholtz equation (2.46) for the moving line source in the half-space $z > 0$, where the coordinate y is replaced by z-h:

$$\Delta_2 \hat{p}_{2h} + \left(\frac{\omega}{c}\right)^2 \hat{p}_{2h} = -\frac{Q}{u} e^{i(\omega-\Omega)\frac{x}{u}} \delta(z-h).$$

The half-space pressure $\hat{p}_{2h}(x,z)$ is thus composed of three terms: the incident pressure $p_{in} = \hat{p}_{2f}(x,z)$, the pressure $p_{ref} = \hat{p}_{2f}(x,-z)$ reflected at a the rigid plane, and the correction term σ_1 or σ_2.

In the case of convergence, the upper limit disappears and both line integrals give the same result since

$$I_1 = \int_0^\infty e^{(\Gamma + i\kappa)\tau} \mathrm{d}\tau = \left. \frac{e^{\tau(\Gamma + i\kappa)}}{(\Gamma + i\kappa)} \right|_0^\infty = \frac{-1}{(\Gamma + i\kappa)} \tag{2.291}$$

or

$$I_2 = \int_0^\infty i e^{i(\Gamma + i\kappa)\tau} \mathrm{d}\tau = \left. \frac{i e^{i\tau(\Gamma + i\kappa)}}{i(\Gamma + i\kappa)} \right|_0^\infty = \frac{-1}{(\Gamma + i\kappa)}. \tag{2.292}$$

The final result is

$$\sigma := \sigma_1 = \sigma_2 = \frac{iQ\Gamma}{u} e^{i(\omega - \Omega)\frac{x}{u}} \frac{e^{i\kappa(z+h)}}{-\kappa(\Gamma + i\kappa)}. \tag{2.293}$$

Thus, we obtain

$$\hat{p}_{2h}(x, z) = p_{\text{in}} + p_{\text{ref}} + \frac{iQ\Gamma}{u} e^{i(\omega - \Omega)\frac{x}{u}} \frac{e^{i\kappa(z+h)}}{-\kappa(\Gamma + i\kappa)}, \; z > 0. \tag{2.294}$$

By introducing the reflection factor

$$R(\omega) = \frac{i\kappa - \Gamma}{i\kappa + \Gamma}, \tag{2.295}$$

we can reformulate Eq. (2.294) in a more natural way

$$\begin{aligned} \hat{p}_{2h} &= p_{\text{in}} + R(\omega) p_{\text{ref}} \\ &= \frac{iQ}{2u\kappa} e^{i(\omega - \Omega)\frac{x}{u}} \left\{ e^{i\kappa|z-h|} + R(\omega) e^{i\kappa(z+h)} \right\}, \; z > 0. \end{aligned} \tag{2.296}$$

The corresponding acoustic potential is given by

$$\hat{\varphi}_{2h} = \frac{Q_1}{2iu\kappa\rho} e^{i(\omega - \Omega)\frac{x}{u}} \left\{ e^{i\kappa|z-h|} + R(\omega) e^{i\kappa(z+h)} \right\}, \; z > 0, \tag{2.297}$$

since $\hat{\varphi}_{2h} = \hat{p}_{2h}(\omega)/(i\omega\rho)$ and $Q = -i\omega Q_1$.

It is very surprising that the half-space solution for a moving source is much simpler than the solution for the corresponding stationary source with $u = 0$ derived in [45, Eqs. (5.51–5.53)] since the latter consists of Hankel functions and an infinite line integral.

If the integrals appearing in (2.288) and (2.289) cannot be solved in closed form, which is the case in 3D (see next section), we must solve the integral numerically and have to choose the convergent form, i.e., either (2.291) or (2.292) or equivalently, the correct integration path in the complex plane. This is explained in detail in [45, Chap. 5] for a stationary source. If we look at the integral I_1 more closely, we find that it converges against $-1/(\Gamma + i\kappa)$ only if the real part of the exponent is negative, i.e.,$\mathrm{Re}\{\Gamma + i\kappa\} < 0$. Here, it should be noted that κ is a real quantity if $\omega \in (\omega_-, \omega_+)$, as we have (see Eq. (2.42))

$$\kappa = \sqrt{\left(\frac{\omega}{c}\right)^2 - \left(\frac{\omega - \Omega}{u}\right)^2} = \frac{1}{u\gamma}\sqrt{(\omega - \omega_-)(\omega_+ - \omega)}$$
$$= \begin{cases} \text{real and positive for } \omega \in (\omega_-, \omega_+) \\ \text{imaginary for } \omega \notin (\omega_-, \omega_+) \end{cases} \tag{2.298}$$

with $\omega_- = \Omega/(1 + M)$, $\omega_+ = \Omega/(1 - M)$ and $\gamma = 1/\sqrt{1 - M^2}$.

Outside the Doppler interval (ω_-, ω_+), the spectrum tends very quickly toward zero (see Sect. 2.2.2.2). So we only look at the spectrum within the Doppler interval (ω_-, ω_+) and assume that $\mathrm{Re}\{\Gamma + i\kappa\} < 0 \Leftrightarrow \mathrm{Re}\{\Gamma\} < 0 \Leftrightarrow \mathrm{Im}\{Z\} < 0$. Therefore, considering the time dependence $\exp(-i\omega t)$, the normal impedance of the impedance plane has to be mass-like, for example, $Z = \rho c Z_0 = -i\omega m''$ for a mass m'' per unit area assigned to the impedance surface, as explained in greater detail in [45, p. 44].

In contrast, the integral I_2 converges against $-1/(\Gamma + i\kappa)$ only if $\mathrm{Re}\{i(\Gamma + i\kappa)\} < 0$. Since

$$i(\Gamma + i\kappa) = \frac{-\omega/c}{Z_0} - \kappa = \frac{-\omega/c(\mathrm{Re}Z_0 - i\mathrm{Im}Z_0)}{|Z_0|^2} - \kappa,$$

we obtain

$$\mathrm{Re}\{i(\Gamma + i\kappa)\} < 0 \Leftrightarrow -\mathrm{Re}\left\{\frac{\omega/c(\mathrm{Re}Z_0)}{|Z_0|^2} + \kappa\right\} < 0 \Leftrightarrow \mathrm{Re}Z_0 + \frac{|Z_0|^2}{\omega/c}\mathrm{Re}\kappa > 0.$$

Hence, due to Eq. (2.298), the integral is convergent for all kinds of surface impedances when $\omega \in (\omega_-, \omega_+)$—even if the impedance Z_0 is purely imaginary, such as a mass-like or spring-like impedance. However, the integral I_2 occurring in the complex representation (IF2) is often more difficult to calculate. Further details can be found in [45, pp. 45–46].

To validate the solution (2.296), we compare it with the time domain solution for the velocity potential $\varphi(x_L, z_L)$ in Lorentz coordinates given by Dragna and Blanc-Benon in [9].

Equations (28–30) of [9] reads in our notation ($k_0 = \Omega/c$)

$$\varphi(x_L, z_L) = \varphi_D(x_L, z_L) + \varphi_R(x_L, z_L)$$
$$= \frac{1}{2\pi}\int_{-\infty}^{\infty} \frac{\gamma}{2i\sqrt{k_0^2 - \lambda^2}} e^{i\lambda x_L}\left\{e^{i\sqrt{k_0^2-\lambda^2}|z_L - \gamma h|}\right.$$
$$\left. + R_{\mathrm{Dra}}(\lambda) e^{i\sqrt{k_0^2-\lambda^2}(z_L+\gamma h)}\right\} \mathrm{d}\lambda\, e^{-i\Omega t_L} \tag{2.299}$$

or by considering that $x_L = \gamma^2(x - Mct)$, $z_L = \gamma z$, $t_L = \gamma^2(t - Mx/c)$

$$\varphi = \frac{\gamma}{4\pi i} \int_{-\infty}^{\infty} \frac{1}{\sqrt{k_0^2 - \lambda^2}} e^{i\gamma^2\lambda(x-ut)} \Big\{ e^{i\gamma\sqrt{k_0^2-\lambda^2}|z-h|}$$
$$+ R_{\mathrm{Dra}}(\lambda) e^{i\gamma\sqrt{k_0^2-\lambda^2}(z+h)} \Big\} \mathrm{d}\lambda\, e^{-i\gamma^2\Omega(t-Mx/c)}. \tag{2.300}$$

Note that Dragna and Blanc-Benon have chosen the source strength $Q_1 = \rho$ (cp. our Eq. (2.45) and Eq. (3) in [9]).

Therefore, we have to prove that the Fourier transform of φ agrees with the velocity potential $\hat{\varphi}_{2h}$ given in Eq. (2.297):

$$\hat{\varphi} := F_t(\varphi) = \int_{-\infty}^{\infty} \varphi e^{i\omega t} \mathrm{d}t = \hat{\varphi}_{2h} \text{ if } Q_1 = \rho. \tag{2.301}$$

Proof of (2.301): Performing the temporal Fourier transform of Eq. (2.300), we get

$$\hat{\varphi} = \frac{\gamma}{4\pi i} \int_{-\infty}^{\infty} \frac{e^{ix\gamma^2(\lambda+k_0M)}}{\sqrt{k_0^2 - \lambda^2}} \Big\{ e^{i\gamma\sqrt{k_0^2-\lambda^2}|z-h|}$$
$$+ R_{\mathrm{Dra}}(\lambda) e^{i\gamma\sqrt{k_0^2-\lambda^2}(z+h)} \Big\} \int_{-\infty}^{\infty} e^{-i\gamma^2(\lambda ut+\Omega t)+i\omega t} \mathrm{d}t\mathrm{d}\lambda. \tag{2.302}$$

The Fourier integral can be easily solved

$$\int_{-\infty}^{\infty} e^{-i\gamma^2(\lambda ut+\Omega t)+i\omega t} dt = \frac{2\pi}{|\gamma^2 u|} \delta(\lambda - \lambda_0) \text{ with } \lambda_0 = \frac{\omega}{\gamma^2 u} - \frac{\Omega}{u}, \tag{2.303}$$

and thus, we obtain from the last two equations

$$\hat{\varphi} = \frac{1}{2\gamma i u} \frac{e^{ix\gamma^2(\lambda_0+k_0M)}}{\sqrt{k_0^2 - \lambda_0^2}} \Big\{ e^{i\gamma\sqrt{k_0^2-\lambda_0^2}|z-h|} + R_{Dra}(\lambda_0) e^{i\gamma\sqrt{k_0^2-\lambda_0^2}(z+h)} \Big\}.$$

A straightforward calculation leads to

$$\gamma^2(\lambda_0 + k_0 M) = \frac{\omega - \Omega}{u} \tag{2.304}$$

and to

$$\gamma\sqrt{k_0^2-\lambda_0^2}=\sqrt{\left(\frac{\Omega}{c}\gamma\right)^2-\frac{1}{u^2\gamma^2}(\omega-\gamma^2\Omega)^2}=\sqrt{\left(\frac{\omega}{c}\right)^2-\left(\frac{\omega-\Omega}{u}\right)^2}=\kappa. \tag{2.305}$$

Therefore, we finally obtain

$$\hat{\varphi}=\frac{1}{2iu}\frac{e^{i\left(\frac{\omega-\Omega}{u}\right)x}}{\kappa}\left\{e^{i\kappa|z-h|}+R_{\mathrm{Dra}}(\lambda_0)e^{i\kappa(z+h)}\right\}. \tag{2.306}$$

The only thing missing is the analysis of the reflection factor. Equation (27) in [9] states that

$$R_{\mathrm{Dra}}(\lambda_0)=\frac{\sqrt{k_0^2-\lambda_0^2}-\frac{(k_0+\lambda_0 M)\gamma}{Z_0\left[\gamma^2(\Omega+\lambda_0 u)\right]}}{\sqrt{k_0^2-\lambda_0^2}+\frac{(k_0+\lambda_0 M)\gamma}{Z_0\left[\gamma^2(\Omega+\lambda_0 u)\right]}}. \tag{2.307}$$

At first glance, this expression appears somewhat complicated, but it simplifies considerably, as

$$Z_0\left[\gamma^2(\Omega+\lambda_0 u)\right]=Z_0\left[\gamma^2\left(\Omega+\left(\frac{\omega}{\gamma^2 u}-\frac{\Omega}{u}\right)u\right)\right]=Z_0(\omega)$$

is nothing else than the normalized impedance at the circular frequency ω.

Together with (2.305), and the identity $k_0+\lambda_0 M=\omega/(\gamma^2 c)$ we obtain—cf. Eq. (2.295) for $R(\omega)$

$$R_{\mathrm{Dra}}(\lambda_0)=\frac{\kappa/\gamma-\frac{\omega/(\gamma c)}{Z_0(\omega)}}{\kappa/\gamma+\frac{\omega/(\gamma c)}{Z_0(\omega)}}=\frac{\kappa-\frac{\omega}{cZ_0}}{\kappa+\frac{\omega}{cZ_0}}=\frac{i\kappa-\Gamma}{i\kappa+\Gamma}=R(\omega).$$

In conclusion, the spectra (2.297) and (2.306) agree when $Q_1=\rho$ is chosen. This proves the validity of the claimed Eq. (2.301). □

It is amazing how much simpler the integral-free frequency solution of the moving source (2.296/2.297) is constructed compared to the solution (2.300) in the time domain.

Using the identity

$$\begin{aligned}\varphi_D&=\frac{\gamma}{4\pi i}\int_{-\infty}^{\infty}\frac{1}{\sqrt{k_0^2-\lambda^2}}e^{i\gamma^2\lambda(x-ut)}e^{i\gamma\sqrt{k_0^2-\lambda^2}|z-h|}\mathrm{d}\lambda\, e^{-i\gamma^2\Omega(t-Mx/c)}\\&=\frac{\gamma}{4i}H_0^{(1)}\left(k_0\gamma^2\sqrt{(x-ut)^2+(z-h)^2/\gamma^2}\right)e^{-i\gamma^2\Omega(t-Mx/c)}.\end{aligned} \tag{2.308}$$

given in [9, Eq. (29)], we obtain the inverse Fourier transform of the direct field $\hat{\varphi}_{2f} = \hat{p}_{2f}(x,z)/(i\omega\rho)$, i.e., the time domain potential of the line source (first term of Eq. (2.297)), as a side result of the above derivation:

$$\varphi_{2f} = F_t^{-1}\left[\hat{\varphi}_{2f}\right] = \frac{1}{2\pi}\int_{-\infty}^{\infty} \frac{Q_1}{2iu\rho} e^{i(\omega-\Omega)\frac{x}{u}} \frac{e^{i\kappa|z-h|}}{\kappa} e^{-i\omega t} d\omega$$

$$= \frac{\gamma Q_1}{4i\rho} H_0^{(1)}\left(k_0\gamma^2\sqrt{(x-ut)^2+(z-h)^2/\gamma^2}\right) e^{-i\gamma^2\Omega(t-Mx/c)}. \tag{2.309}$$

Similar results were derived in a different way in Piscoya and Ochmann [10] for an absorbing ground, where the surface impedance Z is real and frequency independent, and are also described in Sect. 3.1.3 of the present book (see for example Eq. (3.26)). The starting point for these investigations was the solution for the acoustic impulse response of a line source above an absorbing plane (see [46]).

2.8.2 *The Spectral Pressure of a* Point *Source Moving Along a Straight Line*

2.8.2.1 Derivation of the Superposition Integral

The spectrum of a point source pulsating with the angular frequency Ω and moving rectilinearly with constant velocity u in the positive x-direction at height $z = h$ in the three-dimensional space is given by Eq. (2.19) with $r = \sqrt{y^2 + (z-h)^2}$

$$\hat{p}_{3f}(\omega) = \frac{i}{4u} e^{i\left(\frac{\omega-\Omega}{u}\right)x} H_0^{(1)}\left(\kappa\sqrt{y^2+(z-h)^2}\right), \ \kappa = \sqrt{\left(\frac{\omega}{c}\right)^2 - \left(\frac{\omega-\Omega}{u}\right)^2}, \tag{2.310}$$

whereby in this section we set $Q = 1\ \text{kg/s}^2$ for the sake of simplicity.

To fulfill the impedance boundary condition (2.284), we apply either the superposition integral (IF1) or (IF2) and obtain

$$\hat{p}_{3h}(x,y,z) = \hat{p}_{3f}(x,y,z) + \hat{p}_{3f}(x,y,-z) + \sigma_{1,2}, \tag{2.311}$$

where

$$\sigma_1 := 2\Gamma \int_0^{\infty} e^{\Gamma\tau} \hat{p}_{3f}(x, y, -z-\tau)\mathrm{d}\tau$$

$$= \frac{i\Gamma}{2u} e^{i(\omega-\Omega)\frac{x}{u}} \int_0^{\infty} e^{\Gamma\tau} H_0^{(1)}\left(\kappa\sqrt{y^2+(z+h+\tau)^2}\right)\mathrm{d}\tau \tag{2.312}$$

or

$$\sigma_2 := 2i\Gamma \int_0^{\infty} e^{i\Gamma\tau} \hat{p}_{3f}(x, y, -z-i\tau)\mathrm{d}\tau$$

$$= \frac{i\Gamma}{2u} e^{i(\omega-\Omega)\frac{x}{u}} \int_0^{\infty} ie^{i\Gamma\tau} H_0^{(1)}\left(\kappa\sqrt{y^2+(z+h+i\tau)^2}\right)\mathrm{d}\tau \tag{2.313}$$

Appendix E shows that the formulas (2.312) and (2.313) also satisfy the inhomogeneous Helmholtz Eq. (2.13) for the moving point source in the half-space $z > 0$ (with $Q = 1\,kg/s^2$ and $z \to z - h$)

$$\Delta\hat{p} + \left(\frac{\omega}{c}\right)^2 \hat{p} = -\frac{1}{u} e^{i(\omega-\Omega)\frac{x}{u}} \delta(y)\delta(z-h).$$

Here, too, the half-space pressure $\hat{p}_{3h}(x, y, z)$ consists of three terms: the incident pressure $p_{in} = \hat{p}_{3f}(x, y, z)$, the pressure $p_{ref} = \hat{p}_{3f}(x, y, -z)$ reflected at the rigid plane, and the correction term σ_1 or σ_2, respectively. The integral contained in σ_1 or in σ_2 cannot be solved in closed form. But the formulas are valid for all kinds of impedances and are, therefore, a new result.

2.8.2.2 Verification of the Superposition Integral for Moving Sources

In [5], a time domain representation is given for a radiating monopole flying rectilinearly above a plane equipped with a mass-like or alternatively with an absorbing impedance. The inverse Fourier transform of the incident wave p_{in} and of the reflected wave p_{ref} leads to the correct and well-known time domain expressions as shown in Sect. 2.2.1.2 (Eq. 2.36). But the correction terms for a mass-like or an absorbing impedance, respectively, are completely different in the time domain. In the frequency domain, however, the two corresponding spectra must be given by the same correction term (2.312) or (2.313). We start with the case of a mass-like impedance.

Test 1: Plane with mass-like impedance

The correction term in the time domain is given by Eq. (21d) in [5],

$$\sigma_{\text{mass}} = -2\zeta e^{-i\gamma^2\Omega(t-Mx/c)} \int_0^\infty \exp(-\zeta\eta) \frac{1}{4\pi R(-h-\eta)} \exp\left[ik_0\gamma^2 R(-h-\eta)\right]\mathrm{d}\eta, \tag{2.314}$$

where

$$R(-\eta - h) := \left[(x-ut)^2 + (y/\gamma)^2 + ((z_h+\eta)/\gamma)^2\right]^{1/2} \tag{2.315}$$

with $z_h = z + h$ and $k_0 = \Omega/c$. $\zeta = \frac{\rho}{m''}$ is the ratio between the density of the fluid and the mass (per unit area) of the mass-like plane. The corresponding normalized mass impedance is $Z_0 = -i\omega m''/(\rho c)$. Therefore, we have to prove that the Fourier transform of σ_{mass} agrees with σ_1 (or with σ_2):

$$\sigma(\omega) := \int_{-\infty}^{\infty} \sigma_{\text{mass}}(t) e^{i\omega t}\mathrm{d}t = \sigma_1. \tag{2.316}$$

Proof of (2.316): The Fourier transform leads to

$$\sigma(\omega) = -2\zeta e^{i\gamma^2\Omega Mx/c} \int_0^\infty e^{-\zeta\eta} I(\omega)\mathrm{d}\eta, \tag{2.317}$$

which contains the Fourier integral

$$I(\omega) = \int_{-\infty}^{\infty} \frac{\exp\left[ik_0\gamma^2\sqrt{(x-ut)^2 + (y/\gamma)^2 + ((z_h+\eta)/\gamma)^2}\right]}{4\pi\sqrt{(x-ut)^2 + (y/\gamma)^2 + ((z_h+\eta)/\gamma)^2}} e^{-i\gamma^2\Omega t} e^{i\omega t}\mathrm{d}t. \tag{2.318}$$

Using the substitution $\alpha = ut - x$, we obtain

$$I(\omega) = \frac{1}{4\pi u} e^{-ix\left(\frac{\Omega\gamma^2-\omega}{u}\right)} \int_{-\infty}^{\infty} \frac{\exp\left[i\frac{\Omega}{c}\gamma^2\sqrt{\alpha^2 + (y/\gamma)^2 + ((z_h+\eta)/\gamma)^2}\right]}{\sqrt{\alpha^2 + (y/\gamma)^2 + ((z_h+\eta)/\gamma)^2}} e^{-i\alpha\left(\frac{\Omega\gamma^2-\omega}{u}\right)}\mathrm{d}\alpha \tag{2.319}$$

Using formula (2.26), i.e.,

$$\int_{-\infty}^{\infty} \frac{e^{i\beta\sqrt{\alpha^2+r^2}}}{\sqrt{\alpha^2+r^2}} e^{-ik_z\alpha}\mathrm{d}\alpha = \pi i H_0^{(1)}\left(r\sqrt{\beta^2 - k_z^2}\right),$$

the integral I can be written as a Hankel function, and we obtain

$$I(\omega) = \frac{1}{4\pi u} e^{-ix\left(\frac{\Omega\gamma^2-\omega}{u}\right)} \pi i H_0^{(1)}\left(\sqrt{\frac{y^2}{\gamma^2} + \frac{(z_h+\eta)^2}{\gamma^2}} \sqrt{\left(\frac{\Omega}{c}\gamma^2\right)^2 - \left(\frac{\Omega\gamma^2-\omega}{u}\right)^2}\right)$$

or

$$I(\omega) = \frac{i}{4u} e^{-ix\left(\frac{\Omega\gamma^2-\omega}{u}\right)} H_0^{(1)}\left(\sqrt{y^2+(z_h+\eta)^2} \sqrt{\left(\frac{\Omega}{c}\gamma\right)^2 - \frac{1}{u^2\gamma^2}(\omega-\Omega\gamma^2)^2}\right). \tag{2.320}$$

Taking into account Eq. 2.305, we receive

$$I(\omega) = \frac{i}{4u} e^{-ix\left(\frac{\Omega\gamma^2-\omega}{u}\right)} H_0^{(1)}\left(\kappa\sqrt{y^2+(z_h+\eta)^2}\right). \tag{2.321}$$

Introducing (2.321) into (2.317), we get

$$\sigma(\omega) = \frac{-i\zeta}{2u} e^{i\gamma^2\Omega M x/c} e^{-ix\left(\frac{\Omega\gamma^2-\omega}{u}\right)} \int_0^\infty e^{-\zeta\eta} H_0^{(1)}\left(\kappa\sqrt{y^2+(z_h+\eta)^2}\right) \mathrm{d}\eta. \tag{2.322}$$

Since

$$e^{i\gamma^2\Omega M x/c} e^{-ix\left(\frac{\Omega\gamma^2-\omega}{u}\right)} = e^{ix\left(\frac{\omega-\Omega}{u}\right)} \text{ and } \zeta = \frac{\rho}{m''} = \frac{i\omega/c}{i\omega m''/(\rho c)} = \frac{i\omega/c}{-Z_0} = -\Gamma, \tag{2.323}$$

we obtain immediately (cf. Eq. (2.312))

$$\sigma(\omega) = \frac{i\Gamma}{2u} e^{ix\left(\frac{\omega-\Omega}{u}\right)} \int_0^\infty e^{\Gamma\eta} H_0^{(1)}\left(\kappa\sqrt{y^2+(z_h+\eta)^2}\right) \mathrm{d}\eta = \sigma_1, \tag{2.324}$$

which concludes the proof. □

Test 2: Plane with a constant and real absorbing impedance Z_0.

The proof is a little more difficult and elaborate than the one for the mass-like impedance.

The correction term in the time domain is given by Eqs. (23b) and (23c) of Ochmann [5]

$$\sigma_{\text{abs}} = i\frac{\gamma^2}{2\pi}\frac{\gamma}{Z_0}\left(k_0 - iM\frac{\partial}{\partial\left[\gamma^2(x-ut)\right]}\right)e^{-i\gamma^2\Omega(t-Mx/c)}\int\limits_0^\infty \frac{1}{\tilde{R}(\eta)}e^{ik_0\tilde{R}(\eta)}e^{i\frac{\gamma}{Z_0}k_0\eta}\mathrm{d}\eta, \tag{2.325}$$

where

$$\begin{aligned}\tilde{R}(\eta) :&= \left[\left(x_L + \frac{\gamma M}{Z_0}\eta\right)^2 + y_L^2 + (z_L + \gamma h + \eta)^2\right]^{1/2} \\ &= \gamma^2\left[\left((x-ut) + \frac{M}{Z_0}\frac{\eta}{\gamma}\right)^2 + \left(\frac{y}{\gamma}\right)^2 + \left(\frac{z_h}{\gamma} + \frac{\eta}{\gamma^2}\right)^2\right]^{1/2}.\end{aligned} \tag{2.326}$$

Note that a factor γ is missing ($h \to \gamma h$) in Eq. (23c) given in [5].

Therefore, similar to Test 1, we have to prove that the Fourier transform of σ_{abs} is equal to σ_1(or to σ_2):

$$\sigma(\omega) := \int\limits_{-\infty}^{\infty} \sigma_{\text{abs}}(t)e^{i\omega t}\mathrm{d}t = \sigma_1. \tag{2.327}$$

Proof of (2.327): The Fourier transform leads to

$$\sigma(\omega) = \frac{i\gamma}{2\pi Z_0}e^{ix\gamma^2\Omega M/c}\int\limits_0^\infty e^{i\frac{\gamma}{Z_0}k_0\eta}I(\omega)\mathrm{d}\eta, \tag{2.328}$$

which contains the Fourier integral

$$\begin{aligned}I(\omega) = &\int\limits_{-\infty}^{\infty}\left(\frac{\Omega}{c} - \frac{iM}{\gamma^2}\frac{\partial}{\partial(x-ut)}\right) \\ &\times\left\{\frac{\exp\left[i\frac{\Omega}{c}\gamma^2\sqrt{\left((x-ut)+\frac{M}{Z_0}\frac{\eta}{\gamma}\right)^2 + \left(\frac{y}{\gamma}\right)^2 + \left(\frac{z_h}{\gamma}+\frac{\eta}{\gamma^2}\right)^2}\right]}{\sqrt{\left((x-ut)+\frac{M}{Z_0}\frac{\eta}{\gamma}\right)^2 + \left(\frac{y}{\gamma}\right)^2 + \left(\frac{z_h}{\gamma}+\frac{\eta}{\gamma^2}\right)^2}}e^{it(\omega-\gamma^2\Omega)}\right\}\mathrm{d}t\end{aligned} \tag{2.329}$$

We modify the well-known rule that the time derivative of a function $f(t)$ corresponds to the multiplication by the factor $-i\omega$ of the Fourier-transformed function in the following way:

$$\int_{-\infty}^{\infty} \frac{\partial}{\partial t} f(t) e^{it(\omega-\gamma^2\Omega)} \mathrm{d}t = -i(\omega - \gamma^2\Omega) \int_{-\infty}^{\infty} f(t) e^{it(\omega-\gamma^2\Omega)} \mathrm{d}t. \tag{2.330}$$

Thus, together with $\frac{\partial}{\partial(x-ut)} = -\frac{1}{u}\frac{\partial}{\partial t}$, we obtain for the first bracket in the integral

$$\left(\frac{\Omega}{c} - \frac{iM}{\gamma^2}\frac{\partial}{\partial(x-ut)}\right) \rightarrow \frac{\Omega}{c} + \frac{iM}{u\gamma^2}\Big(-i\big(\omega - \gamma^2\Omega\big)\Big) = \frac{\Omega}{c} + \frac{1}{c\gamma^2}\big(\omega - \gamma^2\Omega\big) = \frac{\omega}{c}\frac{1}{\gamma^2}. \tag{2.331}$$

With the help of (2.331) and the substitution

$$\alpha = (x - ut) + \frac{M}{Z_0}\frac{\eta}{\gamma} \text{ or } t = -\frac{1}{u}\left(\alpha - x - \frac{M}{Z_0}\frac{\eta}{\gamma}\right),$$

we obtain for the integral

$$I(\omega) = \frac{\omega}{c} \cdot \frac{1}{u\gamma^2} \int_{-\infty}^{\infty} \left\{ \frac{\exp\left[i\frac{\Omega}{c}\gamma^2 \sqrt{\alpha^2 + \left(\frac{y}{\gamma}\right)^2 + \left(\frac{z_h}{\gamma} + \frac{\eta}{\gamma^2}\right)^2} \right]}{\sqrt{\alpha^2 + \left(\frac{y}{\gamma}\right)^2 + \left(\frac{z_h}{\gamma} + \frac{\eta}{\gamma^2}\right)^2}} \times e^{-i\frac{1}{u}\left(\alpha - x - \frac{M}{Z_0}\frac{\eta}{\gamma}\right)(\omega - \gamma^2\Omega)} \right\} \mathrm{d}\alpha \tag{2.332}$$

or

$$I(\omega) = \frac{\omega}{c}\frac{1}{u\gamma^2} e^{i\frac{1}{u}\left(x + \frac{M}{Z_0}\frac{\eta}{\gamma}\right)(\omega - \gamma^2\Omega)} \int_{-\infty}^{\infty} \left\{ \frac{\exp\left[i\frac{\Omega}{c}\gamma^2 \sqrt{\alpha^2 + (y/\gamma)^2 + \left(\frac{z_h}{\gamma} + \frac{\eta}{\gamma^2}\right)^2} \right]}{\sqrt{\alpha^2 + (y/\gamma)^2 + \left(\frac{z_h}{\gamma} + \frac{\eta}{\gamma^2}\right)^2}} e^{-i\frac{1}{u}\alpha(\omega - \gamma^2\Omega)} \right\} \mathrm{d}\alpha \tag{2.333}$$

According to the formula (2.26)

$$\int_{-\infty}^{\infty} \frac{e^{i\beta\sqrt{\alpha^2 + r^2}}}{\sqrt{\alpha^2 + r^2}} e^{-ik_z\alpha} \mathrm{d}\alpha = \pi i H_0^{(1)}\left(r\sqrt{\beta^2 - k_z^2}\right), \tag{2.334}$$

we can write the integral again as a Hankel function

$$I(\omega) = \frac{\omega}{c}\frac{\pi i}{u\gamma^2} e^{i\frac{1}{u}\left(x + \frac{M}{Z_0}\frac{\eta}{\gamma}\right)(\omega - \gamma^2\Omega)} H_0^{(1)}\left(\sqrt{\frac{y^2}{\gamma^2} + \left(\frac{z_h}{\gamma} + \frac{\eta}{\gamma^2}\right)^2} \sqrt{\left(\frac{\Omega}{c}\gamma^2\right)^2 - \left(\frac{\omega - \Omega\gamma^2}{u}\right)^2} \right). \tag{2.335}$$

By taking into account Eq. 2.305, we obtain

$$I(\omega) = \frac{\omega}{c}\frac{\pi i}{u\gamma^2} e^{i\frac{1}{u}\left(x+\frac{M}{Z_0}\frac{\eta}{\gamma}\right)(\omega-\gamma^2\Omega)} H_0^{(1)}\left(\kappa\sqrt{y^2+\left(z_h+\frac{\eta}{\gamma}\right)^2}\right). \tag{2.336}$$

Furthermore, we get from (2.328) and (2.336)

$$\begin{aligned}\sigma(\omega) &= \frac{i\gamma}{2\pi Z_0} e^{ix\gamma^2\Omega M/c} \int_0^\infty e^{i\frac{\gamma}{Z_0}k_0\eta} I(\omega)\,\mathrm{d}\eta \\ &= \frac{-1}{2Z_0}\frac{\omega}{c}\frac{1}{u\gamma} e^{ix(\gamma^2\Omega M/c+(\omega-\gamma^2\Omega)/u)} \\ &\quad\times \int_0^\infty e^{i\frac{\gamma}{Z_0}k_0\eta} e^{i\frac{1}{Z_0 c}\frac{\eta}{\gamma}(\omega-\gamma^2\Omega)} H_0^{(1)}\left(\kappa\sqrt{y^2+\left(z_h+\frac{\eta}{\gamma}\right)^2}\right)\mathrm{d}\eta. \end{aligned} \tag{2.337}$$

Since (see first part of Eq. (2.323))

$$\frac{\gamma^2\Omega M}{c} + \frac{\omega-\gamma^2\Omega}{u} = \frac{\omega-\Omega}{u},$$

$$\text{and}\quad \frac{i\gamma}{Z_0}\frac{\Omega}{c} + \frac{i}{Z_0 c\gamma}(\omega-\gamma^2\Omega) = \frac{1}{\gamma}\frac{i\omega/c}{Z_0} = \frac{1}{\gamma}\Gamma, \quad \text{with} \quad \Gamma = i\frac{\omega/c}{Z_0}.$$

we obtain

$$\sigma(\omega) = \frac{i\Gamma}{2u}\frac{1}{\gamma} e^{ix\left(\frac{\omega-\Omega}{u}\right)} \int_0^\infty e^{\frac{1}{\gamma}\Gamma\eta} H_0^{(1)}\left(\kappa\sqrt{y^2+\left(z_h+\frac{\eta}{\gamma}\right)^2}\right)\mathrm{d}\eta. \tag{2.338}$$

The substitution $\tilde{\eta} = \eta/\gamma$ and the comparison with Eq. (2.312) immediately shows that

$$\sigma(\omega) = \frac{i\Gamma}{2u} e^{ix\left(\frac{\omega-\Omega}{u}\right)} \int_0^\infty e^{\Gamma\tilde{\eta}} H_0^{(1)}\left(\kappa\sqrt{y^2+(z_h+\tilde{\eta})^2}\right)\mathrm{d}\tilde{\eta} = \sigma_1, \tag{2.339}$$

which concludes the proof. □

We have thus shown that the Fourier transform of the different formulas for σ_{mass} and for σ_{abs}, respectively, leads to the same representation in the frequency domain

$$\sigma_1 = \int_{-\infty}^{\infty} \sigma_{\text{mass}} e^{i\omega t}\,\mathrm{d}t = \int_{-\infty}^{\infty} \sigma_{\text{abs}} e^{i\omega t}\,\mathrm{d}t, \tag{2.340}$$

which is not surprising if one considers that the formula for σ_1(or σ_2 if σ_1 does not converge) is valid for all types of surface impedances.

In a subsequent paper, we will numerically analyze the spectrum of moving point sources above a variety of grounds described, for example, by mass-like, absorbing, and purely spring-like impedance models. Additionally, more realistic impedances will be studied by using a modified Miki model with and without backing by a rigid wall.

For comparison, we can utilize the solution provided by Dragna and Blanc-Benon in [9], which can be readily extended to three dimensions by combining a spatial 2D Fourier transform—instead of a 1D transform—with the modified Lorentz transformation. This solution is valid for arbitrary impedances, too, but is formulated in the time domain. Such an extension was carried out by Wang et al. in [47] and by Wang in [48]. The comparison will also be detailed in the planned separate publication.

Using Eqs. (2.35) and (2.313), the velocity potential can be explicitly described as follows

$$\begin{aligned}\varphi_{3h}(t)/(-Q_1/\rho) = & \frac{1}{4\pi R_+} e^{ik_0\gamma^2 R_+} e^{-ik_0\gamma^2(ct-Mx)} \\ & + \frac{1}{4\pi R_-} e^{ik_0\gamma^2 R_-} e^{-ik_0\gamma^2(ct-Mx)} + \sigma_{3h}(t)\end{aligned}$$

with

$$\sigma_{3h}(t) = \frac{-\Gamma}{4\pi u} \int_{-\infty}^{\infty} e^{i(\omega-\Omega)\frac{x}{u}} e^{-i\omega t} \int_{0}^{\infty} e^{i\Gamma\tau} H_0^{(1)}\left(\kappa(\omega)\sqrt{y^2 + (z+h+i\tau)^2}\right) \mathrm{d}\tau \mathrm{d}\omega \tag{2.341}$$

and

$$R_\pm = \sqrt{(x-ut)^2 - \left(y^2 + (z \mp h)^2/\gamma^2\right)}. \tag{2.342}$$

2.8.3 *The Spectral Pressure of a* “Plane Source” *(1D Case) Moving in a Straight Line*

We will now also examine the one-dimensional case, which, although not of great practical importance, provides further insight into the theoretical foundations. However, if we consider a plane wave moving along the z-direction in the half-space $z > 0$ above the impedance plane, we have to restrict the observation time to a finite interval $[t_0, t_1]$ so that the source does not plunge into the lower half-space $z < 0$.

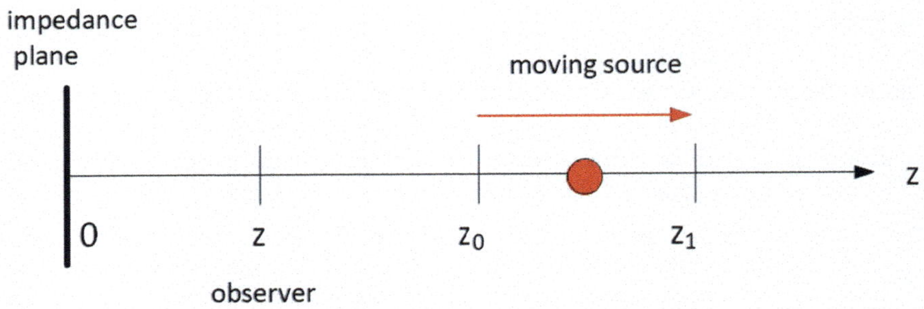

Fig. 2.49 Radiating plane 1D source moving from z_0 to z_1 "above" an impedance plane

Since only one dimension is available, the source can of course not move parallel to the plane at $z = 0$, but only perpendicular to it, see Fig. 2.49.

2.8.3.1 Derivation of the 1D Spectral Pressure for a Time-Limited Source in Free Space

First, we derive the solution in free space without any boundaries. For an infinite time interval, this was done in detail in Sect. 2.2.3. Here, we employ the same method—the Green's function method, but with the difference that we add the characteristic function

$$\chi_{[t_0,t_1]}(t) := \begin{cases} 1, & t \in [t_0, t_1] \\ 0, & t \notin [t_0, t_1] \end{cases}$$

to the right-hand side of the 1D wave equation

$$\frac{\partial^2 \varphi}{\partial z^2} - \frac{1}{c^2}\frac{\partial^2 \varphi}{\partial t^2} = \frac{Q_1}{\rho} e^{-i\Omega t} \delta(z - ut) \chi_{[t_0,t_1]}(t). \tag{2.343}$$

The temporal Fourier transform leads to the inhomogeneous 1D vibration equation for the spectral pressure $\hat{p}(\omega) = i\omega\rho\hat{\varphi}$

$$\frac{\partial^2 \hat{p}}{\partial z^2} + \left(\frac{\omega}{c}\right)^2 \hat{p} = -\frac{Q}{u} e^{i\left(\frac{\omega-\Omega}{u}\right)z} \chi_{[z_0,z_1]}(z) \text{ with } z_0 = ut_0,\ z_1 = ut_1 \text{ and } Q = -i\omega Q_1. \tag{2.344}$$

By convoluting the negative right-hand side of Eq. (2.344) with the 1D Green's function (cf. Eq. (2.67))

$$g(z - \zeta) = \frac{i}{2k} e^{ik|z-\zeta|},\ k = \omega/c, \tag{2.345}$$

we obtain

$$\hat{p}(z) = \frac{iQ}{2ku}\int_{-\infty}^{\infty} e^{ik|z-\zeta|}e^{i\left(\frac{\omega-\Omega}{u}\right)\zeta}\chi_{[z_0,z_1]}(\zeta)\mathrm{d}\zeta = \frac{iQ}{2ku}\int_{z_0}^{z_1} e^{ik|z-\zeta|}e^{i\left(\frac{\omega-\Omega}{u}\right)\zeta}\mathrm{d}\zeta. \quad (2.346)$$

In contrast to the situation with infinite observation times in Sect. 2.2.3, the finite observation time simplifies the further derivation since we do not need to consider the convergence behavior of the integral for $\zeta \to \pm\infty$. Now, we consider the case $0 < z < z_0$(see Fig. 2.49) and therefore $|z - \zeta| = -(z - \zeta)$ for $\zeta \in (z_0, z_1)$. An easy calculation gives

$$\hat{p}(z) = \hat{p}_{\text{in}}(z) = \frac{Qe^{-ikz}}{2ku}\frac{e^{i\left(k+\frac{\omega-\Omega}{u}\right)z_1} - e^{i\left(k+\frac{\omega-\Omega}{u}\right)z_0}}{\left(k+\frac{\omega-\Omega}{u}\right)}. \quad (2.347)$$

Since $\left(\frac{\omega}{c} + \frac{\omega-\Omega}{u}\right) = \frac{1+M}{u}(\omega - \omega_-)$ with $\omega_- = \Omega/(1+M)$, we receive alternatively

$$\hat{p}_{\text{in}}(z) = \frac{Qe^{-ikz}}{2k(1+M)}\frac{e^{i\frac{1+M}{u}(\omega-\omega_-)z_1} - e^{i\frac{1+M}{u}(\omega-\omega_-)z_0}}{\omega-\omega_-}. \quad (2.348)$$

Only the smaller Doppler frequency $\omega_- = \Omega/(1+M)$ appears as the source moves away from the observer.

2.8.3.2 Derivation of the Superposition Integral

Using the time-limited "basis solution" (2.348), we can now apply the superposition integral to obtain the solution in half space.

The source mirrored at the plane $z = 0$ is easily determined by replacing z by -z

$$\hat{p}_{\text{ref}}(z) = \hat{p}(-z) = \frac{Qe^{ikz}}{2k(1+M)}\frac{e^{i\frac{1+M}{u}(\omega-\omega_-)z_1} - e^{i\frac{1+M}{u}(\omega-\omega_-)z_0}}{\omega-\omega_-}. \quad (2.349)$$

By applying the superposition integral (IF1), the pressure over the impedance plane is given by

$$\hat{p}_h(z) = \hat{p}_{\text{in}}(z) + \hat{p}_{\text{ref}}(z) + \sigma, \quad (2.350)$$

with

$$\sigma = 2\Gamma\int_0^{\infty} e^{\Gamma\tau}\hat{p}(-z-\tau)\,\mathrm{d}\tau = \frac{\Gamma A}{ku}\int_0^{\infty} e^{\Gamma\tau}e^{ik(z+\tau)}\,\mathrm{d}\tau,$$

$$A = \frac{Qe^{i\left(k+\frac{\omega-\Omega}{u}\right)z_1} - Qe^{i\left(k+\frac{\omega-\Omega}{u}\right)z_0}}{k+\frac{\omega-\Omega}{u}}.$$

The integration yields

$$\sigma = \frac{\Gamma A}{ku}\int_0^\infty e^{\Gamma\tau}e^{ik(z+\tau)}\,\mathrm{d}\tau = -\frac{\Gamma A}{ku}\cdot\frac{1}{\Gamma+ik}e^{ikz},$$

where the convergence of the integral is assumed. In case of divergence, (IF2) must be used and leads to the same result as described in Sect. 2.8.1. Alternatively, we find that

$$\hat{p}_h(z) = \hat{p}_{\text{in}}(z) + R\hat{p}_{\text{ref}}(z) = \frac{A}{2ku}\left(e^{-ikz} + Re^{ikz}\right) \tag{2.351}$$

with

$$R = \left(1 - \frac{2\Gamma}{\Gamma+ik}\right) = \frac{-\Gamma+ik}{\Gamma+ik} = \frac{ik - \frac{ik}{Z_0}}{ik + \frac{ik}{Z_0}} = \frac{Z_0 - 1}{Z_0+1}. \tag{2.352}$$

As expected, we obtain the reflection coefficient of a plane wave at an impedance plane. This is a further confirmation that the superposition integral also gives the correct result for moving sources over arbitrary impedance planes and arbitrary observation times.

In addition, Appendix E shows that *the two superposition integrals (IF1) and (IF2) also satisfy the Helmholtz equation for a time-limited one-dimensional source.*

2.8.4 Concluding Remark

Assuming that the trajectory of the moving source is completely contained in the half-space above the impedance plane, we can also combine the superposition integral (IF1) or (IF2) with the CCI. For example, the CCI (2.111) for a point source in 3D combined with (IF2) results in a double integral for the correction term, which can only be solved numerically.

Furthermore, all other formulas derived in the preceding chapters for sources moving on various orbits in different coordinate systems can be combined with the superposition integral. However, the corresponding expressions become cumbersome and require numerical solutions.

2.9 The Frequency Spectrum of Moving Rigid Bodies

2.9.1 *Derivation of the Spectral Sound Pressure Emitted into the Outside Space*

In the previous sections, we have analyzed the spectra of moving point sources without considering spatial extension. While this is a theoretical simplification, it often provides reasonably approximate solutions. In this section, we consider as a "real source" a vibrating solid and non-deformable body that has a finite non-zero volume $V(t)$ and a smooth surface $S(t)$ with outward normal vector $\vec{n}$, see Fig. 2.50.

We assume that the body has a solid surface into which the fluid cannot penetrate. Far away from the body, the homogeneous wave equation applies, but in the vicinity of the body, the boundary conditions are not easy to describe due to the vibration and the diversion flow around the body surface.

Lighthill, Curle, and Ffowcs Williams and Hawkings developed a very general theory for analyzing the sound field of moving rigid bodies. If Lighthill's stress tensor and the shear stresses at the body surface are neglected, "*... a reduced form of the Ffowcs Williams-Hawkings equation, which is widely used for subsonic propeller and fan noise ...*" (quoted from Rienstra and Hirschberg [49, p. 271]) is given by (see [49, Eq. (9.28)])

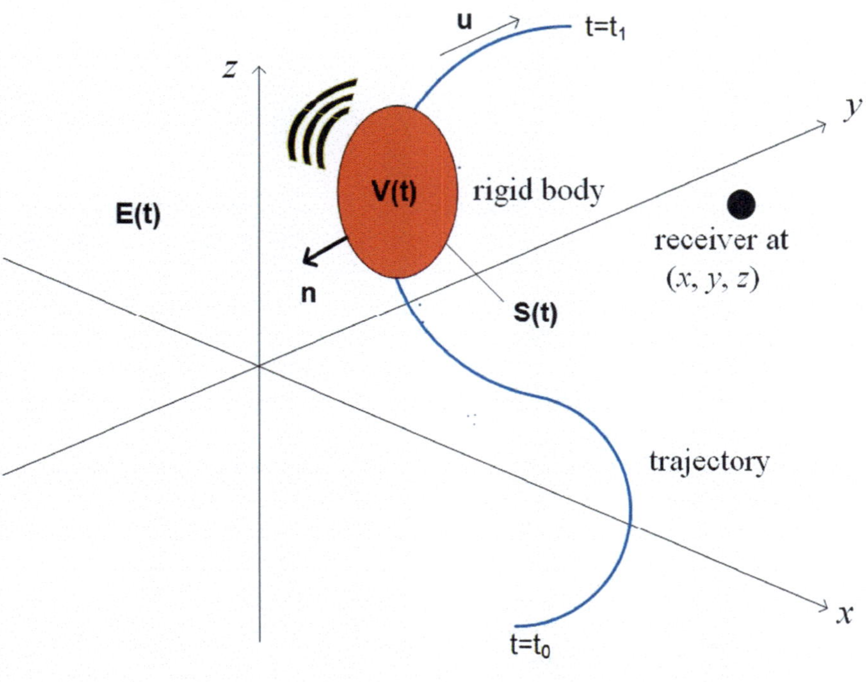

Fig. 2.50 Pulsating body moving with normal velocity $u_n = \vec{u} \cdot \vec{n}$ along a given trajectory

$$\left(\Delta - \frac{1}{c^2}\frac{\partial^2}{\partial t^2}\right)\{pH(f)\} = -q(\vec{x}, t) \tag{2.353}$$

with

$$\begin{aligned} q(\vec{x}, t) &= q_1(\vec{x}, t) + q_2(\vec{x}, t) \\ &= \left\{\frac{\partial}{\partial t}\left[\rho_0 u_n |\nabla f| \delta(f)\right]\right\} + \left\{-\nabla \cdot \left[p\vec{n}|\nabla f|\delta(f)\right]\right\}, \end{aligned} \tag{2.354}$$

where ∇ is the Nabla operator with respect to $\vec{x}$, ρ_0 is the mean density at a great distance from the body, and $u_n = \vec{u} \cdot \vec{n}$ is the normal velocity of the body. The source term $q_1(\vec{x}, t)$, i.e., the first curved bracket, leads to a sound field called thickness noise, and the second term $q_2(\vec{x}, t)$ leads to a sound field called loading noise. To describe the body surface, one utilizes an auxiliary function $f(\vec{x}, t)$ in such a manner that

$$f(\vec{x}, t)\begin{cases} < 0 \; if \; \vec{x} \in V(t) \\ = 0 \; if \; \vec{x} \in S(t) \\ > 0 \; if \; \vec{x} \in E(t) \end{cases}, \tag{2.355}$$

where $V(t)$ is the interior of the body without surface $S(t)$ and $E(t)$ is the space exterior to the body at time t. Thus, we have $\mathbb{R}^3 = V(t) \cup S(t) \cup E(t)$, see Fig. 2.50.

Since H is, as usual, the Heaviside unit step function, we obtain

$$H(f(\vec{x}, t)) = \begin{cases} 0 & \text{if} \; \vec{x} \in V(t) \\ 1/2 & \text{if} \; \vec{x} \in S(t) \\ 1 & \text{if} \; \vec{x} \in E(t) \end{cases}, \tag{2.356}$$

which reminds us of the integral formulation of the boundary element method, abbreviated BEM) (see, e.g., Wu [50, p. 18]). In fact, the Kirchhoff integral for stationary surfaces can be derived from the Ffowcs Williams-Hawkings equation (abbreviation: FW-H equation) as shown in Ehrenfried [51] at p. 319. Furthermore, Farassat and Myers [52] have derived a general Kirchhoff formula for moving surfaces. However, this formula is only valid if the so-called Kirchhoff surface lies entirely within the region where the linear wave equation holds true and encompasses all physical sources and non-linearities.

Presumably, the middle line of expression (2.356) cannot be used because of the more complex boundary condition due to the flow. In the following, we, therefore, assume that the receiver is outside the body at all times. In Sect. 2.9.3, we took a closer look at the meaning of the variable $pH(f)$. The Dirac delta function $\delta(f)$ in Eq. (2.354) causes the source terms to be located exclusively on the body surface.

We assume that the solid surface $S(t)$ moves along the curve $\vec{y}_0(t) = (y_{01}(t), y_{02}(t), y_{03}(t))^T$ for $t \in [t_0, t_1]$.

To illustrate the theory presented here, the example of a rigid sphere moving in a straight line with $\vec{y}_0(t) = (ut, 0, 0)^T$ is studied in parallel.

The surface of the body is described by the vector $\vec{y}_1 = (y_{11}, y_{12}, y_{13})^T$. Thus, the moving surface is given by

$$\vec{y}(t) = \vec{y}_0(t) + \vec{y}_1 \text{ with } t \in [t_0, t_1],\ \vec{y}_1 = (y_{11}, y_{12}, y_{13})^T \in S_1, \tag{2.357}$$

where $\vec{y}_1 = \vec{y}(t) - \vec{y}_0(t)$ is the co-moving coordinate and the surface S_1 is time-independent.

Our example of a rectilinear moving sphere of radius a is described in spherical coordinates by

$$\vec{y}(t) = \begin{pmatrix} ut \\ 0 \\ 0 \end{pmatrix} + \begin{pmatrix} a\cos\vartheta \\ a\cos\phi\sin\vartheta \\ a\sin\phi\sin\vartheta \end{pmatrix} \text{ with } t \in [t_0, t_1],\ \phi \in [0, 2\pi),\ \vartheta \in [0, \pi). \tag{2.358}$$

Here, we have rotated the usual spherical coordinates so that the y_{11}-coordinate is described by $r\cos\vartheta$.

The surface of singularities satisfies the relationship

$$|\nabla f(\vec{x}, t)|\delta(f(\vec{x}, t)) = \iint_{S_1(\vec{y}_1)} \delta(\vec{x} - (\vec{y}_0(t) + \vec{y}_1))\mathrm{d}s(\vec{y}_1). \tag{2.359}$$

This can be demonstrated using the identity

$$\iiint_{\mathbb{R}^3} \delta(h(\vec{x}))\psi(\vec{x})\mathrm{d}\vec{x} = \iint_{S(\vec{y})} \frac{\psi(\vec{y})}{|\nabla h(\vec{y})|}\mathrm{d}s(\vec{y}) \tag{2.360}$$

for sufficiently smooth testing functions ψ, where the surface S is defined by $h(\vec{y}) = 0$ (see [51, Eq. (10.4.8)] or [53, pp. 327–328]), as the surface $S_1(\vec{y}_1)$ in Eq. (2.359) does not depend on time. The auxiliary function $f(\vec{x}, t)$ does not appear on the right-hand side of Eq. (2.359), which shows that the solution depends only on the surface S, but not on the specific choice of $f(\vec{x}, t)$ [51, p. 301]. Thus, in the work of Farassat et al. (see [52, end of p. 452]) $f(\vec{x}, t)$ is always chosen in such a way that $|\nabla f(\vec{x}, t)| = 1$. By means of Eq. (2.359), the source terms (2.354) take the form

$$q_1(\vec{x}, t) = \frac{\partial}{\partial t}\left[\iint_{S_1(\vec{y}_1)} \rho_0 u_n(\vec{y}_1, t)\delta(\vec{x} - (\vec{y}_0(t) + \vec{y}_1))\mathrm{d}s(\vec{y}_1)\chi_{[t_0,t_1]}(t)\right] \tag{2.361}$$

and

$$q_2(\vec{x}, t) = -\nabla \cdot \left[\iint\limits_{S_1(\vec{y}_1)} p(\vec{y}_1, t)\vec{n}(\vec{y}_1)\delta(\vec{x} - (\vec{y}_0(t) + \vec{y}_1))\mathrm{d}s(\vec{y}_1) \right] \chi_{[t_0,t_1]}(t). \quad (2.362)$$

As in Sect. 2.3.2, the characteristic function $\chi_{[t_0,t_1]}(t)$ is inserted to limit the observation time to a specific time interval if necessary.

In addition, we have used that

$$\begin{aligned} u_n(\vec{x}, t)\delta(\vec{x} - (\vec{y}_0(t) + \vec{y}_1)) &= u_n(\vec{y}_0(t) + \vec{y}_1, t)\delta(\vec{x} - (\vec{y}_0(t) + \vec{y}_1)) \\ &= u_n(\vec{y}_1, t)\delta(\vec{x} - (\vec{y}_0(t) + \vec{y}_1)), \end{aligned}$$

since $u_n(\vec{y}_0(t) + \vec{y}_1, t)$ stays the same for all $\vec{y}_0(t)$, as the body is rigid and does not change its shape during the flight. The same applies to the surface pressure and the normal in Eq. (2.362).

In other words, the surface variables pressure and velocity naturally move with the surface and, therefore, depend on the co-moving coordinate $\vec{y}_1 = \vec{y}(t) - \vec{y}_0(t)$ in the surface integral. We assume that both quantities emit sound at a fixed angular frequency Ω

$$u_n(\vec{y}_1, t) = \hat{u}_n(\vec{y}_1)\exp(-i\Omega t), \; p(\vec{y}_1, t) = \hat{p}(\vec{y}_1)\exp(-i\Omega t). \quad (2.363)$$

We now proceed as in the previous sections and carry out a Fourier transform with respect to time in the first step

$$\begin{aligned} \hat{q}_1 &= \int\limits_{-\infty}^{\infty} \frac{\partial}{\partial t} \left[\iint\limits_{S_1(\vec{y}_1)} \rho_0 u_n(\vec{y}_1, t)\delta(\vec{x} - (\vec{y}_0(t) + \vec{y}_1))\mathrm{d}s(\vec{y}_1) \right] \chi_{[t_0,t_1]}(t)e^{i\omega t}\mathrm{d}t \\ &= \int\limits_{-\infty}^{\infty} \left[\iint\limits_{S_1(\vec{y}_1)} -i\omega\rho_0\hat{u}_n(\vec{y}_1)\delta(\vec{x} - (\vec{y}_0(t) + \vec{y}_1))\mathrm{d}s(\vec{y}_1) \right] \chi_{[t_0,t_1]}(t)e^{i(\omega-\Omega)t}\mathrm{d}t. \end{aligned} \quad (2.364)$$

Just as in Sect. 2.3.1, the trajectory of the surface should be given by the parameterized curve

$$\vec{y}_0 = \begin{pmatrix} y_{01}(t) \\ y_{02}(t) \\ y_{03}(t) \end{pmatrix} = \begin{pmatrix} ut \\ \Phi(t) \\ \psi(t) \end{pmatrix} \text{ for } t \in [t_0, t_1]. \quad (2.365)$$

Then, we obtain the three-dimensional Dirac delta function

$$\delta(\vec{x} - (\vec{y}_0(t) + \vec{y}_1)) = \delta(x_1 - y_{11} - ut)\delta(x_2 - y_{12} - \Phi(t))\delta(x_3 - y_{13} - \psi(t)),$$

and thus the Fourier integral can be solved using the sifting property of $\delta(x_1 - y_{11} - ut)$

$$\hat{q}_1 = -\frac{1}{u} \iint\limits_{S_1(\vec{y}_1)} \frac{\partial \hat{p}}{\partial \vec{n}}(\vec{y}_1) e^{i\left(\frac{\omega-\Omega}{u}\right)(x_1-y_{11})} \tilde{\delta}_2(\vec{x}-\vec{y}_1) \chi_{[ut_0,ut_1]}(x_1 - y_{11}) \mathrm{d}s(\vec{y}_1), \quad (2.366)$$

where we have defined the resulting two-dimensional Dirac delta function by

$$\tilde{\delta}_2(\vec{x}-\vec{y}_1) := \delta\left(x_2 - y_{12} - \Phi\left(\frac{x_1 - y_{11}}{u}\right), x_3 - y_{13} - \psi\left(\frac{x_1 - y_{11}}{u}\right)\right). \quad (2.367)$$

$\partial\hat{p}/\partial\vec{n} = i\omega\rho_0\hat{u}_n$ is the pressure gradient in normal direction with respect to the surface $S_1(\vec{y}_1)$ and $\vec{x} = (x_1, x_2, x_3)^T$, $\vec{y}_1 = (y_{11}, y_{12}, y_{13})^T$.

At first glance, this formula for $\hat{q}_1$ appears to be somewhat complicated due to its generality. For the special case of our example of a sphere with radius a, which moves on a straight line with constant velocity u in the x-direction and is observed for all times $t \in (-\infty, \infty)$—abbreviated as *Sphere(a,u)*—the formula looks much simpler, because $\Phi(t) = \psi(t) = 0$ and $\vec{y}_1 = (y_{11}, y_{12}, y_{13})^T = (a\cos\vartheta, a\cos\phi\sin\vartheta, a\sin\phi\sin\vartheta)^T$

$$\hat{q}_1 = -\frac{1}{u} \int_0^{2\pi}\int_0^{\pi} \frac{\partial \hat{p}}{\partial \vec{n}}(\vec{y}_1) e^{i\left(\frac{\omega-\Omega}{u}\right)(x_1 - a\cos\vartheta)}$$
$$\times\, \delta(x_2 - a\cos\phi\sin\vartheta,\ x_3 - a\sin\phi\sin\vartheta) a^2 \sin\vartheta\, \mathrm{d}\vartheta\, \mathrm{d}\varphi. \quad (2.368)$$

Let us now turn to the second source term and first determine the Fourier transform of $\vec{q}_0(\vec{x}, t)$ defined by $q_2(\vec{x}, t) = -\nabla \cdot \vec{q}_0(\vec{x}, t)$. We proceed in the same way as for the first source term and obtain the Fourier transform F_t of $q_2(\vec{x}, t)$

$$\begin{aligned}\hat{q}_2 &= -\nabla \cdot F_t[\vec{q}_0] \\ &= \frac{-1}{u}\nabla \cdot \iint\limits_{S_1(\vec{y}_1)} \vec{n}(\vec{y}_1)\hat{p}(\vec{y}_1) e^{i\left(\frac{\omega-\Omega}{u}\right)(x_1-y_{11})} \tilde{\delta}_2(\vec{x}-\vec{y}_1) \chi_{[ut_0,ut_1]}(x_1 - y_{11}) \mathrm{d}s(\vec{y}_1).\end{aligned} \quad (2.369)$$

Within the integrand, the function

$$J(\vec{x}-\vec{y}_1) = e^{i\left(\frac{\omega-\Omega}{u}\right)(x_1-y_{11})} \tilde{\delta}_2(\vec{x}-\vec{y}_1) \chi_{[ut_0,ut_1]}(x_1 - y_{11}) \quad (2.370)$$

only depends on the difference $\vec{x} - \vec{y}_1$ but not on $\vec{x}$ or $\vec{y}_1$ alone. Therefore,

$$\nabla \cdot \vec{n}(\vec{y}_1) J(\vec{x}-\vec{y}_1) = \vec{n}(\vec{y}_1) \cdot \nabla_{\vec{x}} J(\vec{x}-\vec{y}_1) = -\vec{n}(\vec{y}_1) \cdot \nabla_{\vec{y}_1} J(\vec{x}-\vec{y}_1) = -\partial J(\vec{x}-\vec{y}_1)/\partial n(\vec{y}_1),$$

where $\nabla := \nabla_{\vec{x}}$ and $\nabla_{\vec{y}_1}$ are the Nabla operators with respect to $\vec{x}$ or $\vec{y}_1$, and $\partial/\partial n(\vec{y}_1)$ is the normal derivative with respect to $\vec{y}_1$. Thus, we can write

$$\hat{q}_2 = \frac{1}{u} \iint\limits_{S_1(\vec{y}_1)} \hat{p}(\vec{y}_1) \frac{\partial J(\vec{x} - \vec{y}_1)}{\partial n(\vec{y}_1)} \mathrm{d}s(\vec{y}_1) \tag{2.371}$$

or, explicitly written

$$\hat{q}_2 = \frac{1}{u} \iint\limits_{S_1(\vec{y}_1)} \hat{p}(\vec{y}_1) \frac{\partial}{\partial n(\vec{y}_1)} \left[e^{i\left(\frac{\omega-\Omega}{u}\right)(x_1 - y_{11})} \delta_2(\vec{x} - \vec{y}_1) \chi_{[ut_0, ut_1]}(x_1 - y_{11}) \right] \mathrm{d}s(\vec{y}_1). \tag{2.372}$$

For the *Sphere(a,u)* we obtain

$$\hat{q}_2 = \frac{1}{u} \int\limits_0^{2\pi} \int\limits_0^{\pi} \hat{p}(\vec{y}_1) \frac{\partial}{\partial n(\vec{y}_1)} \Big[e^{i\left(\frac{\omega-\Omega}{u}\right)(x_1 - a\cos\vartheta)} \delta(x_2 \\ -a\cos\phi\sin\vartheta, x_3 - a\sin\phi\sin\vartheta) \Big] a^2 \sin\vartheta \,\mathrm{d}\vartheta \,\mathrm{d}\varphi. \tag{2.373}$$

The quantity $F_t[pH(f)]$—cf. the left-hand side of Eq. (2.353)—becomes complicated for arbitrary points $\vec{x} \in \mathbb{R}^3$ because a jump occurs when the moving body crosses or touches the observer (see Sect. 2.9.3). But we assume that $\vec{x}$ lies in the exterior of the moving surface $E = \bigcup_{t\in[t_0,t_1]} E(t)$ for all times considered. With respect to our example *Sphere(a,u)* the receiver $\vec{x}$ must lie outside of an infinite cylinder around the x-axis of radius a.

The Helmholtz equation for the spectral pressure $\hat{p}$ is given by

$$\left(\Delta - \left(\frac{\omega}{c}\right)^2 \right) \hat{p}(\vec{x}) = -\hat{q}_1(\vec{x}) - \hat{q}_2(\vec{x}), \quad \vec{x} \in E. \tag{2.374}$$

By convoluting the negative right-hand side $\hat{q} = \hat{q}_1 + \hat{q}_2$ with the Green's function

$$\begin{aligned} g(\vec{x} - \vec{\eta}) &= g(x_1 - \eta_1, x_2 - \eta_2, x_3 - \eta_3) \\ &= \frac{e^{i\frac{\omega}{c}\sqrt{(x_1-\eta_1)^2+(x_2-\eta_2)^2+(x_3-\eta_3)^2}}}{4\pi\sqrt{(x_1-\eta_1)^2+(x_2-\eta_2)^2+(x_3-\eta_3)^2}}, \end{aligned} \tag{2.375}$$

we obtain the solution

$$\hat{p} = \hat{p}_1 + \hat{p}_2 = \int\limits_{-\infty}^{\infty} \int\limits_{-\infty}^{\infty} \int\limits_{-\infty}^{\infty} [\hat{q}_1(\vec{\eta}) + \hat{q}_2(\vec{\eta})] \hat{g}(\vec{x} - \vec{\eta}) \mathrm{d}\vec{\eta}. \tag{2.376}$$

We now derive, in detail, the partial solution $\hat{p}_1(\vec{x})$ that corresponds to the source term $\hat{q}_1(\vec{x})$. Using $\hat{q}_1$ from Eq. (2.366), we get

$$\hat{p}_1 = -\frac{1}{u} \iiint\limits_{\mathbb{R}^3} \iint\limits_{S_1(\vec{y}_1)} \frac{\partial p}{\partial \vec{n}}(\vec{y}_1) e^{i\left(\frac{\omega-\Omega}{u}\right)(\eta_1 - y_{11})} \tilde{\delta}_2(\vec{\eta} - \vec{y}_1) \times \chi_{[ut_0,\, ut_1]}(\eta_1 - y_{11}) g(\vec{x} - \vec{\eta})\, \mathrm{d}s(\vec{y}_1)\, \mathrm{d}\vec{\eta}. \tag{2.377}$$

The evaluation of the two-dimensional Dirac delta function $\tilde{\delta}_2(\vec{x} - \vec{y}_1)$ reduces the triple integral to a single integral

$$\hat{p}_1 = -\frac{1}{u} \int\limits_{\eta_1=-\infty}^{\eta_1=\infty} \iint\limits_{S_1(\vec{y}_1)} \frac{\partial \hat{p}}{\partial \vec{n}}(\vec{y}_1) e^{i\left(\frac{\omega-\Omega}{u}\right)(\eta_1 - y_{11})} \chi_{[ut_0,\, ut_1]}(\eta_1 - y_{11}) \times g\left(\vec{x} - \vec{\xi}(\eta_1, y_{11})\right) \mathrm{d}s(\vec{y}_1)\, \mathrm{d}\eta_1. \tag{2.378}$$

where we have used the abbreviation

$$\vec{\xi}(\eta_1, y_{11}) = \left(\eta_1, y_{12} + \Phi\left(\frac{\eta_1 - y_{11}}{u}\right), y_{13} + \psi\left(\frac{\eta_1 - y_{11}}{u}\right)\right)^T \tag{2.379}$$

for the sake of clarity. After interchanging the integrals, the characteristic function determines the limits of integration

$$\hat{p}_1 = -\frac{1}{u} \iint\limits_{S_1(\vec{y}_1)} \frac{\partial \hat{p}}{\partial \vec{n}}(\vec{y}_1) \int\limits_{\eta_1=ut_0+y_{11}}^{\eta_1=ut_1+y_{11}} e^{i\left(\frac{\omega-\Omega}{u}\right)(\eta_1 - y_{11})} g\left(\vec{x} - \vec{\xi}(\eta_1, y_{11})\right) \mathrm{d}\eta_1 \mathrm{d}s(\vec{y}_1). \tag{2.380}$$

Since the integrand only depends on the difference $\eta_1 - y_{11}$, we can substitute $\zeta = \eta_1 - y_{11}$ and obtain finally

$$\hat{p}_1 = -\frac{1}{u} \iint\limits_{S_1(\vec{y}_1)} \frac{\partial \hat{p}}{\partial \vec{n}}(\vec{y}_1) \int\limits_{\zeta=ut_0}^{\zeta=ut_1} e^{i\left(\frac{\omega-\Omega}{u}\right)\zeta} \times g\left(x_1 - \zeta - y_{11},\, x_2 - y_{12} - \Phi\left(\frac{\zeta}{u}\right),\, x_3 - y_{13} - \psi\left(\frac{\zeta}{u}\right)\right) \mathrm{d}\zeta\, \mathrm{d}s(\vec{y}_1)$$

or

$$\hat{p}_1(\vec{x}) = - \iint\limits_{S_1(\vec{y}_1)} \frac{\partial \hat{p}}{\partial \vec{n}}(\vec{y}_1) \Psi(\vec{x}, \vec{y}_1) \mathrm{d}s(\vec{y}_1),\ \vec{x} \in E \tag{2.381}$$

with the core function

$$\Psi(\vec{x},\vec{y}_1) = \frac{1}{u}\int\limits_{\zeta=ut_0}^{\zeta=ut_1} e^{i\left(\frac{\omega-\Omega}{u}\right)\zeta} \times g\left(x_1-\zeta-y_{11},\ x_2-y_{12}-\Phi\left(\frac{\zeta}{u}\right),\ x_3-y_{13}-\psi\left(\frac{\zeta}{u}\right)\right)\mathrm{d}\zeta. \tag{2.382}$$

We derive the second partial solution $\hat{p}_2(\vec{x})$, which belongs to the source term $\hat{q}_2(\vec{x})$, in a completely analogous way by using the expression (2.372) and the Green's function method:

$$\hat{p}_2(\vec{x}) = \iint\limits_{S_1(\vec{y}_1)} \hat{p}(\vec{y}_1)\frac{\partial\Psi(\vec{x},\vec{y}_1)}{\partial\vec{n}(\vec{y}_1)}\mathrm{d}s(\vec{y}_1),\ \vec{x}\in E. \tag{2.383}$$

This gives the final result for the sound pressure spectrum emitted by a moving body into the external space E

$$\hat{p}(\vec{x}) = -\iint\limits_{S_1(\vec{y}_1)} \frac{\partial\hat{p}}{\partial\vec{n}}(\vec{y}_1)\Psi(\vec{x},\vec{y}_1)\mathrm{d}s(\vec{y}_1) + \iint\limits_{S_1(\vec{y}_1)} \hat{p}(\vec{y}_1)\frac{\partial\Psi(\vec{x},\vec{y}_1)}{\partial\vec{n}(\vec{y}_1)}\mathrm{d}s(\vec{y}_1),\ \vec{x}\in E. \tag{2.384}$$

This formula is a significant and novel result of the present section. It looks exactly like the integral representation for a *stationary* sound-emitting body upon which the boundary element method (BEM) is based and which is known to be

$$\hat{p}(\vec{x}) = -\iint\limits_{S(\vec{y})} \frac{\partial\hat{p}}{\partial\vec{n}}(\vec{y})g(\vec{x}-\vec{y})\mathrm{d}s(\vec{y}) + \iint\limits_{S(\vec{y})} \hat{p}(\vec{y})\frac{\partial g(\vec{x}-\vec{y})}{\partial\vec{n}(\vec{y})}\mathrm{d}s(\vec{y}),\ \vec{x}\in E. \tag{2.385}$$

Here $g(\vec{x}-\vec{y})$ is the Green's function (2.375) and S is the stationary surface of the radiating body. Since the normal $\vec{n}$ is usually directed into the acoustic medium when using the FW-H equation as shown in Fig. 2.50, but points into the interior of the body when using the BEM, Eq. (2.385) differs from the formula (34) of Wu [50] at p. 18, for example, by a minus sign on the right-hand side. If we compare the last two equations, we see that only the simple Green's function $g(\vec{x}-\vec{y})$ must be replaced by the core function $\Psi(\vec{x},\vec{y}_1)$ in order to move from the stationary to the moving case. The BEM formulation (2.385) can also be derived directly by starting from the FW-H equation and assuming a stationary surface. In the following, we refer to Eq. (2.384) as the boundary element solution, abbreviated as the BE solution.

The core function $\Psi(\vec{x}, \vec{y}_1)$ can be interpreted as a superposition of monopoles and dipoles, which are located on the trajectory and weighted in a suitable manner. Interestingly, the Green's function for a point source above an impedance plane can be interpreted in a very similar way, because: "*The sound field caused by a monopole source above an impedance plane can be calculated by using a superposition of equivalent point sources located along a line in the mirror space below the plane*" (quoted from the abstract in Ochmann [54]).

2.9.2 *Application to a Sphere Moving in a Straight Line*

As already mentioned in the last section, we now turn to the example of the rigid sphere *Sphere(a,u)* with radius a moving along the x-axis with velocity u (Fig. 2.51).

The moving sphere is described in spherical coordinates by

$$\vec{y}(t) = \vec{y}_0(t) + \vec{y}_1 = \begin{pmatrix} ut \\ 0 \\ 0 \end{pmatrix} + \begin{pmatrix} a\cos\vartheta \\ a\cos\phi\sin\vartheta \\ a\sin\phi\sin\vartheta \end{pmatrix} \text{ with } \phi \in [0, 2\pi),\ \vartheta \in [0, \pi),$$

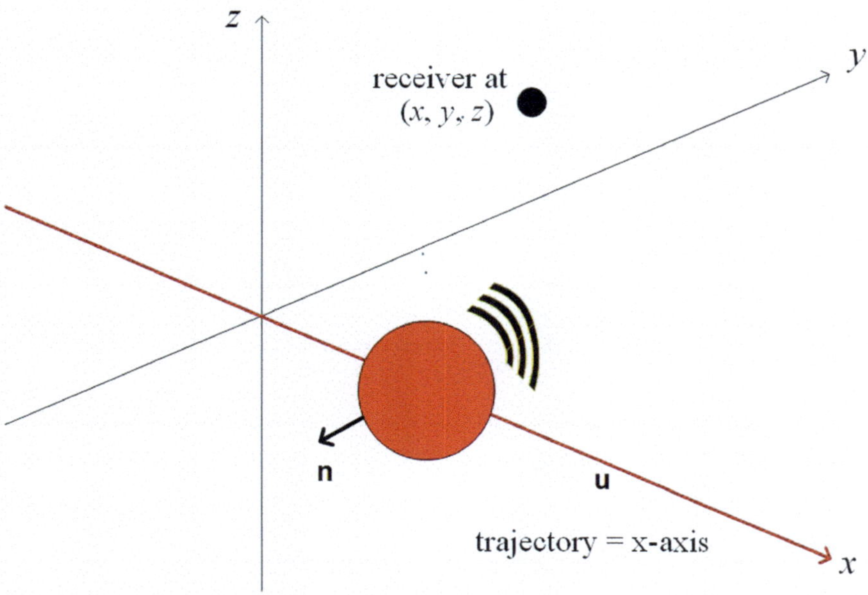

Fig. 2.51 Pulsating sphere moving with constant velocity u along the x-axis

where, for the sake of simplicity, we assume that $t \in (-\infty, \infty)$, which means that the sphere moves from minus infinity to plus infinity without restriction. The time-independent surface S_1 is given by

$$S_1 = S(t=0) = \left\{ \vec{y}_1 = (y_{11}, y_{12}, y_{13})^t \middle| y_{11}^2 + y_{12}^2 + y_{13}^2 = a^2 \right\} =: K_a.$$

Both source terms $q_1(\vec{x}, t)$ and $q_2(\vec{x}, t)$ are known because of the Eqs. (2.368) and (2.373). The BE solution (2.384) is, therefore,

$$\hat{p}(\vec{x}) = -\iint\limits_{K_a(\vec{y}_1)} \frac{\partial \hat{p}}{\partial \vec{n}}(\vec{y}_1)\Psi(\vec{x}, \vec{y}_1)\mathrm{d}s(\vec{y}_1) + \iint\limits_{K_a(\vec{y}_1)} \hat{p}(\vec{y}_1)\frac{\partial \Psi(\vec{x}, \vec{y}_1)}{\partial \vec{n}(\vec{y}_1)}\mathrm{d}s(\vec{y}_1),\ \vec{x} \in E \tag{2.386}$$

or

$$\hat{p}(\vec{x}) = -\int\limits_0^{2\pi}\int\limits_0^{\pi} \frac{\partial \hat{p}}{\partial \vec{n}}(\vec{y}_1)\Psi(\vec{x}, \vec{y}_1)a^2 \sin\vartheta \mathrm{d}\vartheta \mathrm{d}\phi + \iint\limits_{K_a(\vec{y}_1)} \hat{p}(\vec{y}_1)\frac{\partial \Psi(\vec{x}, \vec{y}_1)}{\partial \vec{n}(\vec{y}_1)}a^2 \sin\vartheta \mathrm{d}\vartheta \mathrm{d}\phi,\ \vec{x} \in E \tag{2.387}$$

According to Eq. (2.365) $\Phi(t) = \psi(t) = 0$ and, thus, the core function (2.382) with $\zeta = \eta_1 - y_{11}$ is simplified to

$$\Psi(\vec{x}, \vec{y}_1) = \frac{1}{u}\int\limits_{\eta_1=-\infty}^{\infty} e^{i\left(\frac{\omega-\Omega}{u}\right)(\eta_1 - y_{11})} g(x_1 - \eta_1, x_2 - y_{12}, x_3 - y_{13})\mathrm{d}\eta_1, \tag{2.388}$$

where

$$g(x_1 - \eta_1, x_2 - y_{12}, x_3 - y_{13}) = \frac{e^{i\frac{\omega}{c}\sqrt{(x_1-\eta)^2+(x_2-y_{12})^2+(x_3-y_{13})^2}}}{4\pi\sqrt{(x_1-\eta)^2+(x_2-y_{12})^2+(x_3-y_{13})^2}}. \tag{2.389}$$

Using Eq. (2.26), the infinite integral (2.388) can be solved analytically and results in

$$\Psi(\vec{x}, \vec{y}_1) = \frac{i}{4u} e^{i(\omega-\Omega)\frac{(x_1-y_{11})}{u}} H_0^{(1)}\left(\sqrt{(x_2-y_{12})^2+(x_3-y_{13})^2}\sqrt{\left(\frac{\omega}{c}\right)^2 - \left(\frac{\omega-\Omega}{u}\right)^2}\right). \tag{2.390}$$

For finite observation times, the integral can, of course, only be solved numerically—as already discussed in Sect. 2.3.2. The BE solution does not include any of the usual approximations listed below (see Sim1–Sim3), namely the assumptions that the sphere is compact, that the position of the receiver is in the far field, and that the Mach number is small. The solution for the spectrum is now to be checked both analytically and numerically.

2.9.2.1 Analytical Verification of the BE Solution for the Spectrum of *Sphere(a,u)*

The problem of a pulsating sphere, which is small compared to the wavelength, i.e., $k_0 a = \Omega a/c \ll 1$ is studied in the time domain by Dowling in her paper [55], in Sect. 2.3 of her dissertation [56], and in Sect. 9.3 of the book [8]. In the paper [57], published shortly afterward, Taylor pointed out that the methods developed by Dowling are not straightforward and are limited to acoustically compact bodies. Taylor developed a method for the treatment of vibrating bodies of arbitrary size in homentropic potential flow at low Mach number. He worked in Fourier space together with a special transformation similar to the Lorentz transformation with respect to time. Since he considered a stationary body in a flow, there exists no Doppler effect. In [58] and [59], Lockard analyzed moving bodies but also converted the problem into an equivalent one of a stationary body in a flow, before transforming the FW-H equation into the frequency domain, ensuring that the moving observer does not experience a Doppler effect. Lee et al. [60] used a vibrating sphere in a subsonic flow as a test case for the dual-reciprocity method.

The main difficulty now is to obtain the quantities at the surface of the pulsating sphere in order to evaluate the surface integrals appearing in Eq. (2.387). In the following, we work with the acoustic potential $\hat{\varphi}$ in the frequency domain instead of the pressure.

We assume that the radius r of the sphere oscillates with frequency Ω about its center position a with the amplitude aA

$$r = a\left(1 + Ae^{-i\Omega t}\right). \tag{2.391}$$

Then, we obtain for the normal derivative of the potential at the surface of the sphere

$$\frac{\partial \hat{\varphi}}{\partial n}(\vec{y}_1) = -iaA\Omega + 3A\frac{u}{a}y_{11}, \quad r = a, \tag{2.392}$$

where we used Eq. (30) of Lee et al. [60], who in turn applied the work of Tayler [57] and Myers [61] to derive this boundary condition. Please note that we use the time dependency $e^{-i\Omega t}$ instead of $e^{i\Omega t}$ as in Lee et al. [60], and u should have the opposite sign. The potential itself reads on the surface

$$\hat{\varphi}(\vec{y}_1) = \frac{c}{k_0} A e^{i\frac{3}{2}Mk_0 y_{11}} \left[i(k_0 a)^2 - \frac{3}{2} M k_0 y_{11} \right], \; M = \frac{u}{c}, \; k_0 = \frac{\Omega}{c}, \tag{2.393}$$

where we have used Eq. (47) of Taylor [57] together with $k_0 a \ll 1$(see Appendix F). These two boundary conditions lead to the analytical solution in the far field

$$\varphi_{\text{far}}(\vec{x}, t) = i\Omega a^3 A e^{-i\Omega t} \frac{e^{ik_0\sqrt{(x_1-ut)^2+R_x^2}}}{\sqrt{(x_1-ut)^2+R_x^2}} e^{ik_0 M(x_1-ut)} \left[1 + \frac{3}{2} M \frac{(x_1-ut)}{\sqrt{(x_1-ut)^2+R_x^2}} \right] \tag{2.394}$$

with $R_x = \sqrt{x_2^2 + x_3^2}$. This time domain solution is given by Taylor ([57], (Eq. 52)) and is valid for a compact sphere in the far field at a low Mach number. This means that all three simplifications Sim1–Sim3 (see below) must be fulfilled. Please note the changes from Taylor's to our notation:

$x_1 \to x_1 - ut$, $M \to -M$, $\cos\vartheta \to x_1/a$ and thus $r = \sqrt{x_1^2 + x_2^2 + x_3^2} \to \sqrt{(x_1-ut)^2 + x_2^2 + x_3^2}$, since we are considering a body moving along the positive x-direction in a stationary medium and not a stationary body in a medium flowing in the positive x-direction.

The main aim of this investigation is to show that the BE solution (2.387), applied to the boundary conditions (2.392) and (2.393), yields the solution (2.394) when all three simplifications made are taken into account:

Sim1: *compact sphere* with $k_0 a \ll 1$,
Sim2: the receiver is in the *far field*: $x_j \gg y_{1j}$, $j = 1, 2, 3$,
Sim3: the sphere moves with *low Mach number*, which implies that terms of $O(M^2)$ are ignored (see Taylor [57, p. 126]).

For that reason, we first calculate the thickness noise and afterward the loading noise.

(a) Calculation of the velocity potential for the "thickness noise" in the far field of a compact sphere

From the presentation (2.386), we obtain for the thickness noise

$$\hat{\varphi}_1(\vec{x}) = - \iint\limits_{K_a(\vec{y}_1)} \frac{\partial \hat{\varphi}}{\partial \vec{n}}(\vec{y}_1) \Psi(\vec{x}, \vec{y}_1) \mathrm{d}s(\vec{y}_1). \tag{2.395}$$

Using the formula (2.390) for the core function and the boundary condition (2.392) for the normal derivative of the velocity potential at the surface of the sphere, we obtain

$$\hat{\varphi}_1(\vec{x}) = \iint\limits_{K_a(\vec{y}_1)} \left[iaA\Omega - 3A\frac{u}{a}y_{11}\right]\frac{i}{4u}e^{iW(x_1-y_{11})} \times H_0^{(1)}\left(\kappa\sqrt{(x_2-y_{12})^2+(x_3-y_{13})^2}\right)\mathrm{d}s(\vec{y}_1), \qquad (2.396)$$

where we have used the abbreviations

$$W = \frac{\omega-\Omega}{u} \text{ and } \kappa = \sqrt{\left(\frac{\omega}{c}\right)^2 - \left(\frac{\omega-\Omega}{u}\right)^2}. \qquad (2.397)$$

In the far field, $x_j \gg y_{1j},\ j = 1, 2, 3$ must apply, and thus

$$R_{xy} := \sqrt{(x_2-y_{12})^2+(x_3-y_{13})^2} \approx \sqrt{x_2^2+x_3^2} = R_x. \qquad (2.398)$$

Therefore, we can draw the Hankel function in front of the integral and get

$$\hat{\varphi}_1(\vec{x}) = \frac{i}{4u}H_0^{(1)}(\kappa R_x)e^{iWx_1}\iint\limits_{K_a(\vec{y}_1)}\left[iaA\Omega - 3A\frac{u}{a}y_{11}\right]e^{-iWy_{11}}\mathrm{d}s(\vec{y}_1). \qquad (2.399)$$

The remaining integral over the sphere can be easily solved analytically with the help of

Lemma 2.9.1 The following estimates for the integrals are obtained for a compact sphere.

$$\iint\limits_{K_a(\vec{y}_1)} e^{-iUy_{11}}\mathrm{d}s(\vec{y}_1) \approx 4\pi a^2, \qquad (2.400)$$

$$\iint\limits_{K_a(\vec{y}_1)} e^{-iUy_{11}}y_{11}\mathrm{d}s(\vec{y}_1) \approx -i\frac{4}{3}\pi a^4 U, \qquad (2.401)$$

$$\iint\limits_{K_a(\vec{y}_1)} e^{-iUy_{11}}y_{11}^2\mathrm{d}s(\vec{y}_1) \approx \frac{4}{3}\pi a^4, \qquad (2.402)$$

where $U \cdot a$ is an arbitrary but small quantity.

Proof: We get for an arbitrary function $f(y_{11}) = f(a\cos\vartheta)$.

$$\iint\limits_{K_a(\vec{y}_1)} f(y_{11})\mathrm{d}s(\vec{y}_1) = 2\pi a^2\int_0^\pi f(a\cos\vartheta)\sin\vartheta\,\mathrm{d}\vartheta = 2\pi a^2\int_{-1}^1 f(a\zeta)\mathrm{d}\zeta \qquad (2.403)$$

with $\zeta = \cos\vartheta$ and therefore

$$\iint_{K_a(\vec{y}_1)} e^{-iUy_{11}} \mathrm{d}s(\vec{y}_1) = 2\pi a^2 \int_{-1}^{1} e^{-iUa\zeta} \mathrm{d}\zeta = 4\pi a^2 \frac{\sin(Ua)}{Ua} \underset{Ua \to 0}{\longrightarrow} 4\pi a^2.$$

Using Eq. 2.322/1 of Gradshteyn and Ryzhik [24], we get for small Ua

$$\begin{aligned}
\iint_{K_a(\vec{y}_1)} e^{-iUy_{11}} y_{11} \mathrm{d}s(\vec{y}_1) &= 2\pi a^2 \int_{-1}^{1} e^{-iUa\zeta} \zeta a \mathrm{d}\zeta \\
&= \frac{4\pi i a^3}{(Ua)^2} [-\sin(Ua) + Ua\cos(Ua)] \\
&\approx \frac{4\pi i a^3}{(Ua)^2} \left[-Ua + \frac{1}{3!}(Ua)^3 + Ua\left(1 - \frac{1}{2}(Ua)^2\right) \right] \\
&= -i\frac{4}{3}\pi a^4 U.
\end{aligned}$$

Using Eq. 2.322/2 of Gradshteyn and Ryzhik [24], we obtain in a similar way

$$\begin{aligned}
\iint_{K_a(\vec{y}_1)} e^{-iUy_{11}} y_{11}^2 \mathrm{d}s(\vec{y}_1) &= 2\pi a^4 \int_{-1}^{1} e^{-iUa\zeta} \zeta^2 \mathrm{d}\zeta \\
&= \frac{2\pi a^4}{(Ua)^2} \left[2Ua\sin(Ua) + 4\cos(Ua) - 4\frac{\sin(Ua)}{Ua} \right] \underset{Ua \to 0}{\longrightarrow} \frac{4}{3}\pi a^4,
\end{aligned}$$

since the term in the square brackets tends toward 2/3 for small Ua.

Inserting Eqs. (2.400) and (2.401) of Lemma 2.9.1 into Eq. (2.399) results in

$$\hat{\varphi}_1(\vec{x}) = \frac{i\pi a^3}{u} A H_0^{(1)}(\kappa R_x) e^{iWx_1} (i\Omega + uiW), \tag{2.404}$$

where we have set $U = W$ and assumed that for a compact sphere with $\Omega a/c \ll 1$

$$Wa = \frac{\omega - \Omega}{u} a \ll 1.$$

Obviously, the thickness noise $\hat{\varphi}_1(\vec{x})$ consists of the monopole component

$$\hat{\varphi}_{1M}(\vec{x}) = \frac{i\pi a^3}{u} i\Omega A H_0^{(1)}(\kappa R_x) e^{iWx_1} \tag{2.405}$$

and the dipole component in the x_1-direction

$$\hat{\varphi}_{1D}(\vec{x}) = i\pi a^3 A H_0^{(1)}(\kappa R_x) e^{iWx_1} iW = i\pi a^3 A \frac{\partial}{\partial x_1}\left(H_0^{(1)}(\kappa R_x) e^{iWx_1} \right). \tag{2.406}$$

(b) **Calculation of the velocity potential for the "loading noise" in the far field**

From the presentation (2.386) we obtain for the loading noise potential

$$\hat{\varphi}_2(\vec{x}) = \iint\limits_{K_a(\vec{y}_1)} \hat{\varphi}(\vec{y}_1)\frac{\partial\Psi(\vec{x},\vec{y}_1)}{\partial\vec{n}(\vec{y}_1)}ds(\vec{y}_1). \tag{2.407}$$

A suitable approximation for the normal derivative of the core function for a compact sphere in the far field is provided by Lemma 2.9.2:

Lemma 2.9.2 We obtain the following approximation in the far field

$$\frac{\partial\Psi(\vec{x},\vec{y}_1)}{\partial\vec{n}(\vec{y}_1)} = -\frac{i}{4ua}e^{iW(x_1-y_{11})}\left\{iWy_{11}H_0^{(1)}(\kappa R_x) - \kappa H_1^{(1)}(\kappa R_x)\frac{x_2y_{12}+x_3y_{13}}{R_x}\right\}. \tag{2.408}$$

Proof: The core function is given by Eq. (2.390)

$$\Psi(\vec{x},\vec{y}_1) = \frac{i}{4u}e^{iW(x_1-y_{11})}H_0^{(1)}\left(\kappa R_{xy}\right)$$

with

$$W = (\omega-\Omega)/u,\ \kappa = \sqrt{\left(\frac{\omega}{c}\right)^2 - \left(\frac{\omega-\Omega}{u}\right)^2} \text{ and } R_{xy} = \sqrt{(x_2-y_{12})^2+(x_3-y_{13})^2}.$$

Thus we obtain

$$\frac{\partial\Psi(\vec{x},\vec{y}_1)}{\partial\vec{n}(\vec{y}_1)} = \nabla_{\vec{y}_1}\Psi(\vec{x},\vec{y}_1)\cdot\vec{n}(\vec{y}_1) \tag{2.409}$$

with

$$\vec{n}(\vec{y}_1) = \frac{1}{a}(y_{11},y_{12},y_{13})^T$$

and

$$\nabla_{\vec{y}_1}\Psi(\vec{x},\vec{y}_1) = \frac{i}{4u}\begin{pmatrix} -iWH_0^{(1)}\left(\kappa R_{xy}\right)e^{iW(x_1-y_{11})} \\ -H_1^{(1)}\left(\kappa R_{xy}\right)e^{iW(x_1-y_{11})}\kappa\frac{y_{12}-x_2}{R_{xy}} \\ -H_1^{(1)}\left(\kappa R_{xy}\right)e^{iW(x_1-y_{11})}\kappa\frac{y_{13}-x_3}{R_{xy}} \end{pmatrix},$$

since

$$dH_0^{(1)}(z)/dz = -H_1^{(1)}(z),$$

or

$$\frac{\partial \Psi(\vec{x}, \vec{y}_1)}{\partial \vec{n}(\vec{y}_1)} = -\frac{i}{4ua} e^{iW(x_1 - y_{11})} \Big\{ iWy_{11} H_0^{(1)}(\kappa R_{xy}) \\ + \kappa H_1^{(1)}(\kappa R_{xy}) \frac{(y_{12} - x_2)y_{12} + (y_{13} - x_3)y_{13}}{R_{xy}} \Big\}. \tag{2.410}$$

This is an exact representation of the normal derivation.

In the far field with $x_j \gg y_{1j}, \ j = 1, 2, 3$, we get.

$$R_{xy} := \sqrt{(x_2 - y_{12})^2 + (x_3 - y_{13})^2} \approx \sqrt{x_2^2 + x_3^2} = R_x \text{ and } (y_{1j} - x_j)/R_{xy} \approx -x_j/R_{xy}, \ j = 1, 2.$$

Therefore we obtain

$$\frac{\partial \Psi(\vec{x}, \vec{y}_1)}{\partial \vec{n}(\vec{y}_1)} \approx -\frac{i}{4ua} e^{iW(x_1 - y_{11})} \Big\{ iWy_{11} H_0^{(1)}(\kappa R_x) - \kappa H_1^{(1)}(\kappa R_x) \frac{x_2 y_{12} + x_3 y_{13}}{R_x} \Big\}$$

which proves the assertion. □

Only the second term in the bracket on the right-hand side depends on the azimuthal angle ϕ in the yz-plane, as it contains $y_{12} = a \cos\phi \sin\vartheta$ and $y_{13} = a \sin\phi \sin\vartheta$. Thus, the integration about ϕ results in zero, since

$$\int_0^{2\pi} \cos\phi \mathrm{d}\phi = \int_0^{2\pi} \sin\phi \mathrm{d}\phi = 0.$$

Therefore, we receive for the loading noise with the help of Lemma 2.9.2 and Eq. (2.393)

$$\hat{\varphi}_2(\vec{x}) = \iint_{K_a(\vec{y}_1)} \hat{\varphi}(\vec{y}_1) \frac{\partial \Psi(\vec{x}, \vec{y}_1)}{\partial \vec{n}(\vec{y}_1)} \, \mathrm{d}s(\vec{y}_1) \\ = -\iint_{K_a(\vec{y}_1)} \frac{c}{k_0} A e^{i\frac{3}{2}Mk_0 y_{11}} \Big[i(k_0 a)^2 - \frac{3}{2} Mk_0 y_{11} \Big] \\ \times \frac{i}{4ua} e^{iW(x_1 - y_{11})} \, iWy_{11} \, H_0^{(1)}(\kappa R_x) \, \mathrm{d}s(\vec{y}_1) \tag{2.411}$$

or

$$\hat{\varphi}_2(\vec{x}) = -\frac{icA}{4k_0 ua} iWH_0^{(1)}(\kappa R_x) e^{iWx_1} \iint_{K_a(\vec{y}_1)} e^{-i(W - \frac{3}{2}Mk_0)y_{11}} \\ \times \Big[i(k_0 a)^2 - \frac{3}{2} Mk_0 y_{11} \Big] y_{11} \, \mathrm{d}s(\vec{y}_1). \tag{2.412}$$

The second term in the square bracket leads to the integral

$$I_r = \iint\limits_{K_a(\vec{y}_1)} e^{-i\tilde{W}y_{11}} \frac{3}{2} Mk_0 y_{11}^2 \mathrm{d}s(\vec{y}_1) = 2\pi a^4 Mk_0, \tag{2.413}$$

where we have used Eq. (2.402) of Lemma 2.9.1 and assumed that $Ua = \tilde{W}a = \left(W - \frac{3}{2}Mk_0\right)a$ is a small quantity. Due to Eq. (2.401) of Lemma 2.9.1, the first term in the square brackets leads to

$$I_l = \iint\limits_{K_a(\vec{y}_1)} e^{-i\tilde{W}y_{11}} (k_0 a)^2 y_{11} \mathrm{d}s(\vec{y}_1) = (k_0 a)^2 \left(-i\frac{4}{3}\pi a^4 \tilde{W}\right) = O\left(a^6\right),$$

and can be neglected since $I_l = O\left(a^6\right) \ll O\left(a^4\right) = I_r$ for small a. Thus, the loading noise is given by

$$\hat{\varphi}_2(\vec{x}) = \frac{i\pi a^3 A}{u} H_0^{(1)}(\kappa R_x) e^{iWx_1} \left(\frac{1}{2} uiW\right). \tag{2.414}$$

The comparison with Eq. (2.406) shows that $\hat{\varphi}_2(\vec{x})$ consists of only one dipole component, namely half of the dipole component of the thickness noise

$$\hat{\varphi}_2(\vec{x}) = i\pi a^3 A \frac{1}{2} \frac{\partial}{\partial x_1} \left(H_0^{(1)}(\kappa R_x) e^{iWx_1}\right) = \frac{1}{2} \hat{\varphi}_{1D}(\vec{x}). \tag{2.415}$$

Together with Eqs. (2.405/2.406), the final result for the spectrum of the moving *compact* sphere in the *far field* is

$$\hat{\varphi}(\vec{x}) = \hat{\varphi}_1(\vec{x}) + \hat{\varphi}_2(\vec{x}) = \frac{i\pi a^3}{u} A H_0^{(1)}(\kappa R_x) e^{iWx_1} \left(i\Omega + \frac{3}{2} uiW\right). \tag{2.416}$$

As already mentioned, the second term in the brackets is proportional to a dipole in the x_1-direction because we can write

$$\hat{\varphi}(\vec{x}) = \frac{i\pi a^3}{u} \left(i\Omega e^{iWx_1} + \frac{3}{2} u \frac{\partial}{\partial x_1}\right) A H_0^{(1)}(\kappa R_x) e^{iWx_1}. \tag{2.417}$$

For comparison with the analytical solution (2.394), we have to perform the inverse Fourier transform of Eq. (2.417). From Eq. (2.36), we know that

$$F_t^{-1}\left[\frac{i\pi}{u} e^{iWx_1} H_0^{(1)}(\kappa R_x)\right] = \frac{1}{R_1} e^{ik_0\gamma^2 R_1} e^{-ik_0\gamma^2(ct - Mx_1)} =: e^{-ik_0\gamma^2 ct} E(\vec{x}, t) \tag{2.418}$$

with $R_1 = \sqrt{(x_1 - ut)^2 + \left(x_2^2 + x_3^2\right)/\gamma^2}$ and $E(\vec{x}, t) = \frac{1}{R_1} e^{ik_0\gamma^2 R_1} e^{ik_0\gamma^2 Mx_1}$.

Therefore, we get

$$\begin{aligned}\varphi(\vec{x},t) &= F_t^{-1}\left[\frac{i\pi a^3 A}{u}\left(i\Omega + \frac{3}{2}u\frac{\partial}{\partial x_1}\right)\left\{H_0^{(1)}(\kappa R_x)e^{iWx_1}\right\}\right] \\ &= a^3 A e^{-ik_0\gamma^2 ct}\left(i\Omega + \frac{3}{2}u\frac{\partial}{\partial x_1}\right)\left\{\frac{1}{R_1}e^{ik_0\gamma^2 R_1}e^{ik_0\gamma^2 Mx_1}\right\} \\ &= a^3 A e^{-ik_0\gamma^2 ct}\left(i\Omega + \frac{3}{2}u\frac{\partial E(\vec{x},t)}{\partial x_1}\right). \end{aligned} \tag{2.419}$$

The derivative of $E(\vec{x},t)$ with respect to x_1 is provided by.

Lemma 2.9.3 The derivative of $E(\vec{x},t)$ with respect to x_1 is given by

$$\frac{\partial E(\vec{x},t)}{\partial x_1} = E(\vec{x},t)\left\{ik_0\gamma^2 M + \frac{x_1 - ut}{R_1}\left[ik_0\gamma^2 - \frac{1}{R_1}\right]\right\}. \tag{2.420}$$

In the far field with $R_1 \to \infty$ we obtain

$$\frac{\partial E(\vec{x},t)}{\partial x_1} \underset{R_1\to\infty}{\longrightarrow} E(\vec{x},t)ik_0\gamma^2\left(M + \frac{x_1 - ut}{R_1}\right). \tag{2.421}$$

Proof:

$$\begin{aligned}\frac{\partial E(\vec{x},t)}{\partial x_1} &= \frac{\partial}{\partial x_1}\left\{\frac{1}{R_1}e^{ik_0\gamma^2 R_1}e^{ik_0\gamma^2 Mx_1}\right\} = \frac{1}{R_1}e^{ik_0\gamma^2(R_1+Mx_1)}\left\{ik_0\gamma^2\left(\frac{\partial R_1}{\partial x_1} + M\right) - \frac{1}{R_1}\frac{\partial R_1}{\partial x_1}\right\} \\ &= \frac{1}{R_1}e^{ik_0\gamma^2(R_1+Mx_1)}\left\{ik_0\gamma^2 M + \frac{x_1 - ut}{R_1}\left[ik_0\gamma^2 - \frac{1}{R_1}\right]\right\} \\ &= E(\vec{x},t)\left\{ik_0\gamma^2 M + \frac{x_1 - ut}{R_1}\left[ik_0\gamma^2 - \frac{1}{R_1}\right]\right\} \end{aligned}$$

□

Thus, we obtain in the far field

$$\begin{aligned}\varphi(\vec{x},t) &= a^3 A e^{-ik_0\gamma^2 ct}E(\vec{x},t)\left(i\Omega + \frac{3}{2}uik_0\gamma^2\left(M + \frac{x_1 - ut}{R_1}\right)\right) \\ &= i\Omega a^3 A e^{-ik_0\gamma^2 ct}E(\vec{x},t)\left(1 + \frac{3}{2}M\gamma^2\left(M + \frac{x_1 - ut}{R_1}\right)\right) \\ &= i\Omega a^3 A e^{-ik_0\gamma^2 ct}\frac{1}{R_1}e^{ik_0\gamma^2(R_1+Mx_1)}\left(1 + \frac{3}{2}M\frac{x_1 - ut}{R_1}\right) + O(M^2), \end{aligned} \tag{2.422}$$

since $\gamma^2 = \frac{1}{1-M^2} \approx 1 + M^2$ for small M.

Taking into account the fact that

$$k_0\gamma^2 = \frac{k_0}{1 - M^2} \approx k_0(1 + M^2) = k_0 + O(M^2)$$

and

$$-k_0\gamma^2 ct = \frac{-\Omega t}{1-M^2} \approx -\Omega\left(1+M^2\right)t = -\Omega t - k_0 M u t,$$

if terms of $O\left(M^2\right)$ and less are ignored, we finally obtain

$$\varphi(\vec{x},t) = i\Omega a^3 A e^{-i\Omega t} e^{-ik_0 M u t} \frac{1}{R_1} e^{ik_0(R_1 + M x_1)} \left(1 + \frac{3}{2} M \frac{x_1 - ut}{R_1}\right). \tag{2.423}$$

This expression is identical to the far field solution (2.394)

$$\varphi_{\text{far}}(\vec{x},t) = i\Omega a^3 A e^{-i\Omega t} \frac{e^{ik_0\sqrt{(x_1-ut)^2+R_x^2}}}{\sqrt{(x_1-ut)^2+R_x^2}} e^{ik_0 M(x-ut)} \left(1 + \frac{3}{2} M \frac{(x_1-ut)}{\sqrt{(x_1-ut)^2+R_x^2}}\right),$$

since $R_1 = \sqrt{(x_1-ut)^2 + R_x^2} + O\left(M^2\right)$.

We have thus demonstrated that the BE solution produces the solution (2.394) when all three simplifications *Sim1–Sim3* are taken into account.

2.9.2.2 Numerical Verification of the BE Solution for the Spectrum of *Sphere(a,u)*

For numerical calculations, we do not need to make all three simplifications Sim1–Sim3 mentioned above. As an example, let us consider a sphere moving at a low Mach number. However, it is not necessary for the sphere to be compact and the receiver to be in the far field. The normal derivative of the potential at the surface of the sphere is again given by Eq. (2.392)

$$\frac{\partial\hat{\varphi}}{\partial n}(\vec{y}_1) = -iaA\Omega + 3A\frac{u}{a}y_{11}.$$

The solution for the potential in the time domain at low Mach number $M =" -M_{Taylor}''$ is given by Eq. (47) of Taylor [57]:

$$\varphi = -\frac{c}{k_0} A e^{-i\Omega t + ik_0 M\left(r + \frac{a^3}{2r^2}\right)\frac{(x_1-ut)}{r}} \left[ik_0 a \frac{h_0^{(1)}(k_0 r)}{h_0^{(1)\prime}(k_0 a)} - \frac{3M}{2}\left(2 - k_0^2 a^2\right) \frac{h_1^{(1)}(k_0 r)}{h_1^{(1)\prime}(k_0 a)} \frac{(x_1-ut)}{r} \right] \tag{2.424}$$

with $r = \sqrt{(x_1-ut)^2 + x_2^2 + x_3^2} = \sqrt{(x_1-ut)^2 + R_x^2}$. The prime denotes differentiation with respect to the argument.

Please note that the previous simplified boundary condition (2.393) is only valid for a compact sphere. We must instead consider the solution (2.424) at the surface of the sphere, where we have

$$\vec{x}(t) = \vec{y}(t) = \vec{y}_0(t) + \vec{y}_1 = \begin{pmatrix} ut + y_{11} \\ y_{12} \\ y_{13} \end{pmatrix}$$

and, therefore,
$y_{11} = x_1 - ut$ and thus $r = \sqrt{(x_1 - ut)^2 + x_2^2 + x_3^2} = \sqrt{y_{11}^2 + y_{12}^2 + y_{13}^2} = a.$
So we get

$$\varphi(\vec{x} \to \vec{y}_1) = -\frac{c}{k_0} A e^{-i\Omega t} e^{ik_0 \frac{3M}{2} y_{11}} \left[ik_0 a \frac{h_0^{(1)}(k_0 a)}{h_0^{(1)\prime}(k_0 a)} - \frac{3M}{2}(2 - k_0^2 a^2) \frac{h_1^{(1)}(k_0 a)}{h_1^{(1)\prime}(k_0 a)} \frac{y_{11}}{a} \right]. \tag{2.425}$$

Every point on the surface moves with the sphere; therefore, no Doppler effect occurs. Hence, the spectrum for an observer on the surface of the sphere pulsating with frequency Ω is simply given by

$$\hat{\varphi}(\vec{y}_1) = -\frac{c}{k_0} A e^{ik_0 \frac{3M}{2} y_{11}} \left[ik_0 a \frac{h_0^{(1)}(k_0 a)}{h_0^{(1)\prime}(k_0 a)} - \frac{3M}{2}(2 - k_0^2 a^2) \frac{h_1^{(1)}(k_0 a)}{h_1^{(1)\prime}(k_0 a)} \frac{y_{11}}{a} \right]. \tag{2.426}$$

We have to calculate

$$\hat{\varphi}(\vec{x}) = \hat{\varphi}_1(\vec{x}) + \hat{\varphi}_2(\vec{x}),\ \vec{x} \in E,$$

where the thickness noise is given by using Eq. (2.396) as before

$$\hat{\varphi}_1(\vec{x}) = \iint\limits_{K_a(\vec{y}_1)} \left[iaA\Omega - 3A\frac{u}{a} y_{11} \right] \frac{i}{4u} e^{i\left(\frac{\omega - \Omega}{u}\right)(x_1 - y_{11})} \times H_0^{(1)}\left(\kappa \sqrt{(x_2 - y_{12})^2 + (x_3 - y_{13})^2} \right) \mathrm{d}s(\vec{y}_1), \tag{2.427}$$

and the loading noise now results from Eq. (2.407) in combination with Eq. (2.426)

$$\hat{\varphi}_2(\vec{x}) = \iint\limits_{K_a(\vec{y}_1)} \frac{-c}{k_0} A e^{ik_0 \frac{3M}{2} y_{11}} \left[ik_0 a \frac{h_0^{(1)}(k_0 a)}{h_0^{(1)\prime}(k_0 a)} - \frac{3M}{2}(2 - k_0^2 a^2) \frac{h_1^{(1)}(k_0 a)}{h_1^{(1)\prime}(k_0 a)} \frac{y_{11}}{a} \right] \frac{\partial \Psi(\vec{x}, \vec{y}_1)}{\partial \vec{n}(\vec{y}_1)} \mathrm{d}s(\vec{y}_1).$$

The normal derivative of the core function is given by Eq. (2.410) and, thus, we receive

$$\hat{\varphi}_2(\vec{x}) = \iint\limits_{K_a(\vec{y}_1)} -\frac{c}{k_0} A e^{ik_0 \frac{3M}{2} y_{11}} \left[ik_0 a \frac{h_0^{(1)}(k_0 a)}{h_0^{(1)\prime}(k_0 a)} - \frac{3M}{2}\left(2 - k_0^2 a^2\right) \frac{h_1^{(1)}(k_0 a)}{h_1^{(1)\prime}(k_0 a)} \frac{y_{11}}{a} \right]$$

$$\times \frac{-i}{4ua} e^{iW(x_1 - y_{11})} \left\{ iW y_{11} H_0^{(1)}(\kappa R_x) \right.$$

$$\left. + \kappa H_1^{(1)}(\kappa R_x) \frac{(y_{12} - x_2) y_{12} + (y_{13} - x_3) y_{13}}{R_{xy}} \right\} ds(\vec{y}_1). \quad (2.428)$$

It is not necessary to divide the surface into finite elements, as required in the BEM for general surfaces, since we can conveniently perform the integration over the sphere in spherical coordinates using the known relationship for a function f

$$\iint\limits_{K_a(\vec{y}_1)} f(\vec{x}, \vec{y}_1) ds(\vec{y}_1) = \int\limits_0^{2\pi} \int\limits_0^{\pi} \tilde{f}(\vec{x}, \phi, \vartheta,) a^2 \sin\vartheta \, d\vartheta \, d\phi$$

with $\tilde{f}(\vec{x}, \phi, \vartheta,) = f(\vec{x}, \vec{y}_1) = f\left(\vec{x}, (a\cos\vartheta, a\cos\phi\sin\vartheta, a\sin\phi\sin\vartheta)^T\right)$. We emphasize that the sole limitation of this solution is that the boundary condition for the potential only applies to small Mach numbers.

Results:

First, we study a moving sphere with radius $a = 0.1$ m and $k_0 a = 0.1$. A Mach number of $M = 0.1$ is chosen. The receiver is located in the far field at $(x_1, x_2, x_3) = (0, 10, 10)m$.

In Fig. 2.52, the following frequency curves are displayed:

The blue curve $\text{phi1}_{\text{num}} = |\hat{\varphi}_1|$ is the numerically calculated absolute value of the thickness noise given by Eq. (2.427).

The black curve $\text{phi2}_{\text{num}} = |\hat{\varphi}_2|$ is the numerically calculated absolute value of the loading noise given by Eq. (2.428).

The green curve $\text{phi1} + 2_{\text{num}} = |\hat{\varphi}_1 + \hat{\varphi}_2|$ is the absolute value of the total potential.

The red curve $\text{phi}_{\text{ana-ff}} = |\hat{\varphi}|$ is the absolute value of the approximated analytical potential in the far field given by Eq. (2.417).

We see that the numerically calculated BE solution (green) agrees very well with the far-field approximation (red) since the observer is in the far field and the sphere is compact. For the parameters considered here, the loading noise is significantly smaller than the thickness noise.

The next Fig. 2.53 additionally shows the Fourier transform of the analytical time domain solution (2.424) given by Taylor (blue curve = "Taylor solution") for the same parameters. Here, too, all curves match very well.

But what happens if the sphere is not compact and the observer is not always in the far field? We, therefore, analyze a moving sphere with a larger radius $a = 1$ m and $k_0 a = 1$. The stationary receiver is placed at $(x_1, x_2, x_3) = (0, 2, 2)m$ and thus in the vicinity of the sphere at time $t = 0$ s. The Mach number is again $M = 0.1$.

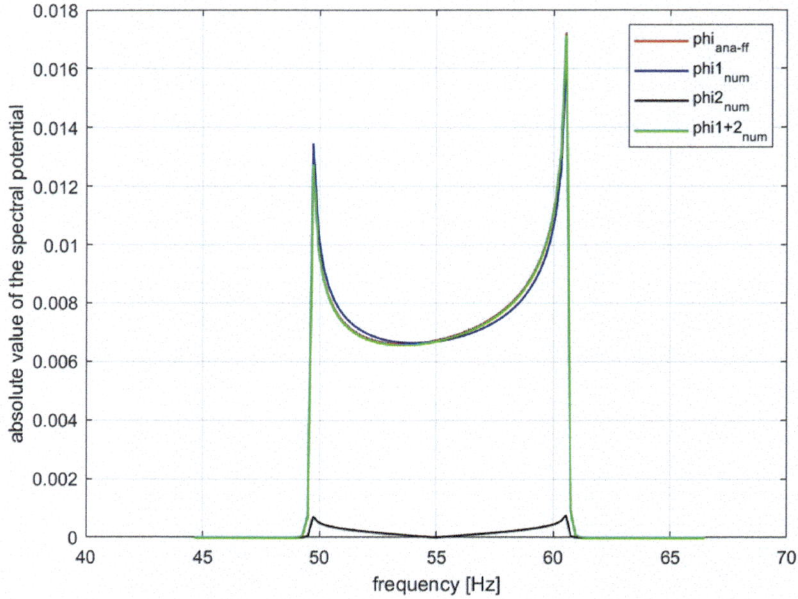

Fig. 2.52 Absolute value of the potential of a compact moving sphere in the far field. Parameters: $a = 0.1$ m, $k_0 a = 0.1$, $M = 0.1$, $(x_1, x_2, x_3) = (0, 10, 10)m$

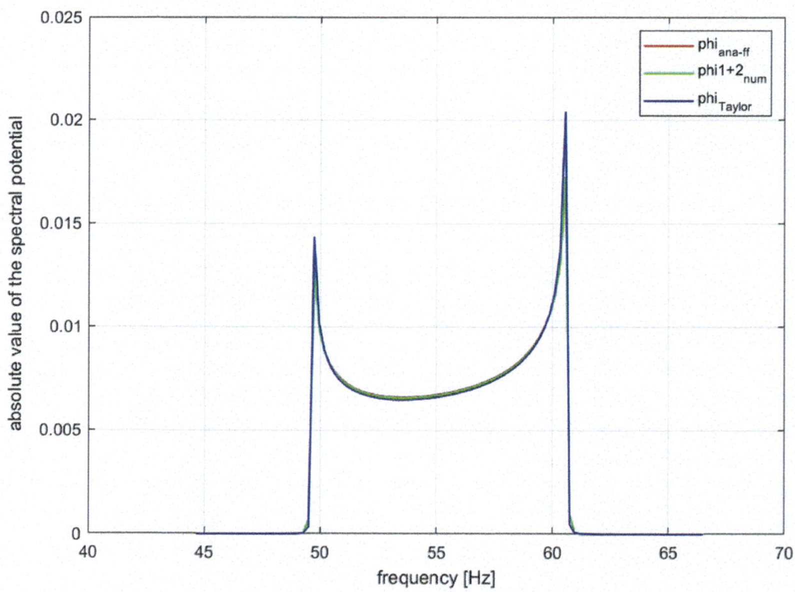

Fig. 2.53 Absolute value of the potential of a moving sphere in the far field. Parameters: $a = 0.1$ m, $k_0 a = 0.1$, $M = 0.1$, $(x_1, x_2, x_3) = (0, 10, 10)m$

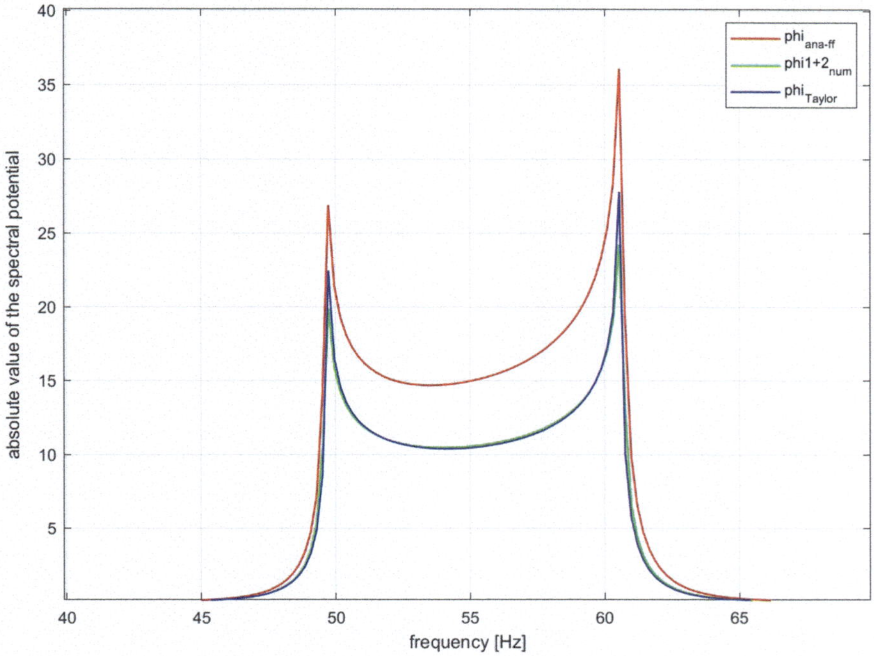

Fig. 2.54 Absolute value of the potential of a moving sphere. Parameters: $a = 1$ m, $k_0 a = 1$, $M = 0.1$, $(x_1, x_2, x_3) = (0, 2, 2)$ m

Figure 2.54 shows that the BE solution (green) and the "Taylor solution" (blue) agree very well, but the solution for a compact sphere in the far field (red) differs considerably. This makes sense, as the BE solution and the "Taylor solution" only require the Mach number to be small.

We cannot omit the requirement of a small Mach number, as it is essential for all solutions utilized here. In principle, the BE solution does not require this assumption, but we take the potential on the spherical surface as an input variable, and this is formulated in Eq. (2.426) of Taylor's theory only for small Mach numbers.

The Fourier transformation of Eq. (2.424) (blue curve) can either be carried out with the Matlab function *integral* or with the fast Fourier transformation (*fft* function of Matlab). Both methods must provide the same result. This has been successfully tested and shown in Fig. 2.55.

The numerical calculations performed thus show that the BE solution corresponds to the analytical approximate solutions for all cases considered here.

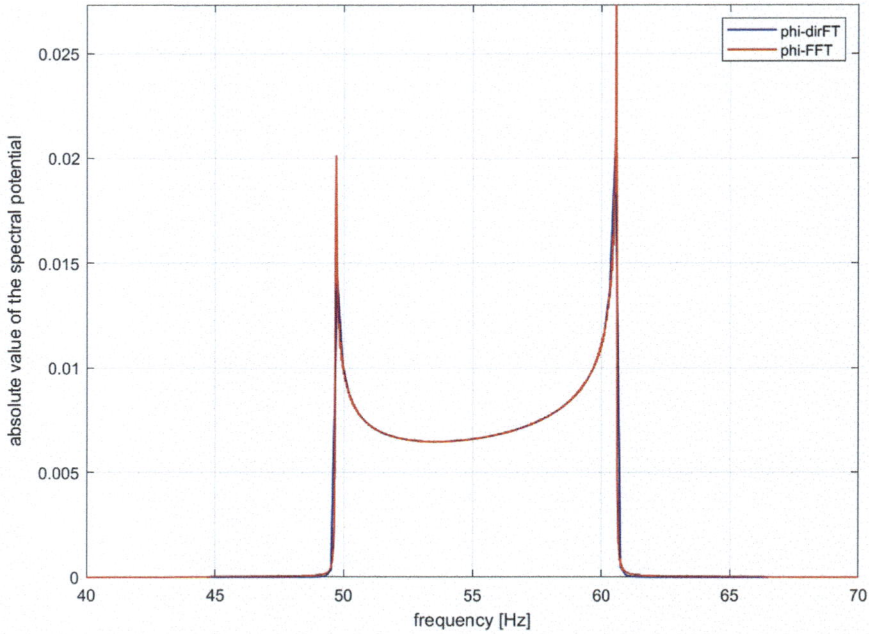

Fig. 2.55 Absolute value of the potential of a moving sphere in the far field. Parameters: $a = 0.1$ m, $k_0a = 0.1$, $M = 0.1$, $(x_1, x_2, x_3) = (0, 10, 10)m$. Blue curve = direct Fourier transform, red curve = FFT

2.9.3 Supplementary Considerations on the Quantity $\bar{p} = pH(f)$

The reduced form of the Ffowcs Williams-Hawkings Eq. (2.353) is formulated for the pressure multiplied by the Heaviside function (2.356). Up to now, we have assumed that the receiver is outside the moving body at all times, so that we obtain $\overline{p}(\vec{x}, t) = p(\vec{x})$. But what happens when this condition is violated?

First, we derive another formula for the Heaviside function. The derivative with respect to time is given by

$$\frac{\partial}{\partial t} H(f(\vec{x}, t)) = \delta(f(\vec{x}, t)) \frac{\partial}{\partial t} f(\vec{x}, t). \tag{2.429}$$

Together with $\frac{\partial}{\partial t} f(\vec{x}, t) = -u_n |\nabla f(\vec{x}, t)|$ (see Rienstra and Hirschberg [49, p. 270]) and Eq. (2.359), this results in

$$\frac{\partial}{\partial t} H(f(\vec{x}, t)) = -u_n(\vec{x}, t) \iint\limits_{S_1(\vec{y}_1)} \delta(\vec{x} - (\vec{y}_0(t) + \vec{y}_1)) \mathrm{d}s(\vec{y}_1)$$

and thus

$$H(f(\vec{x},t)) = -\int_{-\infty}^{t} u_n(\vec{x},\tau) \iint_{S_1(\vec{y}_1)} \delta(\vec{x} - (\vec{y}_0(\tau) + \vec{y}_1))\mathrm{d}s(\vec{y}_1)\mathrm{d}\tau + C. \tag{2.430}$$

For a fixed $\vec{x}$ and t approaching minus infinity, the integral disappears. Thus, the constant of integration must be set to $C = 1$ to ensure that the integral equals 1 when $\vec{x}$ lies outside the body, as required by Eq. (2.356).

We will now analyze, as an example, the case of the rectilinearly moving sphere "*Sphere(a,u)*" in two dimensions $\vec{x} = (x_1, x_2)^T$ in order to avoid overly complex calculations. Such a traveling circular disk is described by

$$\vec{y}(t) = \begin{pmatrix} y_{01} \\ y_{02} \end{pmatrix} + \begin{pmatrix} y_{11} \\ y_{12} \end{pmatrix} = \begin{pmatrix} ut \\ 0 \end{pmatrix} + \begin{pmatrix} a\cos\phi \\ a\sin\phi \end{pmatrix}, \ \phi \in [0, 2\pi),$$

where $S_1 = S(t=0) = \left\{\vec{y}_1 = (y_{11}, y_{12})^T | y_{11}^2 + y_{12}^2 = a^2\right\} =: C_a(\vec{y}_1)$. So, we obtain for the Heaviside function of the disk

$$H(f(\vec{x},t)) = 1 - \int_{-\infty}^{t} u_n(\vec{x},\tau) \int_{C_a(\vec{y}_1)} \delta(x_1 - (y_{11} + u\tau))\delta(x_2 - y_{12})\mathrm{d}s(\vec{y}_1)\mathrm{d}\tau. \tag{2.431}$$

The normal velocity of the body does not depend on time and is given by $u_n(\vec{x},t)|_{\vec{x}\in C_a(\vec{y}_1)} = \vec{u}(\vec{y}_1)\cdot\vec{n}(\vec{y}_1) = (u,0,0)^T\cdot\vec{n}(\vec{y}_1) = un_{y_{11}}$. In contrast to Eq. (2.363), we have neglected the small pulsation of the disk. Based on the first Dirac delta function, the time integral can be solved

$$H(f(\vec{x},t)) = 1 - \int_{C_a(\vec{y}_1)} n_{y_{11}} H(ut - (x_1 - y_{11}))\delta(x_2 - y_{12})\mathrm{d}s(\vec{y}_1). \tag{2.432}$$

The integration over the circle is performed in the polar coordinates $y_{11} = a\cos\phi,\ y_{12} = a\sin\phi$

$$H(f(\vec{x},t)) = 1 - \int_0^{2\pi} \cos\phi H(t - (x_1 - a\cos\phi)/u)\delta(x_2 - a\sin\phi)a\mathrm{d}\phi, \tag{2.433}$$

with $n_{y_{11}} = \cos\phi$. Now, we use the property of the remaining Dirac delta function (cf. Eq. (2.8)) that

$$\delta(g(\phi)) = \sum_j \delta(\phi - \phi_j) / \left| \frac{\partial g}{\partial \phi} \right|_{\phi=\phi_j},$$

where the ϕ_j are all solutions of
$g(\phi_j) := \sin \phi_j - x_2/a = 0$. Since $|\partial g / \partial \phi|_{\phi=\phi_j} = |\cos \phi_j|$, the line integral yields

$$H(f(\vec{x}, t)) = 1 - \sum_j \operatorname{sgn}(\cos \phi_j) H(t - (x_1 - a \cos \phi_j)/u) [H(\phi_j) - H(\phi_j - 2\pi)]. \tag{2.434}$$

We check this expression for $H(f(\vec{x}, t))$:

(a) Exterior space: $\vec{x} \in E(t)$ requires that $x_2 \notin [-a, a] \Leftrightarrow |x_2| > a$. In this case, there exists no real solution of $g(\phi_j) = 0$, and we get, as requested $H(f(\vec{x}, t)) = 1$.
(b) For $x_2 \in (-a, a)$ the situation becomes more complex, as it depends on time whether the receiver is inside or outside the body. For $x_2 \neq 0$, the equation $g(\phi_j) = 0$ possesses the two solutions $a \cos(\phi_{1,2}) = \pm\sqrt{a^2 - x_2^2}$ and, therefore, Eq. (2.434) yields

$$H(f(\vec{x}, t)) = 1 - H\left(t - \left(x_1 - \sqrt{a^2 - x_2^2}\right)/u\right) + H\left(t - \left(x_1 + \sqrt{a^2 - x_2^2}\right)/u\right). \tag{2.435}$$

For $x_2 = 0$, $g(\phi_j) = 0$ has three solution, but the result is again given by Eq. (2.435).
Using the characteristic function χ, we obtain

$$H(f(\vec{x}, t)) = 1 - \chi_{[t_1, t_2]}(t), \quad t_1 = \left(x_1 - \sqrt{a^2 - x_2^2}\right), \quad t_2 = \left(x_1 + \sqrt{a^2 - x_2^2}\right)) \tag{2.436}$$

For $t \in (t_1, t_2)$ it applies that

$$-\sqrt{a^2 - x_2^2} < x_1 - ut < \sqrt{a^2 - x_2^2} \text{ or } (x_1 - ut)^2 + x_2^2 > a^2$$

and thus the receiver $\vec{x} = (x_1, x_2)^T$ is located inside the circular disk, which leads to the required result $H(f(\vec{x}, t)) = 0$.

For $t = t_{1,2} \Leftrightarrow ut - x_1 \pm \sqrt{a^2 - x_2^2}$ the receiver is on the edge of the circular disk. Since

$$H\left(\pm\sqrt{a^2 - x_2^2} + \sqrt{a^2 - x_2^2}\right) = \begin{cases} 1 & \text{for } + \\ 1/2 & \text{for } - \end{cases}$$

and

$$H\left(\pm\sqrt{a^2-x_2^2}-\sqrt{a^2-x_2^2}\right)=\begin{cases}1/2 & \text{for } +\\ 0 & \text{for } -\end{cases}$$

for $x_2 \in (-a, a)$, we obtain from Eq. (2.435) that $H(f(\vec{x}, t)) = 1/2$ as requested.

In summary, we have derived

$$H(f(\vec{x},t))=\begin{cases}1, & |x_2|>a\\ 1-\chi_{[t_1,t_2]}(t), & |x_2|<a\end{cases} \tag{2.437}$$

or

$$\overline{p}(\vec{x},t)=p(\vec{x},t)\begin{cases}1, & |x_2|>a\\ 1-\chi_{[t_1,t_2]}(t), & |x_2|<a\end{cases}. \tag{2.438}$$

This formula for the pressure in the time domain can easily be transformed into the frequency domain by convolution (denoted by *), as the Fourier transform of the signum function is well- known:

$$\begin{aligned}F_t\{p(\vec{x},t)(1-\chi_{[t_1,t_2]}(t))\} &= F_t\{p(\vec{x},t)\} * F_t\{1-\chi_{[t_1,t_2]}(t)\}\\ &= \hat{p}(\vec{x},\omega) * F_t\{1-\chi_{[t_1,t_2]}(t)\}.\end{aligned} \tag{2.439}$$

Since

$$F_t\{1-\chi_{[t_1,t_2]}(t)\}=2\pi\delta(\omega)-\frac{2\sin\left(\omega\sqrt{a^2-x_2^2}/u\right)}{\omega}e^{i\omega x_1/u},$$

the convolution yields

$$F_t\{\overline{p}(\vec{x},t)\}=\hat{p}(\vec{x},\omega)-\frac{1}{\pi}\int\limits_{-\infty}^{\infty}\hat{p}(\vec{x},\omega-\zeta)\frac{\sin\left(\zeta\sqrt{a^2-x_2^2}/u\right)}{\zeta}e^{i\zeta\frac{x_1}{u}}\,\mathrm{d}\zeta,\ |x_2|<a. \tag{2.440}$$

We see that we get a *sinc* function when the receiver crosses or touches the body. However, this is only of interest if the body is permeable to the receiver. If the body is rigid, the receiver is pushed away from the body. Therefore, the BE solution only makes sense for stationary observers who are outside the moving body at all times under consideration.

2.9.4 Symbols

Latin Characters

$arg(z)$	argument (or "phase") of z
c	speed of sound in air
E	exterior of moving surface
$\mathrm{Ei}(x)$	exponential-integral function
$f_\pm$	Doppler frequencies
$F_t(f),\ F_x(f)$	Fourier transform of f with respect to t or x
g, g_2	Green's function
$h_n^{(1,2)}$	spherical Hankel function of the first / second kind and order n
H	Heaviside unit step function
$H_n^{(1)},\ H_n^{(2)}$	Hankel function of the first / second kind and order n
i	imaginary unit
$\mathrm{Im}\{\ \}$	imaginary part
$j_n(z)$	spherical Bessel function of order n
$J_n(z)$	Bessel function of order n
k	wave number ω/c
k_0	wave number Ω/c
$K_0(x)$	modified Bessel function of the second kind and order zero
m''	mass per unit area
M	Mach number
$\vec{n}$	unit surface normal
p	sound pressure in time domain
$P_{\mathrm{stat}}, P_{\mathrm{mov}}$	acoustic power
$P_n^{\|m\|}(\cos\theta)$	associated Legendre polynomials
$q(t),\ q_{1,2},\ Q,\ Q_1$	source strength
$r,\ r_0,\ r_s$	radii in polar, cylindrical or spherical coordinates
$r_<, r_>$	$= \min\{r, r_0\}, = \max\{r, r_0\}$
$r,\ \theta,\ \phi$	spherical coordinates
R	reflection factor
R_x	$= \sqrt{x_2^2 + x_3^2}$
R_{xy}	$= \sqrt{(x_2 - y_{12})^2 + (x_3 - y_{13})^2}$
R_1	$= \sqrt{(x - ut)^2 + (y^2 + z^2)/\gamma^2}$
$\mathbb{R}^3$	field of real 3D numbers
$\mathrm{Re}\{\ \}$	real part
$\mathrm{Res}(\ )$	residue
$\mathrm{sgn}(z)$	signum function
t	time
$\vec{u}(t)$	source velocity
$u_n = \vec{u}\cdot\vec{n}$	normal velocity
u	source velocity in x direction

$u_0(z)$	"basis solution" in free space
$V(t)$	body volume
V_+, V_-	upper and lower half-space
v_n	normal velocity
v_z	source velocity in z direction
w	angular velocity
W	$= (\omega - \Omega)/u$
$\tilde{W}$	$= \left(W - \frac{3}{2}Mk_0\right)$
x, y, z	Cartesian coordinates of the receiver
x_L, y_L, z_L	coordinates in Lorentz space
$\vec{x}_0(t)$	source location $\vec{x}_0(t) = (x_0(t), y_0(t), z_0(t))^T$
$\vec{y}(t)$	$= \vec{y}_0(t) + \vec{y}_1$ (see Eq. (2.357))
Y_0	Neumann function of order zero
$Y_n^m(\theta, \varphi)$	spherical harmonics
z_h	$z + h$
Z	impedance
Z_0	normalized acoustical impedance ($= Z/(\rho c)$)
$\mathbb{Z}$	set of integers

Greek Characters

β	see Eq. (2.121)
Γ	ik/Z_0
δ	Dirac delta function
δ_c	Dirac comb
δ_{nm}	Kronecker delta
Δ, ∇	Laplacian and Nabla operator
Δ_{Dopp}	Doppler interval
ε	damping
ζ	ρ/m''
η	volume viscosity
κ	$+\sqrt{\left(\frac{\omega}{c}\right)^2 - \left(\frac{\omega-\Omega}{u}\right)^2}$ (see Eq. (2.42))
μ	dynamic viscosity
ρ, ρ_0	fluid density
σ, σ_1, σ_2, σ_{mass},σ_{abs}	correction terms for the perfectly rigid case
φ	acoustic velocity potential
ϕ	azimuthal angle in cylindrical or spherical coordinates
Φ, ψ	coordinates of the trajectory
$\Psi(\vec{x}, \vec{y}_1)$	core function
$\chi_{(t_0,t_1)}(t)$	characteristic function of the interval (t_0, t_1)
ω, Ω	angular frequencies
$\omega_\pm$	angular Doppler frequencies

Subscripts

f	quantity in the free field
h	quantity in a half-space
in	incident quantity
L	quantity in Lorenz space
r	quantity in radial direction
ref	reflected quantity
0	coordinate of the position of the source
$1, 2, 3$	quantity in one, two or three dimensions

Superscripts

T transposition of the quantity beneath

Diacritical Signs

^ quantity in frequency domain
~ quantity in Fourier space with respect to x or z

Other Symbols and Abbreviations

□	end of proof
1D, 2D, 3D	one, two, three dimensional
BEM	Boundary Element Method
CCI	Cartesian convolution integral
ESM	Equivalent Source Method
FW-H	Ffowcs Williams and Hawkings equation
GFM	Green's function method
IF	Impedance formula

References

1. F.P. Mechel, *Schallabsorber Band. I (Sound Absorbers Vol. I)* (S. Hirzel Verlag, Stuttgart, 1989)
2. P.M. Morse, K.U. Ingard, *Theoretical Acoustics* (Princeton University Press, 1986)
3. J. Delfs, *Basics of Aeroacoustics*, Lecture notes, WS 2023/2024 at Technische Universität Braunschweig (2023). https://elib.dlr.de/202954/1/Notes_Basics_of_Aeroacoustics_Delfs.pdf
4. G.D. Duffy, *Green's Functions with Applications* (Chapman & Hall and CRC, Boca Raton, London, NewYork, Washington, DC, 2001)
5. M. Ochmann, Exact solutions for sound radiation from a moving monopole above an impedance plane. J. Acoust. Soc. Am. **133**(4), 1911–1921 (2013)
6. F. Ollendorf, Das Frequenzspektrum des Doppler-Effektes (The frequency spectrum of the Doppler effect). ACUSTICA **50**, 99–108 (1982)
7. M. Abramowitz, I.A. Stegun, *Handbook of Mathematical Functions*, 9th edn. (Dover Publications, New York, 1972)
8. A.P. Dowling, J.E. Ffowcs Williams, *Sound and Sources of Sound* (Ellis Horwood Ltd., Chicester, West Sussex, 1983)
9. D. Dragna, P. Blanc-Benon, Sound radiation by a moving line source above an impedance plane with frequency-dependent properties J. Sound. Vib. **349**, 259–275 (2015)

10. R. Piscoya, M. Ochmann, *Das Schallfeld einer sich bewegenden und ein beliebiges Zeitsignal abstrahlenden Linienquelle über absorbierendem Boden (The Sound Field of a Moving Line Source Radiating an Arbitrary Time Signal over Absorbing Ground)* (Fortschritte der Akustik DAGA 2021, Wien)
11. R. Piscoya, M. Ochmann, Sound radiation of a line source with general time dependence moving above an interface between two homogeneous semi-infinite fluid media. J. Theor. Comp. Acoust. **30**(3), 2240003-1–2240003-22 (2022, Open Access)
12. https://en.wikipedia.org/wiki/Residue_(complex_analysis)
13. E. Skudrzyk, *The Foundations of Acoustics, Basis Mathematics and Basis Acoustics* (Springer-Verlag, Wien, New York, 1971)
14. A. Deaño, D. Huybrechs, A. Iserles, *Computing Highly Oscillatory Integrals* (SIAM-Society for Industrial & Applied Mathematics, Philadelphia, 2017)
15. D.G. Crighton, A.P. Dowling, J.E. Ffowcs Williams, M. Heckl, F.G. Leppington, *Modern Methods in Analytical Acoustics*, Lecture Notes (Springer-Verlag, Berlin, Heidelberg GmbH, 1992)
16. ISO 9613-1, *Acoustics—Attenuation of Sound During Propagation Outdoors, Part 1: Calculation of the Absorption of Sound by the Atmosphere* (International Organisation of Standardization, Genève, 1993)
17. L. Cremer, *Vorlesungen über Technische Akustik (Lectures about Technical Acoustics)* (Springer, Berlin, New York, and Heidelberg, 1975)
18. J. Shang, T. Wu, H. Wang, C. Yang, C. Ye, R. Hu, J. Tao, X. He, Measurement of temperature-dependent bulk viscosities of nitrogen, oxygen and air from spontaneous Rayleigh-Brillouin scattering. IEEE Access **7**, 136439–136451 (2019)
19. https://en.wikipedia.org/wiki/Dirac_comb
20. M. A. Poletti, Cylindrical harmonic expansion of the sound field due to a rotating line source, in *Proceedings of 20th International Congress on Acoustics, ICA 2010*, Sydney, Australia, 2010
21. M.A. Poletti, Series expansions of rotating two and three dimensional sound fields. J. Acoust. Soc. Am. **128**, 3363–3374 (2010)
22. https://en.wikipedia.org/wiki/Archimedean_spiral
23. https://dlmf.nist.gov/10.22
24. I.S. Gradshteyn, I.M. Ryzhik, *Table of integrals, series, and products* (Academic Press, New York, 1980)
25. https://en.wikipedia.org/wiki/Inverse_trigonometric_functions
26. A. Erdélyi (ed.), *Tables of Integral Transforms*, vol. I (McGraw-Hill, New York, Toronto, and London, 1954)
27. M. Ochmann, The frequency spectrum of a harmonically radiating monopole moving along arbitrary trajectories—with application to helical orbits. J. Theor. Comp. Acoust. **31**(1), 2350010-1–2350010-19 (2023, Open Access)
28. P.M. Morse, H. Feshbach, *Methods of Theoretical Physics* (McGraw-Hill, New York, 1953)
29. D. Colton, R. Kress, *Inverse Acoustic and Electromagnetic Scattering Theory* (Springer, Berlin, New York, and Heidelberg, 1992)
30. Earl G. Williams, *Fourier Acoustics—Sound Radiation and Nearfield Acoustical Holography* (Academic Press, San Diego, London, Boston, 1999)
31. Y. Mao, Y. Gu, D. Qi, H. Tang, An exact frequency-domain solution of the sound radiated from the rotating dipole point source. J. Acoust. Soc. **132**(3), 1294–1302 (2012)
32. M. Carley, Series expansion for the sound field of rotating sources. J. Acoust. Soc. **120**, 1252–1256 (2006)
33. C.J. Chapman, The spiral Green function in acoustics and electromagnetism. Proc. R. Soc. London Ser. A. **431**, 157–167 (1990)
34. C.J. Chapman, The structure of rotating sound fields. Proc. R. Soc. London Ser. A. **440**, 257–271 (1993)
35. Z. Liu, C. Yi, On acoustic radiation by rotating dipole sources in frequency domain. Mod. Appl. Sci. **5**(5), 80–86 (2011)

36. W. Ma, C. Zhang, A frequency-domain beamforming for rotating sound source identification. J. Acoust. Soc. **148**(3), 1602–1613 (2020)
37. W. Ma, C. Zhang, Doppler effect in the time domain beamforming for rotating sound source identification. J. Acoust. Soc. **148**(1), 430–443 (2020)
38. Y. Mao, C. Xu, Analytical acoustic power spectrum formulations for rotating monopole and dipole point sources. J. Vib. Acoust. **138**, 021009-1–021009-13 (2016)
39. M.A. Poletti, P.D. Teal, Comparison of methods for calculating the sound field due to a rotating monopole. J. Acoust. Soc. Am. **129**(6), 3513–3520 (2011)
40. C.L. Morfey, H.K. Tanna, Sound radiation from a point force in circular motion. J. Sound Vib. **15**(3), 325–351 (1971)
41. W. Pannert, C. Maier, Rotating beamforming—motion-compensation in the frequency domain and application of high-resolution beamforming algorithms J. Sound Vib. **333**, 1899–1912 (2014)
42. N.M. Temme, *Special Functions—An Introduction to the Classical Functions of Mathematical Physics* (Wiley, New York, 1996)
43. F. P. Mechel, *Mathieu Functions—Formulas, Generation Use* (S. Hirzel Verlag, Stuttgart, Leipzig, 1997)
44. R. Barakat, Diffraction of plane waves by an elliptic cylinder. J. Acoust. Soc. **35**(12), 1990–1996 (1963)
45. M. Ochmann, R. Piscoya, *Theory and Application of Acoustic Sources Using Complex Analysis*, 1st edn (Springer Nature, Singapore, 2021)
46. M. Ochmann, Exact solution for the acoustical impulse response of a line source above an absorbing plane. J. Acoust. Soc. Am. **144**(3), 1539–1549 (2018)
47. Y. Wang, B. Tong, K.M. Li, Sound fields generated by a monopole point source moving above a locally reacting surface. in *INTER-NOISE and NOISE-CON Congress, Conference Proceedings*, vol. 252, Institute of Noise Control Engineering, 2016, pp. 683–694
48. Y. Wang, Propagation of en-route aircraft noise, dissertation, Purdue University, 2019
49. S.W. Rienstra, A. Hirschberg, *An Introduction to Acoustics* (Eindhoven University of Technology, 2021). https://www.win.tue.nl/~sjoerdr/papers/boek.pdf
50. T. W. Wu, *Boundary Element Acoustics (ed.)—Fundamentals and Computer Codes (Advances in Boundary Elements Series)* (WITPRESS, Southampton, Boston, 2000)
51. K. Ehrenfried, *Strömungsakustik (Flow acoustics)* (Mensch & Buch Verlag, Berlin, 2004)
52. F. Farassat, M.K. Myers, Extension of Kirchhoff's formula to radiation from moving surfaces. J. Sound Vib. **123**(3), 451–460 (1988)
53. J.E. Ffowcs Williams, D.L. Hawkings, Sound generation by turbulence and surfaces in arbitrary motion. Phil. Trans. R. Soc. Lond. A. Math. Phys. Sci. **264**, 321–342 (1969)
54. M. Ochmann, The complex equivalent source method for sound propagation over an impedance plane. J. Acoust. Soc. Am. **116**(6), 3304–3311 (2004)
55. A.P. Dowling, Convective amplification of real simple sources. J. Fluid Mech. **74**, 529–546 (1976)
56. A. P. Dowling, Acoustic sources in motion, Dissertation, Sidney Sussex College, Cambridge, 1978. https://www.repository.cam.ac.uk/items/31a60961-5455-4557-b8f1-ecdb1ad81e6f
57. K. Taylor, Acoustic generation by vibrating bodies in homentropic potential flow at low Mach number. J. Sound and Vib. **65**(1), 125–136 (1979)
58. D.P. Lockard, An efficient, two-dimensional implementation of the Ffowcs Williams and Hawkings equation. J. Sound and Vib. **229**(4), 897–911 (2000)
59. D.P. Lockard, A comparison of Ffowcs Williams-Hawkings solvers for airframe noise applications, in *8th AIAA/CEAS Aeroacoustics Conference* (2002), pp. 1–11
60. L. Lee, T.W. Wu, P. Zhang, A dual-reciprocity method for acoustic radiation in a subsonic non-uniform flow. Eng. Anal. Boundary Elem. **13**, 365–370 (1994)
61. M.K. Myers, On the acoustic boundary condition in the presence of flow. J. Sound and Vib. **71**(3), 429–435 (1980)

Chapter 3
Sound Field of a Source Moving in Half-Space in Time Domain

In this chapter, we study the sound field produced by a source moving above a semi-infinite medium. We consider two semi-infinite fluid media characterized by the densities and sound speeds ρ, c and ρ_m, c_m, respectively, with a flat interface at $z = 0$. The sound field will be determined only in the upper half-space ($z > 0$), where the source is located. The technique to derive the velocity potential consists of performing a convolution between Green's function of a stationary source and the source function that includes the motion. To obtain an expression in terms of a single integral, it is required that Green's function of the stationary source has a closed formula. For this reason, we will not analyze the sound field of a point source above a homogeneous medium since there is no close formula for that case.

It is assumed that the observer is stationary and located at position $\vec{r}$, and the sound field is determined at time t. For the stationary case (Fig. 3.1a), the source is placed at position $\vec{r}_q$ and the mirror source at $\vec{r}_{qs}$, so that distances to the observer, R and R_s, respectively, are constant. The signals reaching the observer at time t from the source and mirror source were emitted at times τ and τ_s, respectively. For the moving case (Fig. 3.1b), the positions of source and mirror sources are given by $\vec{r}_q = \vec{M} c\tau$ and $\vec{r}_{qs} = \vec{M}_s c\tau$, respectively, with $\vec{M}$ the Mach vector and $\vec{M}_s$ the mirrored Mach vector. The distances from the source and mirror source to the observer (R, R_s) are, therefore, time-dependent.

3.1 Line Source Above a Locally Reacting Medium

A locally reacting medium does not admit wave propagation in the direction parallel to the surface so that the normal velocity at one point on the boundary only depends on the pressure at that point and not at neighboring points. Therefore, there is no need to include the lower domain to determine the sound field in the upper medium and its presence is taken into account through an impedance boundary condition on

M. Ochmann and R. Piscoya, *Moving Acoustic Sources*,
Springer-AAS Acoustics Series, https://doi.org/10.1007/978-981-96-0871-3_3

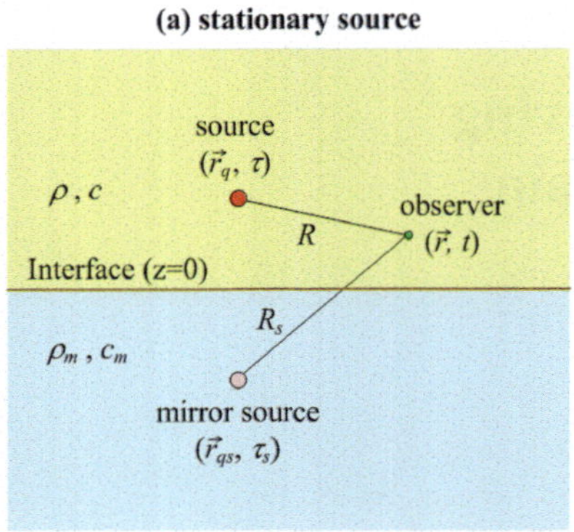

Fig. 3.1 Illustration of the half-space problem for stationary and moving sources

the interface. Consequently, the impedance Z will depend on the position but not on the angle of incidence. The following derivation considers an impedance with a pure absorbing character, i.e., $Im(Z) = 0$.

3.1.1 Stationary Line Source

For the stationary case, the absorbing boundary condition in the time domain is given by

$$\frac{\partial \phi}{\partial z} - \frac{1}{Z_0 c}\frac{\partial \phi}{\partial t} = 0 \text{ at } z = 0, \tag{3.1}$$

where ϕ is Green's function for the velocity potential and Z_0 the normalized impedance ($Z_0 = Z/\rho c$). The approach to obtain ϕ for the line source is to represent the line source as an infinite set of point sources along the y-axis and then integrate three-dimensional Green's function over the source location with respect to the y-axis [6]. In order to obtain a closed formula for 3D Green's function, it was required that the ground impedance be frequency independent. The formula for ϕ for a source sitting at $\vec{r}_q = (0, h)$ emitting a pulse at $t = 0$ is found to be

$$\begin{aligned}\phi(\vec{r}, t; \vec{r}_q, 0) &= \frac{c}{2\pi}\frac{H\left(ct - \sqrt{x^2 + (z-h)^2}\right)}{\sqrt{c^2t^2 - x^2 - (z-h)^2}} \\ &+ \frac{c}{2\pi}\frac{H\left(ct - \sqrt{x^2 + z_h^2}\right)}{\sqrt{c^2t^2 - x^2 - z_h^2}}\left(\frac{(Z_0ct)^2 - z_h^2 - \left(1 + Z_0^2\right)x^2}{(Z_0ct + z_h)^2 + \left(1 - Z_0^2\right)x^2}\right),\end{aligned} \tag{3.2}$$

where $z_h = z + h$, $H(x)$ is the Heaviside step function, and the last term in brackets is the reflection factor. Formula (3.2) can be rewritten considering the location of the source at $\vec{r}_q = (x_q, h)$ and emission time of the pulse at $t = t_q$ as

$$\phi = \phi_{\text{inc}} + \phi_{\text{mrf}} + \phi_{\text{cor}}, \tag{3.3}$$

where ϕ_{inc} is the incident wave, ϕ_{mrf} the mirror-reflected wave and ϕ_{cor} the correction term due to the absorbing character of the ground and are expressed as

$$\phi_{\text{inc}} = \frac{1}{2\pi} \frac{H(t - t_q - t_0)}{\sqrt{(t - t_q)^2 - t_0^2}}, \tag{3.4}$$

$$\phi_{\text{mrf}} = \frac{1}{2\pi} \frac{H(t - t_q - t_{0s})}{\sqrt{(t - t_q)^2 - t_{0s}^2}}, \tag{3.5}$$

$$\phi_{\text{cor}} = -\frac{1}{\pi} \frac{H(t - t_q - t_{0s})}{\sqrt{(t - t_q)^2 - t_{0s}^2}} \left(\frac{Z_0 c(t - t_q) z_h + z_h^2 + (x - x_q)^2}{\left(Z_0 c(t - t_q) + z_h\right)^2 + (1 - Z_0^2)(x - x_q)^2} \right), \tag{3.6}$$

with $t_0 = R/c$, $R = \sqrt{(x - x_q)^2 + (z - h)^2}$, $t_{0s} = R_s/c$, $R_s = \sqrt{(x - x_q)^2 + z_h^2}$.

Equation (3.4) indicates that $\phi_{\text{inc}} = 0$ for $t < t_q + t_0$ and Eqs. (3.5) and (3.6) that $\phi_{\text{mrf}} = \phi_{\text{cor}} = 0$ for $t < t_q + t_{0s}$.

3.1.2 *Moving Line Source*

The velocity potential generated by a moving line source with general time dependence $Q(\vec{r}, t)$ is obtained by using Green's function method with the source strength Q and Green's function ϕ [3]

$$\Phi(\vec{r}, t) = \int_{-\infty}^{+\infty} \int_{-\infty}^{+\infty} Q(\vec{\rho}, \tau) \phi(\vec{r}, t; \vec{\rho}, \tau) d^2\rho d\tau. \tag{3.7}$$

A moving source is described as $Q(\vec{r}, t) = q(t)\delta\left(\vec{r} - \vec{r}_q(t)\right)$; therefore, the velocity potential is written as

$$\Phi(\vec{r}, t) = \int_{-\infty}^{+\infty} q(\tau) \phi\left(\vec{r}, t; \vec{r}_q(\tau), \tau\right) d\tau. \tag{3.8}$$

Here and in the rest of the chapter, we will treat only the case of a source moving with a constant velocity Mc parallel to the ground at a height $z = h$, i.e., $\vec{r}_q(t) = (Mct, h)$. The distances from the source and mirror source to the observer are not constant. At time $t = \tau$, the source and mirror source are located at $(Mc\tau, h)$ and $(Mc\tau, -h)$ and R and R_s are given by

$$R(\tau) = \sqrt{(x - Mc\tau)^2 + (z - h)^2}, \tag{3.9}$$

$$R_s(\tau) = \sqrt{(x - Mc\tau)^2 + z_h^2}, \tag{3.10}$$

and the times of arrival of the signal to the observer from the source and mirror source are functions of the emission time τ. Inserting (3.9) and (3.10) into (3.4)–(3.6) and considering (3.3), (3.8) can be rewritten as

$$\Phi_{inc} = \frac{1}{2\pi}\int_{-\infty}^{+\infty} q(\tau)\frac{H(t - \tau - t_0(\tau))}{\sqrt{(t-\tau)^2 - t_{0s}^2(\tau)}}d\tau, \tag{3.11}$$

$$\Phi_{mrf} = \frac{1}{2\pi}\int_{-\infty}^{+\infty} q(\tau)\frac{H(t - \tau - t_{0s}(\tau))}{\sqrt{(t-\tau)^2 - t_{0s}^2(\tau)}}d\tau, \tag{3.12}$$

$$\Phi_{cor} = -\frac{1}{\pi}\int_{-\infty}^{+\infty} q(\tau)\frac{H(t - \tau - t_{0s}(\tau))}{\sqrt{(t-\tau)^2 - t_{0s}^2(\tau)}}\left(\frac{Z_0c(t-\tau)z_h + z_h^2 + (x - Mc\tau)^2}{(Z_0c(t-\tau) + z_h)^2 + \left(1 - Z_0^2\right)(x - Mc\tau)^2}\right)d\tau. \tag{3.13}$$

In order to simplify Formulas (3.11)–(3.13), it is useful to introduce the inverse Prandtl-Lorentz identities

$$t = t_L + \frac{M}{c}x_L,\ x = x_L + Mct_L,\ z = \frac{z_L}{\gamma},\ h = \frac{h_L}{\gamma},\ \text{with } \gamma = \frac{1}{\sqrt{1 - M^2}}, \tag{3.14}$$

so that

$$c(t - \tau) = Mx_L + c(t_L - \tau) \text{ and } x - Mc\tau = x_L + Mc(t_L - \tau). \tag{3.15}$$

By taking into account (3.15), we find that

$$t - \tau - t_0(\tau) = 0 \rightarrow t_L - \tau - t_{0L} = 0, \tag{3.16}$$

$$t - \tau - t_{0s}(\tau) = 0 \rightarrow t_L - \tau - t_{0sL} = 0, \tag{3.17}$$

where $t_{0L} = R_L/c$, $R_L = \sqrt{x_L^2 + (z_L - h_L)^2}$, $t_{0sL} = R_{sL}/c$, $R_{sL} = \sqrt{x_L^2 + (z_L + h_L)^2}$ and the square roots can be simplified to

$$\sqrt{(t-\tau)^2 - t_0^2(\tau)} = \sqrt{(t_L-\tau)^2 - t_{0L}^2}/\gamma, \tag{3.18}$$

$$\sqrt{(t-\tau)^2 - t_{0s}^2(\tau)} = \sqrt{(t_L-\tau)^2 - t_{0sL}^2}/\gamma. \tag{3.19}$$

Inserting (3.16)–(3.19) into (3.11)–(3.13) and using the commutative property of the convolution, we have

$$\Phi_{inc} = \frac{\gamma}{2\pi}\int_{-\infty}^{+\infty} q(t_L-\tau)\frac{H(\tau - t_{0L})}{\sqrt{\tau^2 - t_{0L}^2}}d\tau, \tag{3.20}$$

$$\Phi_{mrf} = \frac{\gamma}{2\pi}\int_{-\infty}^{+\infty} q(t_L-\tau)\frac{H(\tau - t_{0sL})}{\sqrt{\tau^2 - t_{0sL}^2}}d\tau, \tag{3.21}$$

and introducing the variables $\alpha = \gamma/Z_0$, $\varepsilon = \sqrt{1+\alpha^2M^2}$ and $\nu = z_{hL} + \alpha M x_L$, with $z_{hL} = z_L + h_L$, the correction term is written as

$$\Phi_{cor} = -\frac{\gamma}{\pi}\int_{-\infty}^{+\infty} q(t_L-\tau)\frac{H(\tau - t_{0sL})}{\sqrt{\tau^2 - t_{0sL}^2}}\frac{\varepsilon^2\left(\alpha^2(x_L+Mc\tau)^2 + \alpha z_{hL}(Mx_L+c\tau) + \alpha^2 z_{hL}^2/\gamma^2\right)}{\left(\varepsilon^2 c\tau + \alpha\nu\right)^2 + (\alpha^2-\varepsilon^2)\left(\varepsilon^2 R_{sL}^2 - \nu^2\right)}d\tau. \tag{3.22}$$

Green's function of the moving line source can be obtained by setting $q(t) = \delta(t)$ in (3.22).

3.1.3 Verification Example: Harmonic Time Dependence

As a verification example, we will compare the solution (3.20)–(3.22), applied to a harmonic time dependence $q(t) = e^{-j\omega_s t}$, with the solution obtained by Dragna and Blanc-Benon [4, Eqs. (27), (29), (30)], deduced using the Fourier transform. They solve the Helmholtz equation with a positive right-hand side term. Therefore, we should compare our solution Φ with their negative solution $-\varphi$, given by:

$$-\varphi_1 = \frac{j\gamma}{4}H_0^{(1)}(k_s R_L)e^{-j\omega_s t_L}, \tag{3.23}$$

$$-\varphi_2 = \frac{j\gamma}{4}H_0^{(1)}(k_s R_{sL})e^{-j\omega_s t_L}, \tag{3.24}$$

$$-\varphi_3 = -\frac{j\gamma}{2\pi} e^{-j\omega_s t_L} \int_{-\infty}^{+\infty} \frac{e^{jL_x x_L} e^{j\sqrt{k_s^2 - L_x^2} z_{hL}}}{\sqrt{k_s^2 - L_x^2}} \frac{\beta(\omega_D)(k_s + L_x M) dL_x}{\sqrt{k_s^2 - L_x^2} + (k_s + L_x M)\beta(\omega_D)}, \quad (3.25)$$

where $\omega_D = \gamma^2 \omega_s \left(1 + \frac{ML_x}{k_s}\right)$ is the Doppler-shifted frequency and β is the ground admittance.

Next, we find the solution for the present approach. Inserting $q(t) = e^{-j\omega_s t}$ into our solution (3.20), we have

$$\Phi_{inc} = \frac{\gamma}{2\pi} e^{-j\omega_s t_L} \int_{t_{0L}}^{+\infty} \frac{e^{j\omega_s \tau}}{\sqrt{\tau^2 - t_{0L}^2}} d\tau,$$

and making a change of variable $t_{0L} u = \tau$ we obtain

$$\Phi_{inc} = \frac{j\gamma}{4} e^{-j\omega_s t_L} \left(\frac{2}{j\pi} \int_{1}^{+\infty} \frac{e^{jk_s R_L u}}{\sqrt{u^2 - 1}} du \right).$$

The integral in brackets is an integral representation of $H_0^{(1)}$ [11]; therefore,

$$\Phi_{inc} = \frac{j\gamma}{4} e^{-j\omega_s t_L} H_0^{(1)}(k_s R_L), \quad (3.26)$$

and similarly

$$\Phi_{mrf} = \frac{j\gamma}{4} e^{-j\omega_s t_L} H_0^{(1)}(k_s R_{sL}). \quad (3.27)$$

As expected, both approaches provide the same closed-form solution of direct and mirror-reflected potentials.

The third components of both solutions, Φ_{cor} and $-\varphi_3$, have no closed-form solution and need to be computed numerically.

In order to compute Φ_{cor}, Eq. (3.20) has to be rewritten since it is not suitable for numerical computation. For that purpose, the change of variable $t_{0sL} \cosh u = \tau$ is performed, and the integral becomes

$$\Phi_{cor} = -\frac{\gamma \varepsilon^2 e^{-j\omega_s t_L}}{\pi} \int_{0}^{+\infty} \Lambda(\cosh u) e^{jk_s R_{sL} \cosh u} du, \quad (3.28)$$

with $\Lambda(s) = \frac{\alpha^2 (x_L + MR_{sL} s)^2 + \alpha z_{hL}(Mx_L + R_{sL} s) + \alpha^2 z_{hL}^2/\gamma^2}{\left(\varepsilon^2 R_{sL} s + \alpha \nu\right)^2 + (\alpha^2 - \varepsilon^2)\left(\varepsilon^2 R_{sL}^2 - \nu^2\right)}$.

The term $\Lambda(\cosh u)$ tends to the constant value $(\alpha M/\varepsilon)^2$ for large values of u. The exponential term is an oscillatory function so that the integral will not converge

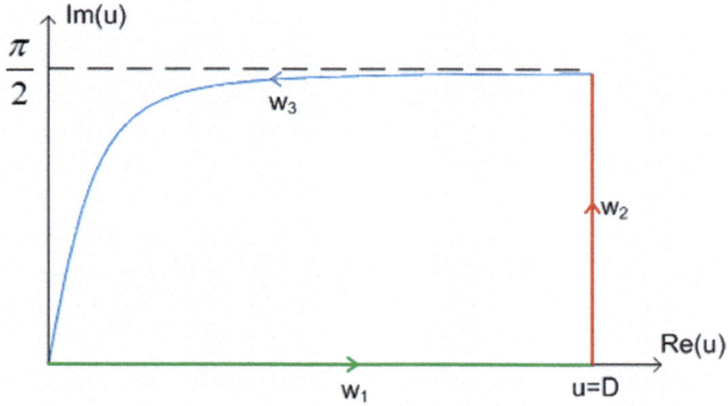

Fig. 3.2 Closed contour $w = w_1 + w_2 + w_3$ for applying Cauchy's integral theorem

numerically. But if the contour of integration for u is moved into the complex plane so that it goes toward the line: $Im(u) = \pi/2$, the exponential term will now decay at large distances. One possible contour can be $u = \sinh\eta + j\frac{\pi}{2}\tanh\eta$, where η is the new real integration variable, and the integration is performed using the trapezoidal rule [8]. To show that this change is possible, Cauchy's integral theorem is applied to the closed contour $w = w_1 + w_2 + w_3$, which is sketched in Fig. 3.2.

The contours are defined as:

$$\begin{aligned} w_1 &: u = \eta + j0 \quad /\eta \in [0, D], \\ w_2 &: u = D + j\eta \quad /\eta \in [0, \pi/2], \\ w_3 &: u = \sinh\eta + j\frac{\pi}{2}\tanh\eta \quad /\eta \in [0, \operatorname{asinh} D]. \end{aligned}$$

Since no singularities or branch cuts exist in the first quadrant, the integral theorem gives

$$\int_w \chi(u)du = \int_{w_1} \chi(u)du + \int_{w_2} \chi(u)du + \int_{w_3} \chi(u)du = 0,$$

where χ is the integrand in (3.28). Thus,

$$\int_{w_1} \chi(u)du = -\int_{w_2} \chi(u)du - \int_{w_3} \chi(u)du. \tag{3.29}$$

To solve (3.28), we have to solve the two integrals on the right-hand side of (3.29) in the limit $D \to +\infty$

$$\Phi_{cor} = -\lim_{D\to+\infty}\left(\int\limits_{w_2}\chi(u)du + \int\limits_{w_3}\chi(u)du\right).$$

The first integral is written as

$$\lim_{D\to+\infty}\int\limits_{w_2}\chi(u)du = \lim_{D\to+\infty}\int\limits_{0}^{\pi/2} j\chi(D+j\eta)d\eta.$$

For large values of u, $\cosh u \approx 0.5e^u$, and $\cosh(D+j\eta) \approx 0.5e^D(\cos\eta + j\sin\eta)$; therefore, the exponential term is approximately

$$e^{jk\rho_{0sL}\cosh u} \approx e^{-0.5k\rho_{0sL}e^D\sin\eta}e^{j0.5k\rho_{0sL}e^D\cos\eta}.$$

In the limit $D \to +\infty$, and for $0 < \eta < \pi/2$, $\chi(D+j\eta) = 0$, so that $\lim\limits_{D\to+\infty}\int_{w_2}\chi(u)du = 0$ and

$$\Phi_{cor} = -\lim_{D\to+\infty}\left(\int\limits_{w_3}\chi(u)du\right) = \int\limits_{0}^{+\infty}\chi(u(\eta))\frac{du}{d\eta}d\eta. \tag{3.30}$$

To calculate $-\varphi_3$, the two integrable singularities at $L_x = -k_s$ and $L_x = k_s$ of Eq. (3.25) have to be considered. It can be overcome by avoiding the singularities going around it along a small semicircle with the center at the singularity. The path is chosen so that $Im\left(\sqrt{k_s^2 - L_x^2}\right) > 0$ and is shown in Fig. 3.3

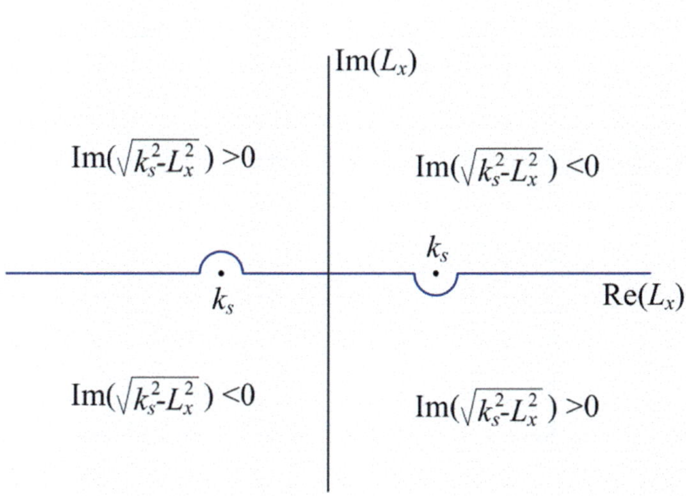

Fig. 3.3 Path of integration for integral (3.25)

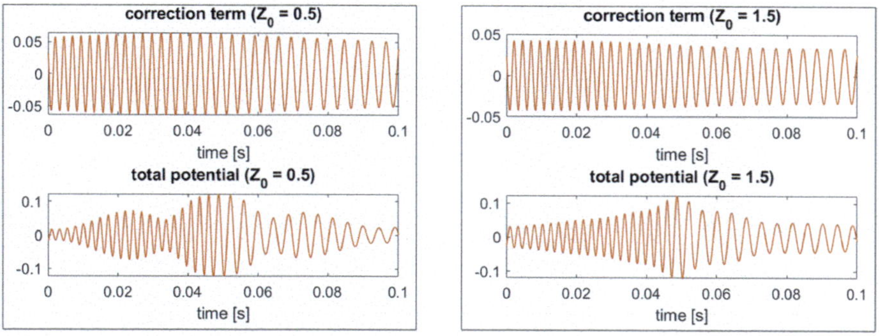

Fig. 3.4 Correction term and total pressure of a moving line source above impedance

Figure 3.4 shows a comparison between Formulas (3.23)–(3.25) and Formulas (3.26)–(3.3–3.28) for two values of ground impedance, $Z_0 = 0.5$ and $Z_0 = 1.5$. The source moves at a speed $M = 0.3$ and a height $h = 2$ m and emits at a frequency $f = 300$ Hz. The observation point is located at $P = (5, 1.5)$ m. The upper plots compare the correction terms (3.25) and (3.28), and the lower plots the total potentials. The curves cannot be distinguished since the difference is very small. Therefore, the results serve to validate both formulas.

3.1.4 Example: Pulse of Short Duration

The expressions (3.20)–(3.22) are suitable for any time signal, especially for signals of short duration. In this example, we consider a short pulse with the form of a Ricker wavelet, which is defined by the formula

$$q(t) = \begin{cases} \left(1 - \left(\frac{t-\mu}{\sigma}\right)^2\right) e^{-\frac{1}{2}\left(\frac{t-\mu}{\sigma}\right)^2}, & |t-\mu| < T/2 \\ 0 & , \textit{elsewhere} \end{cases},$$

with $\mu = 0.01$ s, $\sigma = 0.002$ s and $T = 10\sigma$. Figure 3.5a shows the signal emitted by the source. Figure 3.5b shows the components of the velocity potential and the total potential at the receiving position for a source moving with velocity $M = 0.25$. The amplitude of the correction term is smaller than the direct and mirror-reflected components, and it becomes less significant as the receiver moves away from the ground. Figure 3.5c plots the total potential at the receiving position for different velocities of the source. The shape and amplitude of the received signals do not change much, they appear just to be "pushed" to the left side in the time axis. The ground has an impedance $Z_0 = 10$, the height of the source is $h = 1.5$ m, and the receiver is located at $(x, z) = (5, 1)$ m.

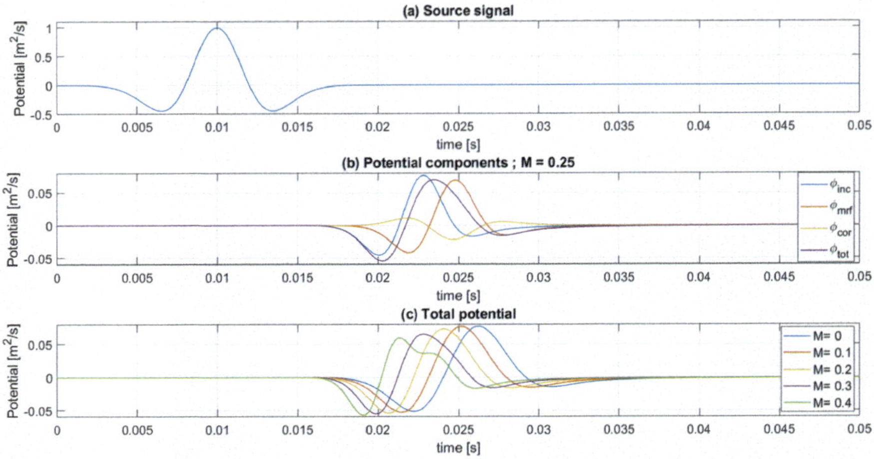

Fig. 3.5 Time series of emitted and received signals

3.1.5 Example: Sequence of Short Pulses

In the third example, we consider a line source moving with velocity Mc and emitting a sequence of short pulses $q_n(t)$ that have the form of real Morlet wavelets

$$q_n(t) = \begin{cases} \cos(\varepsilon(t-\mu_n))e^{-\frac{1}{2}\left(\frac{t-\mu_n}{\sigma}\right)^2}, & |t-\mu_n| < T/2 \\ 0, & elsewhere \end{cases},$$

with $\mu_n = \frac{T}{2} + (n-1)\frac{3T}{2}$, $\varepsilon = 2000$, $\sigma = 0.002$ s and $T = 10\sigma$.

The pulses have a duration of T seconds and are separated by an interval of $T/2$ seconds. Figure 3.6 shows the amplitude of the velocity potential in the half-space at four different times. The first frame shows the pulse shortly after its emission. In the second frame, the pulse reaches the ground. The third and fourth frames show direct and reflected waves propagating.

Figure 3.7 shows the velocity potential emitted by the source (upper plot) and the potential at the position $(x, z) = (8.85, 2.7)$ m (lower plot). The pulses emitted by the source are equally spaced in time but arrive unequally spaced due to the motion. The amplitude of the fourth pulse is larger because it is emitted when the source is nearest to the receiver.

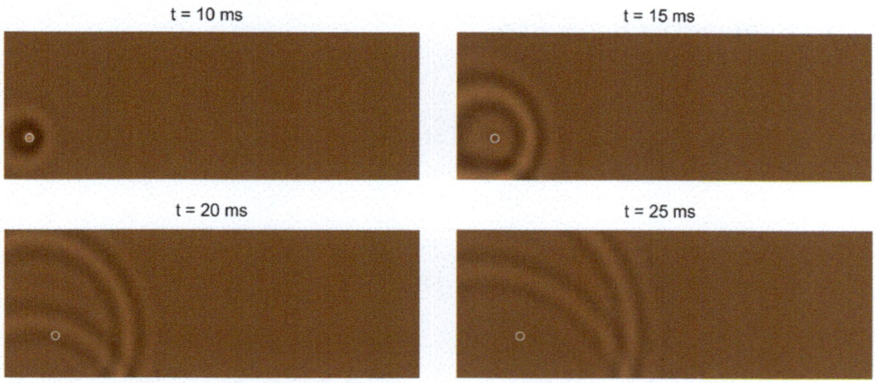

Fig. 3.6 Four snapshots of the emission of the series of short pulses

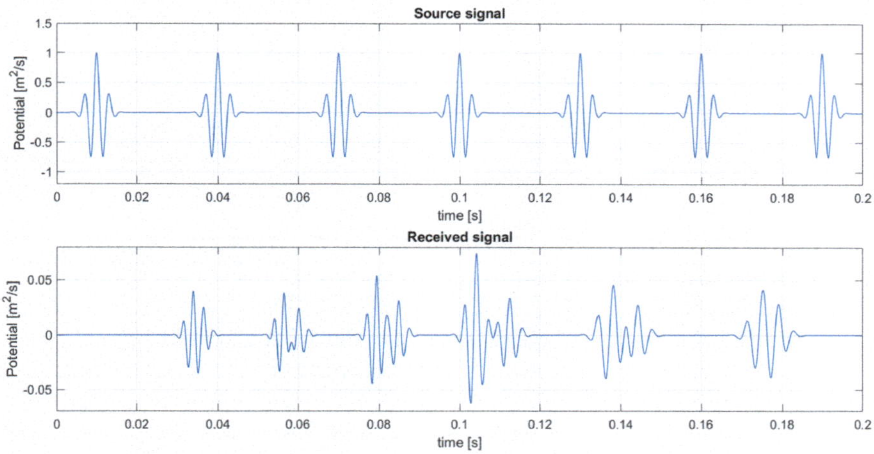

Fig. 3.7 Velocity potential: **a** sequence of pulses; **b** signal captured by the observer

3.2 Line Source Above a Homogeneous Medium

In a homogeneous medium, waves can propagate in the direction parallel to the surface, therefore, the lower medium has to be included to determine the sound field in the upper medium. The direction of the propagation of waves in the lower medium depends on the incident angle, so the impedance on the interface is also dependent on that angle.

3.2.1 Stationary Line Source

A line source is placed at the position $\vec{r}_q = (x_q, h)$ and emits a pulse at time $t = t_q$. A corresponding image source is located at $(x_q, -h)$, and a field point is located at position $\vec{r} = (x, z)$. We denote Green's function in upper and lower media as ϕ and ϕ_m, respectively, and they are the solutions of the differential equations:

$$\frac{1}{c^2}\frac{\partial^2\phi}{\partial t^2} - \Delta\phi = \delta\left(x - x_q\right)\delta(z - h)\delta\left(t - t_q\right), z > 0, \tag{3.31}$$

$$\frac{1}{c_m^2}\frac{\partial^2\phi_m}{\partial t^2} - \Delta\phi_m = 0, z < 0, \tag{3.32}$$

with continuity of sound pressure and normal velocity at the interface as boundary conditions

$$-\rho\left.\frac{\partial\phi}{\partial t}\right|_{z=0} = -\rho_m\left.\frac{\partial\phi_m}{\partial t}\right|_{z=0}, \tag{3.33}$$

$$\left.\frac{\partial\phi}{\partial z}\right|_{z=0} = \left.\frac{\partial\phi_m}{\partial z}\right|_{z=0}. \tag{3.34}$$

A detailed description of the solution procedure is given in Chap. 10.1 of [7], here only the main results will be shown.

To solve the problem, the Cagniard-de Hoop method is used. In order to obtain a closed-form solution, it was required that the densities and sound speeds were frequency independent. In a first step, a Fourier transform in x and a Laplace transform in t are performed. To do that, Eqs. (3.31) and (3.32) are multiplied by $e^{-jk_x x}e^{-st}$ and integrated over $\int_{-\infty}^{+\infty} dk_x \int_0^{+\infty} dt$. After incorporating the boundary conditions, the solution in the (k_x, z, s)-space can be obtained and is given by

$$\phi(k_x, z, s) = \frac{e^{-jk_x x_q}e^{-st_q}}{2n}\left[e^{-\lambda|z-h|} + \left(\frac{\rho_m\lambda - \rho\lambda_m}{\rho_m\lambda + \rho\lambda_m}\right)e^{-\lambda(z+h)}\right], \tag{3.35}$$

where $\lambda = \sqrt{\frac{s^2}{c^2} + k_x^2}$, $\lambda_m = \sqrt{\frac{s^2}{c_m^2} + k_x^2}$.

The second step is to perform the inverse Fourier transform over (3.35)

$$\phi(x, z, s) = \frac{1}{2\pi}\int_{-\infty}^{+\infty}\phi(k_x, z, s)e^{jk_x x}dk_x. \tag{3.36}$$

The last step is to introduce a new parameter p defined as $k_x = jsp$ and find paths in the complex p-plane to convert (3.36) into a Laplace transform

$$\phi(x, z, s) = \int_{0}^{+\infty} F(x, z, p(t))e^{-st}\frac{dp}{dt}dt, \tag{3.37}$$

so that sought Green's function is deduced to be $\phi(x, z, t) = F(x, z, p(t))\frac{dp}{dt}$.

After following the three steps, the solution is found and can also be written as a sum of three terms like (3.3).

The incident and mirror-reflected waves are already given in (3.4) and (3.5) since they are independent of the presence of the lower medium, and the correction term is written as

$$\phi_{cor} = \frac{1}{2\pi}\begin{cases} -Re\left(\Psi\left(t, t_q\right)\right)\frac{H\left(t-t_q-t_{0s}\right)}{\sqrt{\left(t-t_q\right)^2-t_{0s}^2}} & c > c_m \\ Im\left(\Gamma\left(t, t_q\right)\right)\frac{H\left(t-t_q-t_H\right)-H\left(t-t_q-t_{0s}\right)}{\sqrt{t_{0s}^2-\left(t-t_q\right)^2}} - Re\left(\Psi\left(t, t_q\right)\right)\frac{H\left(t-t_q-t_{0s}\right)}{\sqrt{\left(t-t_q\right)^2-t_{0s}^2}} & c < c_m \end{cases}, \tag{3.38}$$

with $\Gamma\left(t, t_q\right) = \frac{\rho_m\eta\left(t,t_q\right)-\rho\eta_m\left(t,t_q\right)}{\rho_m\eta\left(t_q,t\right)+\rho\eta_m\left(t,t_q\right)}$, $\Psi\left(t, t_q\right) = \frac{2\rho\eta_m\left(t,t_q\right)}{\rho_m\eta\left(t,t_q\right)+\rho\eta_m\left(t,t_q\right)}$,
$\eta = \sqrt{\frac{1}{c^2} - p^2\left(t, t_q\right)}$, $\eta_m = \sqrt{\frac{1}{c_m^2} - p^2\left(t, t_q\right)}$,

$$p\left(t, t_q\right) = \begin{cases} \frac{\left|x-x_q\right|\left(t-t_q\right)-(z+h)\sqrt{t_{0s}^2-\left(t-t_q\right)^2}}{R_s^2}, & 0 < t - t_q < t_{0s} \\ \frac{\left|x-x_q\right|\left(t-t_q\right)+j(z+h)\sqrt{\left(t-t_q\right)^2-t_{0s}^2}}{R_s^2}, & t - t_q > t_{0s} \end{cases}, \tag{3.39}$$

where θ_c is a critical angle is found defined as $\sin\theta_c = \frac{c}{c_m}$.

In the case of $c < c_m$, the first term of the correction term in (3.38) is the contribution of the so-called head waves that arrive at an observer at time t_H, before the mirror reflection ($t_H \leq t_{0s}$). The time of arrival of the head waves is obtained by setting $p\left(t - t_q = t_H\right) = 1/c_2$ and is found to be

$$t_H = \frac{\left|x - x_q\right|\sin\theta_c + (z + h)\cos\theta_c}{c}, \ \left|x - x_q\right| > (z + h)\tan\theta_c. \tag{3.40}$$

Equation (3.40) tells us that these head waves reach only the observers situated at certain positions away from the source. Those observers are located in a region denoted as a "wide-angle" reflection [1]. The boundary of this region is obtained by setting $t_H = t_{0s}$. Observers located nearer to the source are placed in the region of "narrow"-angle reflection. At some regions, the head wave arrives even before the direct sound. The limit of these regions can be analytically found by setting $t_H = t_0$ and is given by

$$\left|x - x_q\right| = (z + h)\tan\theta_c + \frac{2\sqrt{zh}}{\cos\theta_c}. \tag{3.41}$$

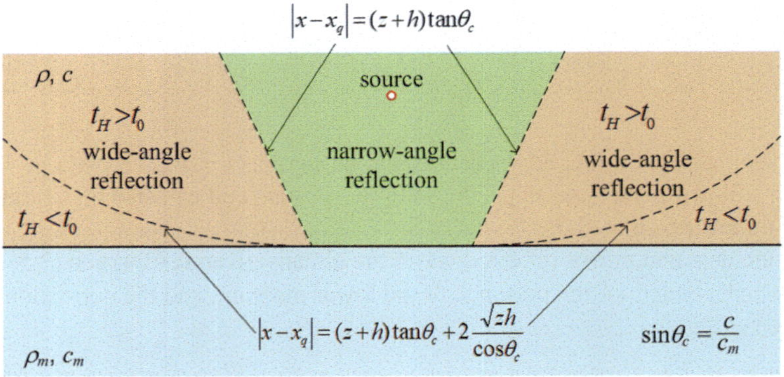

Fig. 3.8 Narrow-angle and wide-angle reflection zones for the case $c < c_m$

Figure 3.8 shows the two reflection zones, and the dotted lines the boundaries where $t_H = t_0$ and $t_H = t_{0s}$.

As can be observed in (3.38), the Cagniard-de Hoop method provides different solutions for the correction term depending on the relation between c and c_m. Grounds in typical outdoor propagation are modeled as porous materials and have, in most cases, a sound speed smaller than in air ($c > c_m$) [2]. If only those cases were studied, we would only need the first solution. However, we do not want to restrict the analysis to that case but also include the case that $c < c_m$ so that the solution can be used, for example, in the air–water interface.

3.2.2 *Moving Line Source*

Now, we only need to find the expression for the correction term, since the direct and mirror-reflected components have already been obtained in Sect. 3.1 and are given by (3.11) and (3.12). Applying again Green's function method

$$\Phi_{cor}(\vec{r}, t) = \int_{-\infty}^{+\infty} q(\tau)\phi_{cor}\big(\vec{r}, t; \vec{r}_q(\tau), \tau\big)d\tau.$$

Setting $x_q = Mc\tau$ and $t_q = \tau$, t_0 and t_{0s} are time-dependent and (3.38) becomes

$$\Phi_{cor} = \frac{1}{2\pi}\int_{-\infty}^{+\infty} q(\tau)Im(\Gamma(t, \tau))\frac{H(t - \tau - t_H(\tau)) - H(t - \tau - t_{0s}(\tau))}{\sqrt{t_{0s}^2(\tau) - (t - \tau)^2}}d\tau$$

$$-\frac{1}{2\pi}\int_{-\infty}^{+\infty} q(\tau)Re(\Psi(t,\tau))\frac{H(t-\tau-t_{0s}(\tau))}{\sqrt{(t-\tau)^2-t_{0s}^2(\tau)}}d\tau, \tag{3.42}$$

with $\Gamma(t,\tau)=\frac{\rho_m\eta(t,\tau)-\rho\eta_m(t,\tau)}{\rho_m\eta(t,\tau)+\rho\eta_m(t,\tau)}$, $\Psi(t,\tau)\frac{2\rho\eta_m(t,\tau)}{\rho_m\eta(t,\tau)+\rho\eta_m(t,\tau)}$,

$$p(t,\tau)=\begin{cases}\frac{|x-Mc\tau|(t-\tau)-z_h\sqrt{t_{0s}^2(\tau)-(t-\tau)^2}}{R_s^2(\tau)} & t-\tau<t_{0s}(\tau)\\ \frac{|x-Mc\tau|(t-\tau)+jz_h\sqrt{(t-\tau)^2-t_{0s}^2(\tau)}}{R_s^2(\tau)} & t-\tau>t_{0s}(\tau)\end{cases}, \tag{3.43}$$

$$t_H(\tau)=\frac{|x-Mc\tau|\sin\theta_c+z_h\cos\theta_c}{c}. \tag{3.44}$$

In Sect. 3.1.2, the values of τ that satisfy $t-\tau-t_0(\tau)>0$ and $t-\tau-t_{0s}(\tau)>0$ were determined. Here, we must additionally find the values of τ that makes $t-\tau-t_H(\tau)>0$.

Relations (3.16)–(3.19) can help us again to simplify the formulas.

After some calculations, (see Appendix G), we find that

$$t-\tau-t_H(\tau)=0\rightarrow\tau=t_L-t_{HL}.$$

with

$$t_{HL}=\frac{1}{\gamma c}\begin{cases}\frac{-\gamma x_L(\sin\theta_c+M)+z_{hL}\cos\theta_c}{1+M\sin\theta_c}, & x_L<x_{HL}^{(-)}\\ \frac{\gamma x_L(\sin\theta_c-M)+z_{hL}\cos\theta_c}{1-M\sin\theta_c}, & x_L>x_{HL}^{(+)}\end{cases}, \tag{3.45}$$

where

$$x_{HL}^{(-)}=-\gamma\frac{(\sin\theta_c+M)}{\cos\theta_c}(z_L+h_L), x_{HL}^{(+)}=\gamma\frac{(\sin\theta_c-M)}{\cos\theta_c}(z_L+h_L). \tag{3.46}$$

For fixed z_{hL}, (3.45) represents a straight line in the plane t_{HL} versus x_L. Next, the positions where the head waves arrive at the same time that the direct waves (x_{H0L}) can be obtained by setting $t_{HL}=t_{0L}$(Appendix G). We obtain then

$$x_{H0L}=\gamma\begin{cases}-\frac{z_{hL}(\sin\theta_c+M)+2(1+M\sin\theta_c)\sqrt{z_Lh_L}}{\cos\theta_c}, & x_L<x_{HL}^{(-)}\\ \frac{z_{hL}(\sin\theta_c-M)+2(1-M\sin\theta_c)\sqrt{z_Lh_L}}{\cos\theta_c}, & x_L>x_{HL}^{(+)}\end{cases}. \tag{3.47}$$

Finally, using the commutative property of the convolution, $(t_L-\tau)\leftrightarrow\tau$, the correction term can be expressed as

$$\phi_{cor}=\frac{\gamma}{2\pi}\left(\int_{t_{HL}}^{t_{0sL}} Im(\Gamma(\tau))\frac{q(t_L-\tau)}{\sqrt{t_{0sL}^2-\tau^2}}d\tau-\int_{t_{0sL}}^{+\infty} Re(\Psi(\tau))\frac{q(t_L-\tau)}{\sqrt{\tau^2-t_{0sL}^2}}d\tau\right), \tag{3.48}$$

$$\text{with } p_L(\tau) = \begin{cases} \dfrac{|x_L + Mc\tau|\left(\tau + \frac{M}{c}x_L\right) - \left(\frac{z_{hL}}{\gamma^2}\right)\sqrt{t_{0sL}^2 - \tau^2}}{(x_L + Mc\tau)^2 + \frac{z_{hL}^2}{\gamma^2}}, & \tau < t_{0sL} \\ \dfrac{|x_L + Mc\tau|\left(\tau + \frac{M}{c}x_L\right) + j\left(\frac{z_{hL}}{\gamma^2}\right)\sqrt{\tau^2 - t_{0sL}^2}}{(x_L + Mc\tau)^2 + \frac{z_{hL}^2}{\gamma^2}}, & \tau > t_{0sL} \end{cases}.$$

Figure 3.9 shows the variation of the integration interval of the first integral in (3.48), $\tau \in [t_{HL}, t_{0sL}]$ (red colored area) with respect to x_L, for fixed z_L and h_L. Within the narrow-angle region, there are no head waves; i.e., the first term in (3.48) is zero. The head waves start at the values $x_{HL}^{(-)}$ and $x_{HL}^{(+)}$ where $t_{HL} = t_{0sL}$. In the plot on the left, we have the case $M < \sin\theta_c$ (one straight line with a positive slope and the other with a negative slope), and in the plot on the right, the case $M > \sin\theta_c$ (both straight lines with negative slope); i.e., the right branch of t_{HL} can have negative values.

In Fig. 3.10, we can see how the regions of narrow- and wide-angle reflection, $t_{HL} = t_{0sL}$, and the curves $t_{HL} = t_{0L}$, change by increasing source velocity v. The curves seem to shift slightly against the direction on movement of the source and to rotate counterclockwise by a small angle that increases with the velocity.

One can easily see that if we set $M = 0(\gamma = 1)$ in (3.45)–(3.47), we recover (3.40) and (3.41) with $x_q = 0$.

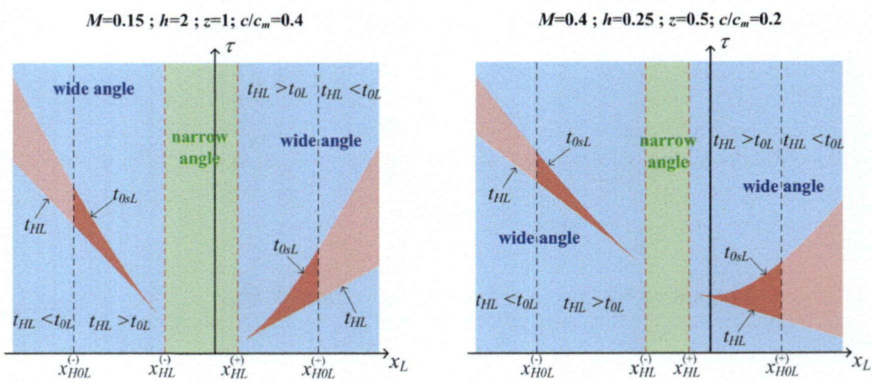

Fig. 3.9 Parameters of the head waves for a moving source

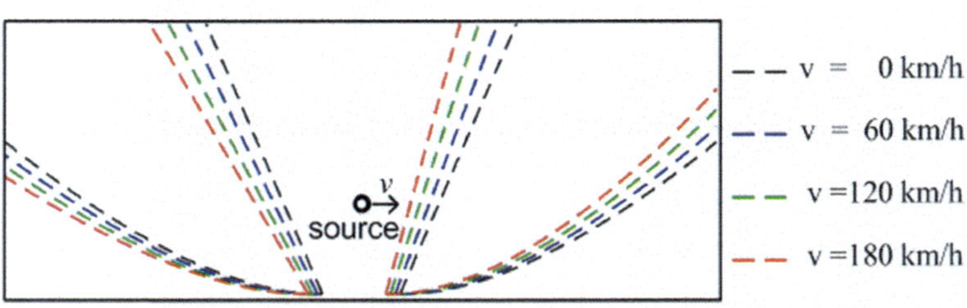

Fig. 3.10 Narrow- and wide-angle reflection zones of a moving source

3.2.3 Verification Example: Harmonic Time Dependence

In order to validate our formulas, we can compare them to the results of the solution given by Wang et al. [10] for a line source emitting a harmonic signal with circular frequency ω_s. In their work, the parameters for the second medium are assumed to be frequency-dependent and complex; i.e., damping is also considered. They solve the differential equation by performing a Fourier transform in x and a Lorentz transformation.

The correction term is given by

$$\phi_{cor}^{W} = -\frac{j\gamma}{2\pi} e^{-j\omega_s t_L} \int\limits_{-\infty}^{+\infty} \frac{k_s \Omega(L_x)\beta(L_x)/\gamma}{\sqrt{k_s^2 - L_x^2} + k_s \Omega(L_x)\beta(L_x)/\gamma} \frac{e^{j\left[L_x x_L + \sqrt{k_s^2 - L_x^2} z_{hL}\right]}}{\sqrt{k_s^2 - L_x^2}} dL_x, \tag{3.49}$$

and

$$\beta(L_x) = \frac{\rho}{\rho_m(\omega_D(L_x))} \sqrt{\frac{c^2}{c_m^2(\omega_D(L_x))} - \left[\frac{\Upsilon(L_x)}{\Omega(L_x)}\right]^2}, \; \omega_D(L_x) = \Omega(L_x)\omega_s,$$

$$\Omega(L_x) = \gamma^2\left(1 + \frac{M L_x}{k_s}\right), \; \Upsilon(L_x) = \gamma^2\left(M + \frac{L_x}{k_s}\right).$$

The correction term has to be computed numerically, but it needs to be rewritten in an efficient way to perform the calculation. In order to have a convergent integral, the proper branch of the square root must be considered; i.e., it will be defined as

$$\sqrt{k_s^2 - L_x^2} = \begin{cases} \sqrt{k_s^2 - L_x^2}, & |L_x| < k_s \\ j\sqrt{L_x^2 - k_s^2}, & |L_x| > k_s \end{cases},$$

which leads to split the integral (3.49) into two integrals. To eliminate the integrable singularity, two changes of variables are used: $L_x = k_s \sin\varepsilon$ for $|L_x| < k_s$ and $L_x = k_s \cosh\varepsilon$ for $|L_x| > k_s$. The correction term becomes

$$\phi_{cor}^{W} = -\frac{j\gamma}{2\pi} e^{-j\omega_s t_L} \left\{ \int\limits_{0}^{\frac{\pi}{2}} e^{jk_s z_{hL}\cos\varepsilon} \left(\frac{e^{jk_s x_L \sin\varepsilon} \gamma \Omega_s(\sin\varepsilon)\beta_s(\sin\varepsilon)}{\cos\varepsilon + \gamma\Omega_s(\sin\varepsilon)\beta_s(\sin\varepsilon)} \right.\right.$$

$$\left. + \frac{e^{-jk_s x_L \sin\varepsilon} \gamma \Omega_s(-\sin\varepsilon)\beta_s(-\sin\varepsilon)}{\cos\varepsilon + \gamma\Omega_s(-\sin\varepsilon)\beta_s(-\sin\varepsilon)} \right) d\varepsilon$$

$$- j \int\limits_{0}^{+\infty} e^{-k_s z_{hL}\sinh\varepsilon} \left(\frac{e^{jk_s x_L \cosh\varepsilon} \gamma \Omega_s(\cosh\varepsilon)\beta_s(\cosh\varepsilon)}{j\sinh\varepsilon + \gamma\Omega_s(\cosh\varepsilon)\beta_s(\cosh\varepsilon)} \right.$$

$$+\frac{e^{-jk_s x_L \cosh\varepsilon}\gamma\Omega_s(-\cosh\varepsilon)\beta_s(-\cosh\varepsilon)}{j\sinh\varepsilon+\gamma\Omega_s(-\cosh\varepsilon)\beta_s(-\cosh\varepsilon)}\Bigg)d\varepsilon, \tag{3.50}$$

with $\Omega_s(\mu)=1+M\,\mu$ and $\beta_s(\mu)=\frac{\rho}{\rho_m(\gamma^2\omega_s\Omega_s(\mu))}\sqrt{\frac{c^2}{c_m^2(\gamma^2\omega_s\Omega_s(\mu))}-\left[\frac{M+\mu}{1+M\,\mu}\right]^2}$.

To find the solution with our formulas, we insert the source amplitude $q(t)=e^{-j\omega_s t}$.

The correction term is given by

$$\phi_{cor}=\frac{\gamma e^{-j\omega_s t_L}}{2\pi}\left(\int_{t_{HL}}^{t_{0sL}}\frac{Im(\Gamma(\tau))e^{j\omega_s\tau}}{\sqrt{t_{0sL}^2-\tau^2}}d\tau-\int_{t_{0sL}}^{+\infty}\frac{Re(\Psi(\tau))e^{j\omega_s\tau}}{\sqrt{\tau^2-t_{0sL}^2}}d\tau\right). \tag{3.51}$$

The integrals in (3.51) need to be computed numerically. The singularities are integrable. To avoid them, we introduce a change of variable.

For the first integral, we consider the change of variable $\tau=t_{0sL}\sin u$ so that the integration range becomes $]u_0,\pi/2[$ with $u_0=\text{asin}\left(\frac{t_{HL}}{t_{0sL}}\right)$. The integral is expressed as

$$I_1=\frac{\gamma e^{-j\omega_s t_L}}{2\pi}\int_{u_0}^{+\pi/2}Im(\Gamma(t_{0sL}\sin u))e^{jk_s R_{0sL}\sin u}du. \tag{3.52}$$

Since the integration interval is finite, (3.52) can be performed numerically using a quadrature scheme.

For the second integral, we consider the change of variable $\tau=t_{0sL}\cosh u$ so that the integration range becomes $[0,+\infty[$. The integral is expressed as

$$I_2=-\frac{\gamma e^{-j\omega_s t_L}}{2\pi}\int_0^{+\infty}Re(\Psi(t_{0sL}\cosh u))e^{jk_s R_{0sL}\cosh u}du. \tag{3.53}$$

Integral (3.53) has a form similar to (3.28), with the integrand having a regular factor and the same highly oscillating exponential factor, which makes the integral numerically non-convergent. The solution procedure is the same, i.e., shifting the integration path into the complex plane so that it goes toward the line $Im(u)=\pi/2$. Along the new integration path, w_3, the factor accompanying the exponential term, $\text{Re}(\Psi)$, will become complex, therefore, we need to express it explicitly before changing the integration path. For real values of u, we can write $\eta=\alpha_\eta+j\beta_\eta$ and $\eta_m=\mu_\eta+j\nu_\eta$ where $\alpha_\eta,\beta_\eta,\mu_\eta,\nu_\eta$ are real functions. Hence, I_2 can be written as

$$I_2=-\frac{\gamma e^{-j\omega_s t_L}}{2\pi}\int_{w_3}\Pi(t_{0sL}\cosh u)e^{jk_s R_{0sL}\cosh u}du, \tag{3.54}$$

$$\Pi = -\frac{2\rho\left(\mu_\eta^2 + \nu_\eta^2\right) + 2\rho\rho_m\left(\alpha_\eta\mu_\eta + \beta_\eta\nu_\eta\right)}{\rho^2\left(\mu_\eta^2 + \nu_\eta^2\right) + \rho_m^2\left(\alpha_\eta^2 + \beta_\eta^2\right) + 2\rho\rho_m\left(\alpha_\eta\mu_\eta + \beta_\eta\nu_\eta\right)}.$$

Expressions η and η_m contain square roots of complex quantities, so there will be branch cuts where Π will have a discontinuity. Therefore, care shall be taken to choose a contour that does not cross the branches. Due to the complexity of the argument of all square roots, the branch points can only be computed numerically.

Figure 3.11 shows two integration paths C_1 and C_2, and the values of Π in a section of the u complex plane for a source emitting at 100 Hz at a height $h = 2$ m. The Mach number is $M = 0.15$, and the lower medium has properties $\frac{\rho_m}{\rho} = 2.5$ and $\frac{c_m}{c} = 4$. The observer is located at $(x, z) = (5, 1)$ m, and the integral is evaluated at $t = 0.102$ s. The paths are given by

$$C_1 : u = \sinh\eta + j\frac{\pi}{2}\tanh\eta \text{ and } C_2 : u = e^{\sqrt{\eta+1}} - e + j\frac{\pi}{2}\tanh\frac{2\eta}{5}. \qquad (3.55)$$

We can see that the contour C_1 crosses a discontinuity of the function Π that arises from the multivalued character of the square root function, while C_2 does not. Therefore, C_2 is an appropriate integration path to calculate (3.54).

Figure 3.12 shows the two terms of (3.51) in the time interval [0, 0.2], computed with a time resolution $\Delta t = 4 \times 10^{-4}$ s. Since the source is moving, the regions of narrow-angle reflection (NA) and wide-angle reflection (WA) move along with it. Therefore, a stationary observer will, certain times, be in the NA region and at other times be in the WA region. One can verify in Fig. 3.12 that the contribution of the head waves vanishes when the observer is located in the NA region.

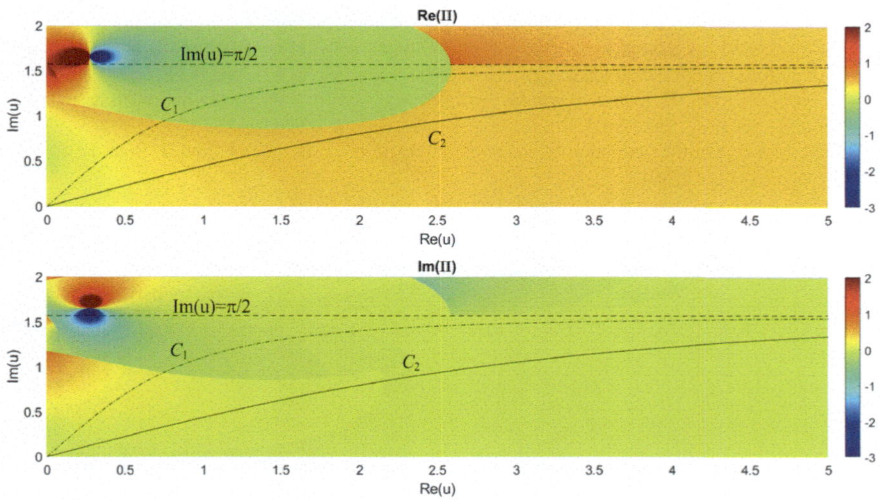

Fig. 3.11 Contour curves w_3 to compute integral (3.54). *Above*: real part of Π; *below*: imaginary part of Π

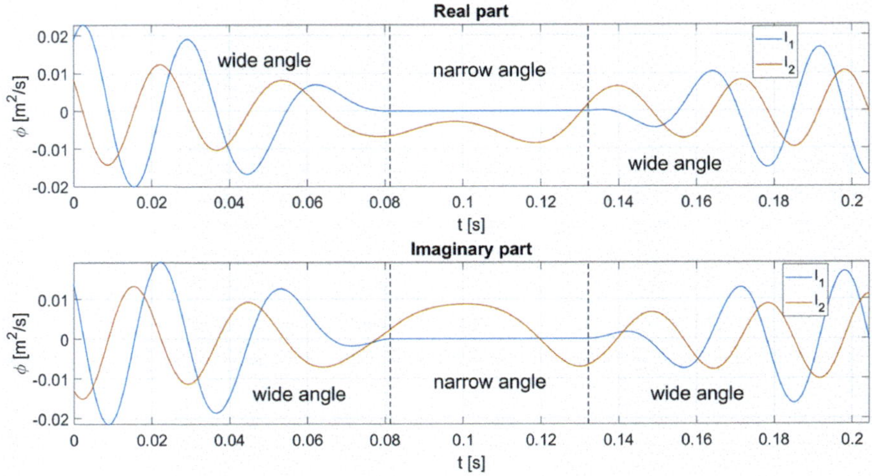

Fig. 3.12 Two components of the correction term

Figure 3.13 shows the comparison of the results obtained with both formulations. In the upper plots, the correction term is presented. The curves lie practically on top of each other since the agreement is very good. The lower plots show the relative error ε,

$$\varepsilon = \frac{\left|Re\left(\phi_{cor} - \phi_{cor}^{W}\right)\right|}{\left|Re\left(\phi_{cor}^{W}\right)\right|} + j\frac{\left|Im\left(\phi_{cor} - \phi_{cor}^{W}\right)\right|}{\left|Im\left(\phi_{cor}^{W}\right)\right|}, \tag{3.56}$$

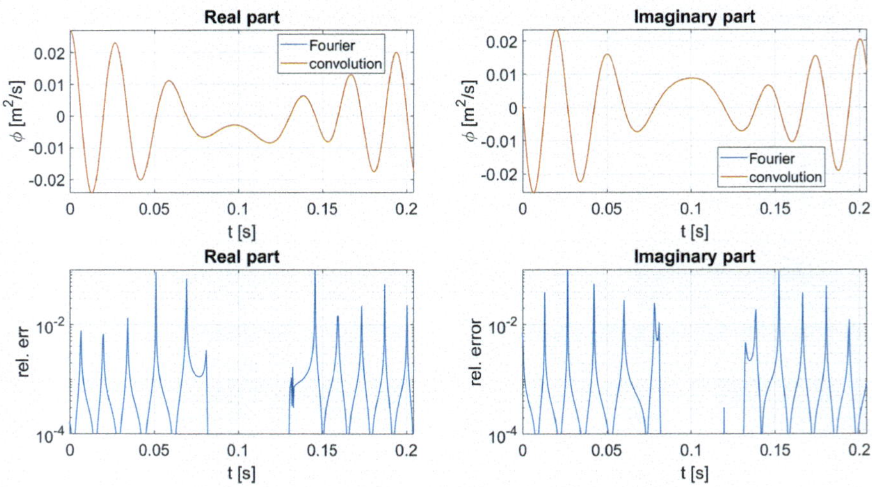

Fig. 3.13 Comparison of the correction term: Fourier = Eq. (3.50); convolution = Eq. (3.51)

which mostly remains below 1%. At some time steps, the error is bigger than 1%, but at these time steps the values of the correction term are very small and have the same order of magnitude of the difference between both formulations. Therefore, the formulation developed in this section is validated.

3.2.4 *Example: Source Emitting a Single, Infinitely Thin Pulse (Delta Function)*

The pulse is defined as $q(t) = \delta(t)$. When inserted in the previous formulas, we obtained closed-form expressions for the three components; i.e., no integration is required:

$$\phi_{dir} = \frac{\gamma}{2\pi} \frac{H(t_L - t_{0L})}{\sqrt{t_L^2 - t_{0L}^2}}, \tag{3.57}$$

$$\phi_{mr} = \frac{\gamma}{2\pi} \frac{H(t_L - t_{0sL})}{\sqrt{t_L^2 - t_{0sL}^2}}, \tag{3.58}$$

$$\phi_{cor} = \frac{\gamma}{2\pi}\left(Im(\Gamma(t_L)) \frac{H(t_L - t_{HL}) - H(t_L - t_{0sL})}{\sqrt{t_{0sL}^2 - t_L^2}} - Re(\Psi(t_L)) \frac{H(t_L - t_{0sL})}{\sqrt{t_L^2 - t_{0sL}^2}} \right). \tag{3.59}$$

The first term in (3.59) describes the head wave that propagates only in the wide-angle region. Figure 3.14 shows four snapshots of the propagation of that head wave. We can check that the head wave propagates in the wide-angle reflection zone, which travels along with the source (the dotted lines accompany the source).

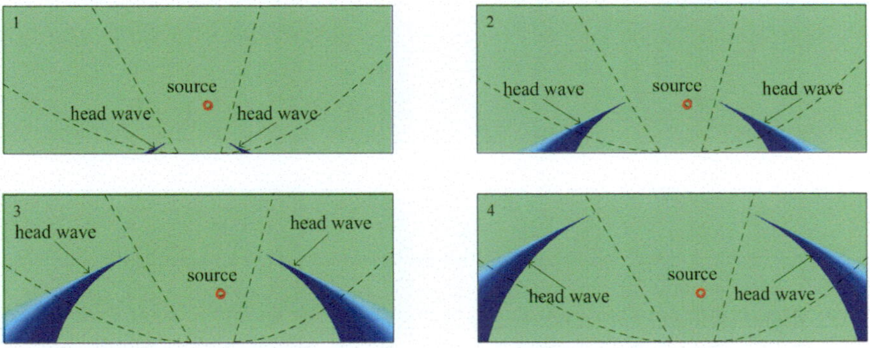

Fig. 3.14 Snapshots of the propagation of a head wave that corresponds to I_1 (Eq. 3.52)

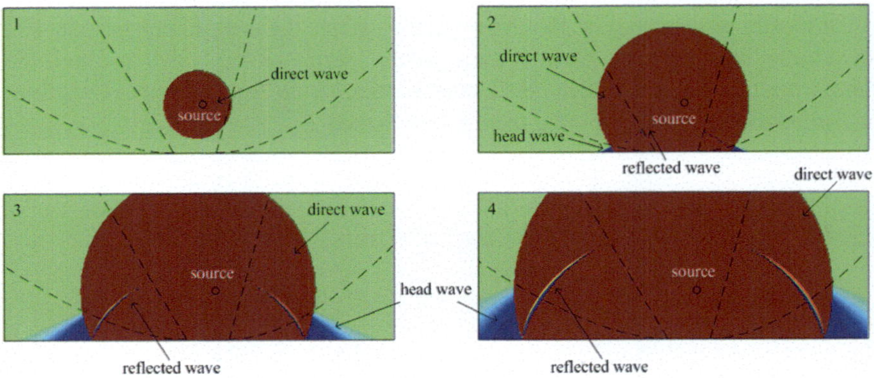

Fig. 3.15 Snapshots of the sound propagation of a pulse, direct, head and mirror-reflected waves can be well-distinguished

Figure 3.15 shows the propagation of the total sound. One can verify that the intersection between the wavefronts of the direct sound and the head wave follows the dotted curve $t_{HL} = t_{0L}$.

3.2.5 Example: Source Sending an "Expcos" Pulse

The source signal is defined in the time interval $[t_a, t_a + T]$ with a time dependence given by

$$q(t) = e^{-\alpha(t-t_a)^2/T^2}\cos\left[\frac{3\pi}{2T}(t-t_a)\right][H(t-t_a) - H(t-t_a-\mathrm{T})], \tag{3.60}$$

with $\alpha = \frac{\ln 2}{(2/3)^2}$ and $T = 5 \text{ x } 10^{-3}$

The direct sound will be given by

$$\phi_{dir} = \frac{\gamma}{2\pi}\int\limits_{t_{0L}}^{+\infty}\frac{e^{\frac{-\alpha(t_L-t_a-\tau)^2}{T^2}}}{\sqrt{\tau^2 - t_{0L}^2}}\cos\left[\frac{3\pi}{2T}(t_L - t_a - \tau)\right]$$

$$[H(t_L - t_a - \tau) - H(t_L - t_a - T - \tau)]d\tau.$$

The integration range is the intersection of two intervals, a semi-infinite fixed one $[t_{0L}, +\infty[$ and a finite moving one $[t_L - t_a - T, t_L - t_a]$; see (Fig. 3.16):

The direct sound can then be written as

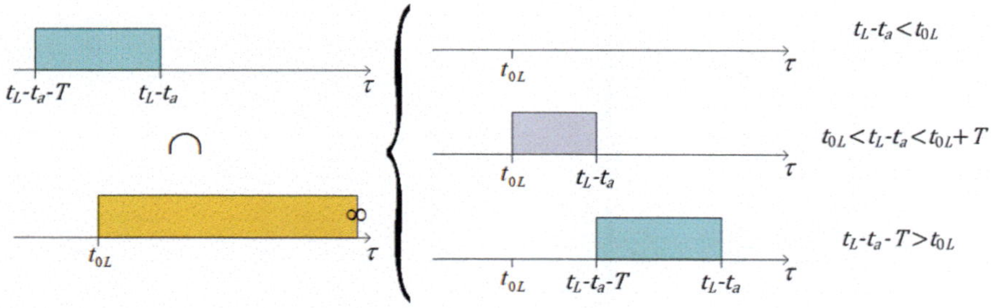

Fig. 3.16 Illustration of the integration range for the direct sound

$$\phi_{dir} = \frac{\gamma}{2\pi} \int_{\tau_1}^{\tau_2} \frac{e^{-\alpha(t_L - t_a - \tau)^2/T^2}}{\sqrt{\tau^2 - t_{0L}^2}} \cos\left[\frac{3\pi}{2T}(t_L - t_a - \tau)\right] d\tau, \tag{3.61}$$

with $\tau_1 = \max(t_L - t_a - T, t_{0L})$ and $\tau_2 = \max(t_L - t_a, t_{0L})$ for $\tau_2 > \tau_1$.

To avoid the singularity in (3.61), we use the change of variable $\tau = t_{0L} \cosh u$ and obtain

$$\phi_{dir} = \frac{\gamma}{2\pi} \int_{u_1}^{u_2} e^{-\alpha(t_L - t_a - t_{0L}\cosh u)^2/T^2} \cos\left(\frac{3\pi}{2T}(t_L - t_a - t_{0L}\cosh u)\right) du, \tag{3.62}$$

where $u_1 = \mathrm{acosh}(\tau_1/t_{0L})$ and $u_2 = \mathrm{acosh}(\tau_2/t_{0L})$.

In a similar way, the mirror-reflected wave can be obtained and is given by

$$\phi_{mr} = \frac{\gamma}{2\pi} \int_{u_{1s}}^{u_{2s}} e^{-\alpha(t_L - t_a - t_{0sL}\cosh u)^2/T^2} \cos\left(\frac{3\pi}{2T}(t_L - t_a - t_{0sL}\cosh u)\right) du, \tag{3.63}$$

where $u_{1s} = \mathrm{acosh}(\tau_{1s}/t_{0sL})$ and $u_{2s} = \mathrm{acosh}(\tau_{2s}/t_{0sL})$, with $\tau_{1s} = \max(t_L - t_a - T, t_{0sL})$ and $\tau_2 = \max(t_L - t_a, t_{0sL})$ for $\tau_{2s} > \tau_{1s}$.

The correction term has two integrals. The integration range is finite in both integrals since the source term is also finite. The first integral is given by

$$I_1 = \frac{\gamma}{2\pi} \int_{t_{HL}}^{t_{0sL}} Im(\Gamma(\tau)) \frac{e^{-\alpha(t_L - t_a - \tau)^2/T^2}}{\sqrt{t_{0sL}^2 - \tau^2}} \cos\left[\frac{3\pi}{2T}(t_L - t_a - \tau)\right]$$
$$[H(t_L - t_a - \tau) - H(t_L - t_a - T - \tau)] d\tau.$$

The integration range is the intersection of two finite intervals, a fixed one $[t_{HL}, t_{0sL}]$ and a moving one $[t_L - t_a - T, t_L - t_a]$ (Fig. 3.17).

The contribution of the head waves reads then

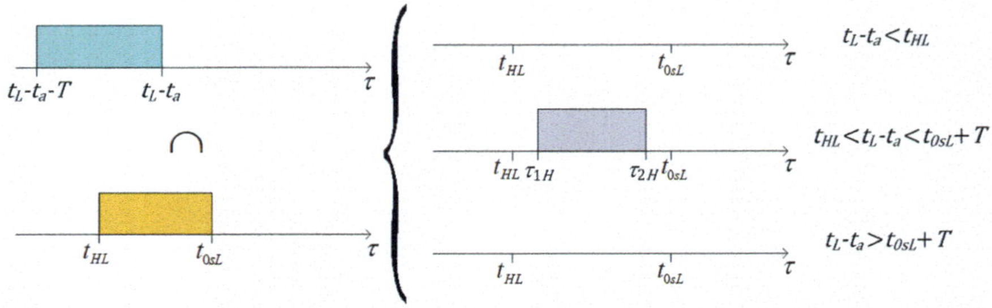

Fig. 3.17 Illustration of the integration range of I_1

$$I_1 = \frac{\gamma}{2\pi} \int_{\tau_{1H}}^{\tau_{2H}} Im(\Gamma(x_L, z_L, \tau)) \frac{e^{-(t_L-\mu-\tau)^2/2\sigma^2}}{\sqrt{t_{0sL}^2 - \tau^2}} d\tau,$$

with $\tau_{1H} = \max(t_L - \mu - \Delta, t_{HL})$ and $\tau_{2H} = \min(t_L - \mu + \Delta, t_{0sL})$ for $\tau_{2H} > \tau_{1H}$.

After introducing the change of variable $\tau = t_{0sL} \sin u$, the integral becomes

$$I_1 = \frac{\gamma}{2\pi} \int_{u_{1H}}^{u_{2H}} Im(\Gamma(t_{0sL} \sin u)) e^{-(t_L-\mu-t_{0sL} \sin u)^2/2\sigma^2} du, \tag{3.64}$$

where $u_{1H} = \text{asin}(\tau_{1H}/t_{0sL})$ and $u_{2H} = \text{asin}(\tau_{2H}/t_{0sL})$.

The second integral I_2 follows the same analysis as for the mirror-reflected wave, so it is written as

$$I_2 = \frac{\gamma}{2\pi} \int_{u_{1s}}^{u_{2s}} \text{Re}(\Psi(t_{0sL} \cosh u)) e^{-\alpha(t_L-t_a-t_{0sL} \cosh u)^2/T^2} \cos\left(\frac{3\pi}{2T}(t_L - t_a - t_{0sL} \cosh u)\right) du \tag{3.65}$$

Figure 3.18 shows the three components of the velocity potential produced by a line source moving with $M = 0.3$ at a height $h = 2.5$ m above a homogeneous medium with parameters $\rho_m = 4\rho$ and $c_m = 2.5c$. The signal of the source is plotted in Fig. 3.18a. The observer in field point 1 is located in the narrow-angle region so there is no head wave reaching it (Fig. 3.18b). The field point 2 is in the wide-angle region so that a head wave arrives before the mirror reflection (Fig. 3.18c). The last field point is also located in the wide-angle region but now the head wave arrives even before the direct wave. From the results, one can notice that the amplitude of the correction term is smaller the further away the observer is from the interface.

Figure 3.19 compares the velocity potential for different Mach numbers, the upper plot shows the curves for field point 1 and the lower plot for field point 3. With increasing Mach number, the curves are compressed in the time domain.

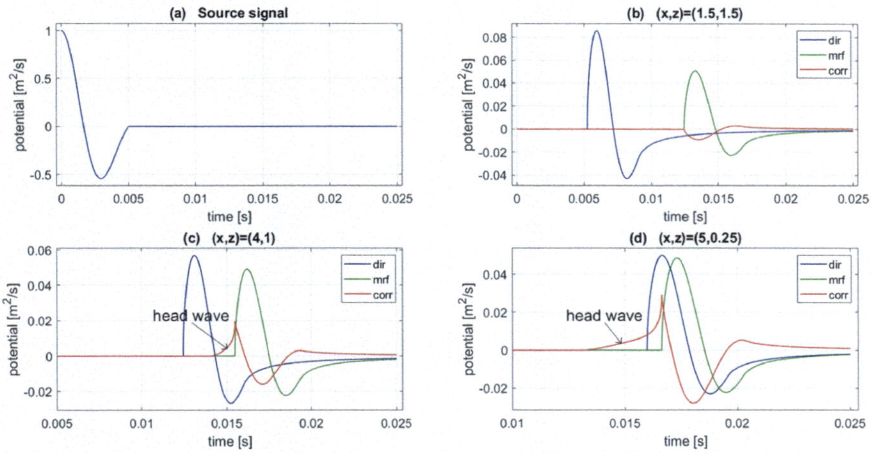

Fig. 3.18 Components of the velocity potential produced by the "expcos" source at three different field points

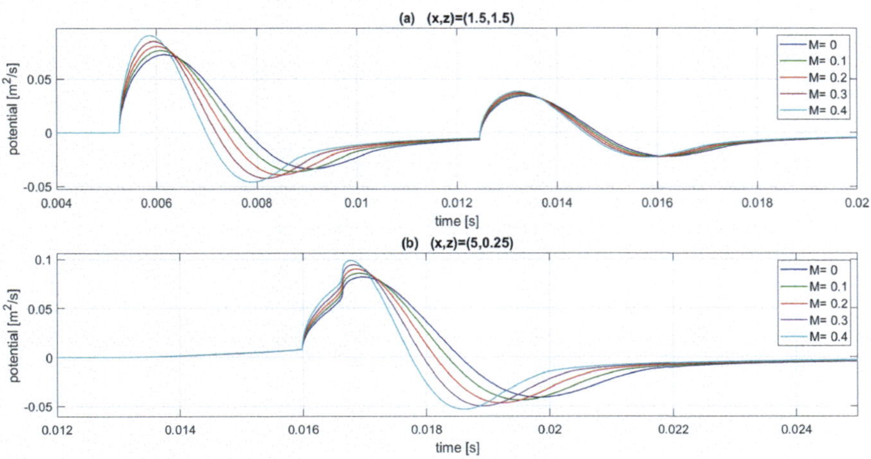

Fig. 3.19 Velocity potential for different values of the Mach number

3.3 Transition from Homogeneous to Locally Reacting Interface

It can be shown that both solutions agree if the sound speed and density of the lower medium are changed in a way that the sound speed is reduced to zero and the density is increased to infinity so that $\rho_2 c_2$ remains constant and the ratio $\rho_2 c_2/\rho_1 c_1$ equals the normalized impedance Z_0.

The correction term for the case of an impedance surface is given by Formula (3.22). After the change of variable $\tau = t_{0sL} \cosh u$, we obtain

$$\phi_{cor}^{Z} = -\frac{\gamma}{\pi}\int_{0}^{+\infty} q(t_L - t_{0sL}\cosh u)\alpha\varepsilon^2\left(\frac{\alpha[(x_L + MR_{0sL}\cosh u)^2 + z_{hL}^2/\gamma^2] + z_{hL}(Mx_L + R_{0sL}\cosh u)}{\left(\varepsilon^2 R_{0sL}^2\cosh u + \alpha v\right)^2 + (\alpha^2 - \varepsilon^2)\left(\varepsilon^2 R_{0sL}^2 - v^2\right)}\right)du. \quad (3.66)$$

For the homogeneous medium, the correction term for the case $c > c_m$ is given by the second term in (3.48). After the same change of variable $\tau = t_{0sL}\cosh u$

$$\phi_{cor} = -\frac{\gamma}{\pi}\int_{t_{0sL}}^{+\infty} q(t_L - t_{0sL}\cosh u)\mathrm{Re}(\Psi_L(x_L, z_L, t_{0sL}\cosh u))du \quad (3.67)$$

By extracting the sound speed from the square roots η and η_m, we define $\delta = \sqrt{1 - c^2p^2}$ and $\delta_m = \sqrt{1 - c_m^2p^2}$ and write $Re(\Psi_L)$ as

$$Re(\Psi_L) = Re\left(\frac{\rho c\delta_m}{\rho_m c_m\delta + \rho c\delta_m}\right).$$

By denoting $\beta = \frac{\rho_1 c_1}{\rho_2 c_2}$, we have

$$Re(\Psi_L) = Re\left(\frac{\beta\delta_m}{\delta + \beta\delta_m}\right) = \frac{\beta\left(\Re(\delta_m\delta^*) + \beta|\delta_m|^2\right)}{|\delta|^2 + 2\beta\Re(\delta_m\delta^*) + \beta^2|\delta_m|^2},$$

where $()^*$ denotes the conjugate complex. By setting $c_m = 0$, we have $\delta_m = 1$ and

$$\phi_{cor} = -\frac{\gamma}{\pi}\int_{0}^{+\infty} q(t_L - t_{0sL}\cosh u)\frac{\beta(Re(\delta^*) + \beta)}{|\delta|^2 + 2\beta\Re(\delta^*) + \beta^2}du, \quad (3.68)$$

δ can be written explicitly as

$$\delta = \frac{1}{\gamma}\frac{z_{hL}(Mx_L + R_{0sL}\cosh u) - j|x_L + MR_{0sL}\cosh u|R_{0sL}\sinh u}{(x_L + MR_{0sL}\cosh u)^2 + z_{hL}^2/\gamma^2}. \quad (3.69)$$

By using (3.69), the numerator of the second term in the integrand in (3.68) can be rewritten as

$$\beta\left(Re(\delta^*) + \beta\right) = \frac{\beta z_{hL}(Mx_L + R_{0sL}\cosh u) + \gamma\beta^2\left[(x_L + MR_{0sL}\cosh u)^2 + z_{hL}^2/\gamma^2\right]}{\gamma\left[(x_L + MR_{0sL}\cosh u)^2 + z_{hL}^2/\gamma^2\right]}.$$

We calculate $|\delta|^2$ as

$$|\delta|^2 = \frac{z_{hL}^2(Mx_L + R_{0sL}\cosh u)^2 + |x_L + MR_{0sL}\cosh u|^2 R_{0sL}^2 \sinh^2 u}{\gamma^2\left[(x_L + MR_{0sL}\cosh u)^2 + z_{hL}^2/\gamma^2\right]^2}.$$

After some arrangement, $|\delta|^2$ can be simplified to

$$|\delta|^2 = \frac{R_{0SL}^2\cosh^2 u - x_L^2}{\gamma^2\left[(x_L + MR_{0sL}\cosh u)^2 + z_{hL}^2/\gamma^2\right]}. \tag{3.70}$$

By using (3.70), the denominator of the second term in the integrand in (3.68) can be rewritten as

$$|\delta|^2 + 2\beta\Re(\delta^*) + \beta^2 = \frac{R_{0SL}^2\cosh^2 u - x_L^2}{\gamma^2\left[(x_L + MR_{0sL}\cosh u)^2 + z_{hL}^2/\gamma^2\right]}$$
$$+ \frac{2\gamma\beta z_{hL}(Mx_L + R_{0sL}\cosh u) + \gamma^2\beta^2\left[(x_L + MR_{0sL}\cosh u)^2 + z_{hL}^2/\gamma^2\right]}{\gamma^2\left[(x_L + MR_{0sL}\cosh u)^2 + z_{hL}^2/\gamma^2\right]},$$

and after some rearrangement, we obtain

$$|\delta|^2 + 2\beta\Re(\delta^*) + \beta^2 = \frac{\left[\varepsilon^2 R_{0sL}\cosh u + \gamma\beta v\right]^2 - \left(1-\beta^2\right)(x_L - \gamma\beta M z_{hL})^2}{\varepsilon^2\gamma^2\left[(x_L + MR_{0sL}\cosh u)^2 + z_{hL}^2/\gamma^2\right]}.$$

The solution for the homogeneous medium reads then

$$\phi_{cor} = -\frac{\gamma}{\pi}\int_0^{+\infty} q(t_L - t_{0sL}\cosh u)\varepsilon^2\gamma\beta$$
$$\times \frac{z_{hL}(Mx_L + R_{0sL}\cosh u) + \gamma\beta\left[(x_L + MR_{0sL}\cosh u)^2 + z_{hL}^2/\gamma^2\right]}{\left[\varepsilon^2 R_{0sL}\cosh u + \gamma\beta v\right]^2 - \left(1-\beta^2\right)(x_L - \gamma\beta M z_{hL})^2}\,du. \tag{3.71}$$

Finally, it is easy to show that
$1-\beta^2 = \varepsilon^2 - \alpha^2$ and $(x_L - \gamma\beta M z_{hL})^2 = \varepsilon^2 R_{0sL}^2 - v^2$, so both solutions agree.

Figure 3.20 shows the correction term due to the Ricker wavelet of Example 3.1.4 for the ratio c_m/c decreasing from 0.9 to 0.1 (dashed color lines), but the constant value of $Z_0 = \rho_m c_m/\rho c$. We can observe that by decreasing values of c_m, the curves tend to the solution for the impedance case ($c_m/c = 0$, black solid line). The arrows show the direction of change with decreasing c_m/c.

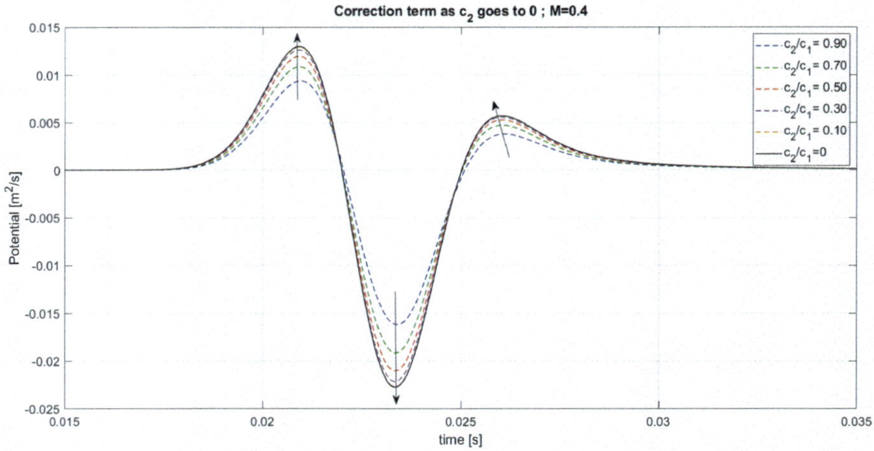

Fig. 3.20 Correction term for different values of c_m

3.4 Point Source Above a Locally Reacting Medium

We continue with the previous approach of using Green's method with the motion of the source included in the source function. We need, therefore, closed-form solutions for 3D half-space Green's function in the time domain.

3.4.1 Stationary Point Source

The source is located at position $\vec{r}_q = (x_q, y_q, h)$ and emits a pulse at the time t_q. As in the 2D case, the velocity potential can be written as the sum of three terms: direct sound, mirror reflection and a correction term. The first two are independent of the type of interface and are given by

$$\phi_{inc} = \frac{\delta(t - t_q - t_0)}{4\pi R}, \tag{3.72}$$

$$\phi_{mrf} = \frac{\delta(t - t_q - t_{0s})}{4\pi R_s}, \tag{3.73}$$

with $t_0 = R/c$, $R = \sqrt{(x - x_q)^2 + (y - y_q)^2 + (z - h)^2}$,

$$t_{0s} = R_s/c, R_s = \sqrt{(x - x_q)^2 + (y - y_q)^2 + z_h^2}.$$

Ochmann deduced in [5] closed-form solutions for two types of impedances, (a) mass character and (b) absorbing character.

3.4.1.1 Impedance with Mass Character

The expression for the correction term is given by

$$\phi_{cor} = -\frac{c\xi e^{\xi z_h}}{2\pi} \frac{e^{-\xi\sqrt{c^2(t-t_q)^2-\Upsilon^2}}}{\sqrt{c^2(t-t_q)^2-\Upsilon^2}} H(t-t_q-t_{0s}), \tag{3.74}$$

with $\Upsilon = \sqrt{(x-x_q)^2+(y-y_q)^2}$, $\xi = \rho/m''$ and m'' is the mass per unit area of the interface.

3.4.1.2 Impedance with Absorbing Character

The correction term is expressed as

$$\phi_{cor} = -\frac{1}{2\pi}\frac{\partial}{\partial t}\left[\frac{H(t-t_q-t_{0s})}{\sqrt{(Z_0c(t-t_q)+z_h)^2+(1-Z_0^2)\Upsilon^2}}\right], \tag{3.75}$$

where Z_0 is the normalized impedance of the interface.

3.4.2 Moving Point Source

The moving source is described as $Q(\vec{r},t) = q(t)\delta(\vec{r}-\vec{r}_q(t))$, with $\vec{r}_q(t) = (Mct, 0, h)$. Therefore, applying Green's function method, the velocity potential is written as

$$\Phi(\vec{r},t) = \int_{-\infty}^{+\infty} q(\tau)\phi(\vec{r},t;\vec{r}_q(\tau),\tau)d\tau. \tag{3.76}$$

Direct and mirror reflection terms can be obtained inserting (3.72) and (3.73) into (3.76). The expressions are given then by

$$\Phi_{inc} = \int_{-\infty}^{+\infty} q(\tau)\frac{\delta(t-\tau-t_0(\tau))}{4\pi R(\tau)}d\tau, \tag{3.77}$$

$$\Phi_{mrf} = \int_{-\infty}^{+\infty} q(\tau)\frac{\delta(t-\tau-t_{0s}(\tau))}{4\pi R_s(\tau)}d\tau, \tag{3.78}$$

with $t_0(\tau) = R(\tau)/c$, $R(\tau) = \sqrt{(x-Mc\tau)^2 + y^2 + (z-h)^2}$,

$$t_{0s}(\tau) = R_s(\tau)/c, R_s(\tau) = \sqrt{(x-Mc\tau)^2 + y^2 + z_h^2}.$$

To solve integrals (3.77) and (3.78), we consider the property of the delta function

$$\delta(f(\eta)) = \sum_i \frac{\delta(\eta-\eta_i)}{|\partial f/\partial \eta|_{\eta=\eta_i}}, \tag{3.79}$$

where η_i are the solutions of the equation $f(\eta) = 0$.

Applying (3.79) to (3.77) and (3.78), we obtain

$$\Phi_{inc} = \sum_i \frac{q\left(\tau_i^{inc}\right)}{4\pi R\left(\tau_i^{inc}\right)\left|\partial f^{inc}/\partial\tau\right|_{\tau=\tau_i^{inc}}}, \tag{3.80}$$

$$\Phi_{mrf} = \sum_i \frac{q\left(\tau_i^{mrf}\right)}{4\pi R_s\left(\tau_i^{mrf}\right)\left|\partial f^{mrf}/\partial\tau\right|_{\tau=\tau_i^{mrf}}}, \tag{3.81}$$

where $f^{inc} = t-\tau-t_0(\tau)$ and $f^{mrf} = t-\tau-t_{0s}(\tau)$.

The derivatives of f^{inc} and f^{mrf} are found to be

$$\left|\frac{\partial f^{inc}}{\partial\tau}\right| = \frac{\sqrt{(x-Mc\tau)^2+y^2+(z-h)^2} - M(x-Mc\tau)}{\sqrt{(x-Mc\tau)^2+y^2+(z-h)^2}}, \tag{3.82}$$

$$\left|\frac{\partial f^{mrf}}{\partial\tau}\right| = \frac{\sqrt{(x-Mc\tau)^2+y^2+z_h^2} - M(x-Mc\tau)}{\sqrt{(x-Mc\tau)^2+y^2+z_h^2}}, \tag{3.83}$$

and inserting the results into (3.80) and (3.81)

$$\Phi_{inc} = \sum_i \frac{q\left(\tau_i^{inc}\right)}{4\pi\left|\sqrt{\left(x-Mc\tau_i^{inc}\right)^2+y^2+(z-h)^2} - M\left(x-Mc\tau_i^{inc}\right)\right|}, \tag{3.84}$$

$$\Phi_{mrf} = \sum_i \frac{q\left(\tau_i^{mrf}\right)}{4\pi \left| \sqrt{\left(x - Mc\tau_i^{mrf}\right)^2 + y^2 + (z+h)^2} - M\left(x - Mc\tau_i^{mrf}\right) \right|}. \tag{3.85}$$

We proceed now to find the solutions τ_i^{inc} and τ_i^{mrf}.

$$f^{inc} = 0 \rightarrow c(t - \tau) = \sqrt{(x - Mc\tau)^2 + y^2 + (z - h)^2},$$

$$f^{mrf} = 0 \rightarrow c(t - \tau) = \sqrt{(x - Mc\tau)^2 + y^2 + z_h^2}.$$

Once again, we apply the inverse Prandtl-Lorentz identities (3.14) plus the relation $y = y_L/\gamma$ that accounts for the third spatial dimension. The previous two equations become

$$Mx_L + c(t_L - \tau) = \sqrt{(x_L + Mc(t_L - \tau))^2 + y_L^2/\gamma^2 + (z_L - h_L)^2/\gamma^2},$$

$$Mx_L + c(t_L - \tau) = \sqrt{(x_L + Mc(t_L - \tau))^2 + y_L^2/\gamma^2 + z_{hL}^2/\gamma^2}.$$

The solutions of the first equations are $\tau_{\pm}^{inc} = t_L \pm R_L$ and of the second equations are $\tau_{\pm}^{mrf} = t_L \pm R_{sL}$. For subsonic motion, only the solutions with negative signs are physically meaningful. After some calculations, one can find that

$$\sqrt{\left(x - Mc\tau_i^{inc}\right)^2 + y^2 + (z - h)^2} - M\left(x - Mc\tau_i^{inc}\right) = \frac{R_L}{\gamma^2},$$

$$\sqrt{\left(x - Mc\tau_i^{mrf}\right)^2 + y^2 + (z + h)^2} - M\left(x - Mc\tau_i^{mrf}\right) = \frac{R_{sL}}{\gamma^2},$$

and the final expressions for direct wave and mirror reflection are written as

$$\Phi_{inc} = \gamma^2 \frac{q(t_L - t_{0L})}{4\pi R_L}, \tag{3.86}$$

$$\Phi_{mrf} = \gamma^2 \frac{q(t_L - t_{0sL})}{4\pi R_{SL}}, \tag{3.87}$$

with $t_{0L} = R_L/c$, $R_L = \sqrt{x_L^2 + y_L^2 + (z_L - h_L)^2}$, $t_{0sL} = R_{sL}/c$, $R_{sL} = \sqrt{x_L^2 + y_L^2 + z_{hL}^2}$

3.4.2.1 Impedance of Mass Character

The correction term for the mass character interface is written as

$$\Phi_{cor} = -\frac{c\xi e^{\xi z_h}}{2\pi} \int_{-\infty}^{+\infty} q(\tau) \frac{e^{-\xi\sqrt{c^2(t-\tau)^2 - \Upsilon^2(\tau)}}}{\sqrt{c^2(t-\tau)^2 - \Upsilon^2(\tau)}} H(t - \tau - t_{0s}(\tau)) d\tau, \tag{3.88}$$

with $\Upsilon(\tau) = \sqrt{(x - Mc\tau)^2 + y^2}$. Using the Prandtl-Lorentz identities, we can rewrite the denominator in (3.88) as

$$\sqrt{c^2(t-\tau)^2 - \Upsilon^2(\tau)} = \sqrt{c^2(t_L - \tau)^2 - \Upsilon_L^2}/\gamma,$$

and the correction term becomes

$$\Phi_{cor} = -\gamma^2 \frac{c\xi_L e^{\xi_L z_{hL}}}{2\pi} \int_{-\infty}^{+\infty} q(\tau) \frac{e^{-\xi_L\sqrt{c^2(t_L-\tau)^2 - \Upsilon_L^2}}}{\sqrt{c^2(t_L-\tau)^2 - \Upsilon_L^2}} H(t_L - \tau - t_{0sL}) d\tau, \tag{3.89}$$

with $\xi_L = \xi/\gamma$. Applying the commutative property of convolution, we rewrite (3.89) as

$$\Phi_{cor} = -\gamma^2 \frac{c\xi_L e^{\xi_L z_{hL}}}{2\pi} \int_{t_{0sL}}^{+\infty} q(t_L - \tau) \frac{e^{-\xi_L\sqrt{c^2\tau^2 - \Upsilon_L^2}}}{\sqrt{c^2\tau^2 - \Upsilon_L^2}} d\tau. \tag{3.90}$$

Finally, we perform the change of variable $\tau = \frac{\Upsilon_L}{c} \cosh u$ to obtain

$$\Phi_{cor} = -\gamma^2 \frac{\xi_L e^{\xi_L z_{hL}}}{2\pi} \int_{u_0}^{+\infty} q\left(t_L - \frac{\Upsilon_L}{c}\cosh u\right) e^{-\xi_L \Upsilon_L \sinh u} du, \tag{3.91}$$

with $u_0 = \mathrm{acosh}\left(\frac{R_{0sL}}{\Upsilon_L}\right)$. For the special case $z_{hL} = 0$, i.e., when source and receiver are sitting on the interface, $u_0 = 1$

3.4.2.2 Impedance of Absorbing Character

The correction term for the absorbing character interface is written as

$$\Phi_{cor} = -\frac{1}{2\pi}\frac{\partial}{\partial t}\left(\int_{-\infty}^{\infty} q(\tau) \frac{H(t - \tau - t_{0s}(\tau))}{\sqrt{(Z_0 c(t-\tau) + z_h)^2 + (1 - Z_0^2)\Upsilon^2(\tau)}} d\tau \right). \tag{3.92}$$

Here, we have taken the time derivative out of the integral since the first term does not depend on the time. Using the Prandtl-Lorentz identities and the variables α, ε and ν defined in Sect. 3.1.2, we can rewrite the denominator in (3.92) as

$$\begin{aligned}&\sqrt{(Z_0c(t-\tau)+z_h)^2+\left(1-Z_0^2\right)\Upsilon^2(\tau)}\\&=\frac{1}{\alpha\varepsilon}\sqrt{\left(\varepsilon^2c(t_L-\tau)+\alpha\nu\right)^2+\left(\alpha^2-\varepsilon^2\right)\left(\varepsilon^2R_{0sL}^2-\nu^2\right)},\end{aligned}\tag{3.93}$$

and the correction term becomes

$$\Phi_{cor}=-\frac{\alpha\varepsilon}{2\pi}\frac{\partial}{\partial t}\left(\int_{-\infty}^{\infty}q(\tau)\frac{H(t_L-\tau-t_{0sL})}{\sqrt{\left(\varepsilon^2c(t_L-\tau)+\alpha\nu\right)^2+\left(\alpha^2-\varepsilon^2\right)\left(\varepsilon^2R_{0sL}^2-\nu^2\right)}}d\tau\right).\tag{3.94}$$

Applying the commutative property of convolution, we rewrite (3.94) as

$$\Phi_{cor}=-\frac{\alpha\varepsilon}{2\pi}\frac{\partial}{\partial t}\left(\int_{-\infty}^{\infty}q(t_L-\tau)\frac{H(\tau-t_{0sL})}{\sqrt{\left(\varepsilon^2c\tau+\alpha\nu\right)^2+\left(\alpha^2-\varepsilon^2\right)\left(\varepsilon^2R_{0sL}^2-\nu^2\right)}}d\tau\right).\tag{3.95}$$

The last step is to perform the time derivative over the expression inside

$$\begin{aligned}\Phi_{cor}=-\frac{\alpha\varepsilon}{2\pi}&\left\{\int_{-\infty}^{\infty}q(t_L-\tau)\frac{\frac{\partial H}{\partial t}(\tau-t_{0sL})}{\sqrt{\left(\varepsilon^2c\tau+\alpha\nu\right)^2+\left(\alpha^2-\varepsilon^2\right)\left(\varepsilon^2R_{0sL}^2-\nu^2\right)}}d\tau\right.\\&+\int_{-\infty}^{\infty}\frac{\partial q}{\partial t}(t_L-\tau)\frac{H(\tau-t_{0sL})}{\sqrt{\left(\varepsilon^2c\tau+\alpha\nu\right)^2+\left(\alpha^2-\varepsilon^2\right)\left(\varepsilon^2R_{0sL}^2-\nu^2\right)}}d\tau\\&\left.+\int_{-\infty}^{\infty}q(t_L-\tau)H(\tau-t_{0sL})\frac{\partial}{\partial t}\left(\frac{1}{\sqrt{\left(\varepsilon^2c\tau+\alpha\nu\right)^2+\left(\alpha^2-\varepsilon^2\right)\left(\varepsilon^2R_{0sL}^2-\nu^2\right)}}\right)d\tau\right\}.\end{aligned}$$

The time derivatives can be expressed explicitly

$$\frac{\partial q}{\partial t}(t_L-\tau)=\frac{\partial q}{\partial t_L}(t_L-\tau)\frac{\partial t_L}{\partial t}=\gamma^2\frac{\partial q}{\partial t_L}(t_L-\tau),$$

$$\frac{\partial\nu}{\partial t}=\frac{\partial\nu}{\partial x_L}\frac{\partial x_L}{\partial t}=-\gamma^2\alpha M^2c,$$

$$\frac{\partial R_{0sL}}{\partial t} = \frac{x_L}{R_{0sL}}\frac{\partial x_L}{\partial t} = -\frac{\gamma^2 Mcx_L}{R_{0sL}},$$

$$\frac{\partial H}{\partial t}(\tau - t_{0sL}) = -\delta(\tau - t_{0sL})\frac{1}{c}\frac{\partial R_{0sL}}{\partial t} = \frac{\gamma^2 Mx_L}{R_{0sL}}\delta(\tau - t_{0sL}),$$

$$\begin{aligned}
&\frac{\partial}{\partial t}\left(\frac{1}{\sqrt{\left(\varepsilon^2 c\tau + \alpha\nu\right)^2 + \left(\alpha^2 - \varepsilon^2\right)\left(\varepsilon^2 R_{0sL}^2 - \nu^2\right)}}\right) \\
&= -\frac{\left(\varepsilon^2 c\tau + \alpha\nu\right)\alpha\frac{\partial \nu}{\partial t} + \left(\alpha^2 - \varepsilon^2\right)\left(\varepsilon^2 R_{0sL}\frac{\partial R_{0sL}}{\partial t} - \nu\frac{\partial \nu}{\partial t}\right)}{\left(\left(\varepsilon^2 c\tau + \alpha\nu\right)^2 + \left(\alpha^2 - \varepsilon^2\right)\left(\varepsilon^2 R_{0sL}^2 - \nu^2\right)\right)^{3/2}} \\
&= \frac{\gamma^2 Mc\varepsilon^2\left[(\alpha c\tau + \nu)\alpha M + \left(\alpha^2 - \varepsilon^2\right)x_L\right]}{\left[\left(\varepsilon^2 c\tau + \alpha\nu\right)^2 + \left(\alpha^2 - \varepsilon^2\right)\left(\varepsilon^2 R_{0sL}^2 - \nu^2\right)\right]^{3/2}}.
\end{aligned}$$

Therefore, the correction terms becomes

$$\begin{aligned}
\Phi_{cor} = -\frac{\alpha\varepsilon\gamma^2}{2\pi}\Bigg\{&\frac{Mx_L}{R_{0sL}}\frac{q(t_L - t_{0sL})}{\varepsilon(\alpha R_{0sL} + \nu)} + \int_{t_{0sL}}^{\infty}\frac{\frac{\partial q}{\partial t_L}(t_L - \tau)d\tau}{\sqrt{\left(\varepsilon^2 c\tau + \alpha\nu\right)^2 + \left(\alpha^2 - \varepsilon^2\right)\left(\varepsilon^2 R_{0sL}^2 - \nu^2\right)}} \\
&+ Mc\varepsilon^2\int_{t_{0sL}}^{\infty}\frac{q(t_L - \tau)\left[(\alpha c\tau + \nu)\alpha M + \left(\alpha^2 - \varepsilon^2\right)x_L\right]}{\left[\left(\varepsilon^2 c\tau + \alpha\nu\right)^2 + \left(\alpha^2 - \varepsilon^2\right)\left(\varepsilon^2 R_{0sL}^2 - \nu^2\right)\right]^{3/2}}d\tau\Bigg\}. \qquad (3.96)
\end{aligned}$$

3.4.3 Verification Example: Harmonic Time Dependence

As has been done in the 2D cases, the formulas derived in this section can be validated by comparing their results with other solutions for the harmonic case, $q(t) = e^{-j\omega t}$. A solution was obtained by Wang et al. [9] by performing a spatial Fourier transform in the Lorentz space. Direct wave, mirror reflection and correction terms are given by

$$\Phi_{inc}^W = \gamma^2 e^{-j\omega t_L}\frac{e^{jkR_L}}{4\pi R_L}, \qquad (3.97)$$

$$\Phi_{mrf}^W = \gamma^2 e^{-j\omega t_L}\frac{e^{jkR_{sL}}}{4\pi R_{SL}}, \qquad (3.98)$$

$$\Phi_{cor}^{W} = \frac{\gamma^2 e^{-j\omega t_L}}{4\pi^2} \int_{-\infty}^{+\infty}\int_{-\infty}^{+\infty} \frac{U(k_x)}{k_z + U(k_x)} \frac{e^{j\left(k_x x_L + k_y y_L + k_z z_{hL}\right)}}{jk_z} dk_x dk_y, \tag{3.99}$$

where $k_z = \sqrt{k^2 - k_x^2 - k_y^2}$, $k = \frac{\omega}{c}$,

$$U(k_x) = \gamma k\left(1 + M\frac{k_x}{k}\right)\beta\left(\gamma^2\omega\left(1 + M\frac{k_x}{k}\right)\right),$$

with $\beta(\omega) = \frac{\rho c}{Z(\omega)}$, the normalized frequency-dependent admittance.

Now, we compare (3.97)–(3.99) with the solution using the present approach. By considering $q(t) = e^{-j\omega t}$, we can easily see that direct wave and mirror reflection are the same in both approaches. It remains only to show that (3.91) and (3.96) provide the same results as (3.99). These three equations can only be computed numerically.

To compute (3.99) numerically, it will be rewritten in a more convenient form. By performing the transformations

$$k_x = ku\cos\varepsilon,\ k_y = ku\sin\varepsilon \rightarrow dk_x dk_y = k^2 u du d\varepsilon,\ k_z = k\sqrt{1-u^2} \tag{3.100}$$

$x_L = r_L\cos\delta_L$, $y_L = r_L\sin\delta_L$ with

$$\delta_L = \operatorname{atan}(y_L/x_L), \tag{3.101}$$

the correction term is written as

$$\Phi_{cor}^{W} = \int_0^{+\infty} \tilde{I}(u) e^{jk_z z_{hL}} \frac{udu}{jk_z},\ \tilde{I}(u) = \int_0^{2\pi} F(u,\varepsilon) e^{jukr_L\cos(\varepsilon - \delta_L)} d\varepsilon, \tag{3.102}$$

with $F(u,\varepsilon) = \frac{\gamma^2 e^{-j\omega t_L}}{4\pi^2}\frac{U(u,\varepsilon)}{k_z(u)+U(u,\varepsilon)}$, $U(u,\varepsilon) = \gamma k(1 + Mu\cos\varepsilon)\beta\left(\gamma^2\omega(1 + Mu\cos\varepsilon)\right)$.

To perform the integrations properly, we need to choose the square root as

$$k_z = \begin{cases} k\sqrt{1-u^2},\ u < 1 \\ jk\sqrt{u^2-1},\ u > 1 \end{cases}, \tag{3.103}$$

and the correction term reads

$$\Phi_{cor}^{W} = -jk\int_0^1 \tilde{I}^{<}(u) e^{j\sqrt{1-u^2}kz_{hL}} \frac{udu}{\sqrt{1-u^2}} - k\int_1^{+\infty} \tilde{I}^{>}(u) e^{-\sqrt{u^2-1}kz_{hL}} \frac{udu}{\sqrt{u^2-1}}, \tag{3.104}$$

where $\tilde{I}(u) = \begin{cases} \tilde{I}^{<}(u) = \int_0^{2\pi} F^{<}(u,\varepsilon)e^{jukr_L\cos(\varepsilon-\delta_L)}d\varepsilon,\ u<1 \\ \tilde{I}^{>}(u) = \int_0^{2\pi} F^{>}(u,\varepsilon)e^{jukr_L\cos(\varepsilon-\delta_L)}d\varepsilon,\ u>1 \end{cases}$ and

$$F^{<}(u,\varepsilon) = \frac{\gamma^2 e^{-j\omega t_L}}{4\pi^2}\frac{\hat{U}(u,\varepsilon)}{\sqrt{1-u^2}+\hat{U}(u,\varepsilon)},\ F^{>}(u,\varepsilon) = \frac{\gamma^2 e^{-j\omega t_L}}{4\pi^2}\frac{\hat{U}(u,\varepsilon)}{j\sqrt{u^2-1}+\hat{U}(u,\varepsilon)},$$

$$\hat{U}(u,\varepsilon) = \gamma(1+Mu\cos\varepsilon)\beta\big(\gamma^2\omega(1+Mu\cos\varepsilon)\big).$$

We introduce the following change of variables:

$$u = \sin\alpha \rightarrow du = \cos\alpha d\alpha,\ \sqrt{1-u^2} = \cos\alpha,\ \alpha \in [0,\pi/2], \tag{3.105a}$$

$$u = \cosh\beta \rightarrow du = \sinh\beta d\beta,\ \sqrt{u^2-1} = \sinh\beta,\ \beta \in [0,+\infty[, \tag{3.105b}$$

and the correction term finally becomes

$$\Phi^W_{cor} = -jk\int_0^{\frac{\pi}{2}} \tilde{I}^{<}(\sin\alpha)e^{j\cos\alpha k z_{hL}}\sin\alpha d\alpha - k\int_0^{+\infty} \tilde{I}^{>}(\cosh\beta)e^{-\sinh\beta k z_{hL}}\cosh\beta d\beta. \tag{3.106}$$

The correction term for the mass character impedance is given by

$$\Phi_{cor} = -\frac{\gamma^2\xi_L}{2\pi}e^{\xi_L z_{hL}}e^{-j\omega t_L}\int_{u_0}^{+\infty} e^{jk\Upsilon_L\cosh u}e^{-\xi_L\Upsilon_L\sinh u}du. \tag{3.107}$$

The second term in the integrand in (3.107) makes the integral numerically convergent.

The correction term for the absorbing character impedance becomes

$$\Phi_{cor} = -\frac{\alpha\varepsilon\gamma^2}{2\pi}\left\{\frac{Mx_L}{R_{0sL}}\frac{e^{-j(\omega t_L - kR_{0sL})}}{\varepsilon(\alpha R_{0sL}+\nu)} + \int_{t_{0sL}}^{\infty}\frac{-j\omega e^{-j\omega(t_L-\tau)}d\tau}{\sqrt{\left(\varepsilon^2 c\tau+\alpha\nu\right)^2+\left(\alpha^2-\varepsilon^2\right)\left(\varepsilon^2R_{0sL}^2-\nu^2\right)}} + Mc\varepsilon^2\int_{t_{0sL}}^{\infty}\frac{e^{-j\omega(t_L-\tau)}\left[(\alpha c\tau+\nu)\alpha M+\left(\alpha^2-\varepsilon^2\right)x_L\right]}{\left[\left(\varepsilon^2 c\tau+\alpha\nu\right)^2+\left(\alpha^2-\varepsilon^2\right)\left(\varepsilon^2R_{0sL}^2-\nu^2\right)\right]^{3/2}}d\tau\right\}. \tag{3.108}$$

By performing the change of variable $c(\tau - t_{0sL}) = u$, (3.108) can be rewritten as

$$\Phi_{cor} = -\frac{\alpha\varepsilon\gamma^2}{2\pi}e^{-j(\omega t_L - kR_{0sL})}\left\{\frac{Mx_L}{R_{0sL}}\frac{1}{\varepsilon(\alpha R_{0sL}+\nu)}\right.$$
$$-\int_0^\infty \frac{jke^{jku}du}{\sqrt{\left(\varepsilon^2(u+R_{0sL})+\alpha\nu\right)^2+\left(\alpha^2-\varepsilon^2\right)\left(\varepsilon^2R_{0sL}^2-\nu^2\right)}}$$
$$\left.+M\varepsilon^2\int_0^\infty \frac{e^{jku}\left[(\alpha(u+R_{0sL})+\nu)\alpha M+\left(\alpha^2-\varepsilon^2\right)x_L\right]}{\left[\left(\varepsilon^2(u+R_{0sL})+\alpha\nu\right)^2+\left(\alpha^2-\varepsilon^2\right)\left(\varepsilon^2R_{0sL}^2-\nu^2\right)\right]^{3/2}}du\right\}. \tag{3.109}$$

The integrals in (3.109) are numerically not convergent due to the oscillatory term e^{jku}. Similarly to the 2D cases, the integral can be calculated if the integration path is shifted into the complex u-plane. We consider the closed contour $w = w_1 + w_2 + w_3$ shown in Fig. 3.21

The contours are defined as:

$$\begin{aligned} w_1 &: u = \eta + j0 \quad /\eta \in [0, D], \\ w_2 &: u = De^{j\eta} \quad /\eta \in [0, \pi/2], \\ w_3 &: u = 0 + j\eta \quad /\eta \in [D, 0] \end{aligned}$$

Since there are no singularities in the integrands in (3.109) in the first quadrant of the complex u-plane, Cauchy's integral theorem indicates that

$$\int_w \chi(u)du = \int_{w_1} \chi(u)du + \int_{w_2} \chi(u)du + \int_{w_3} \chi(u)du = 0,$$

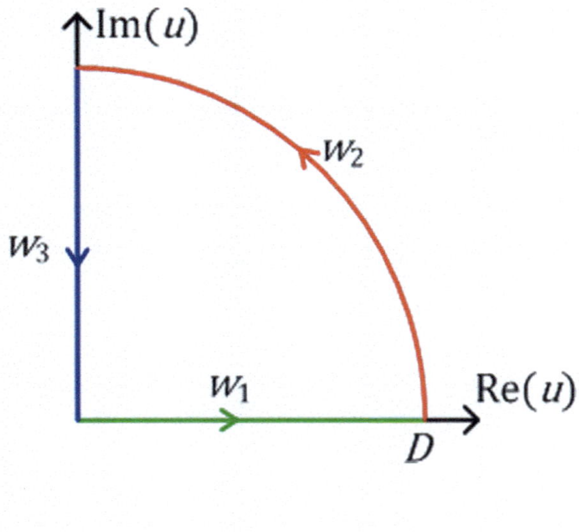

Fig. 3.21 Integration contours for (3.107)

where χ is any integrand in (3.109). Thus,

$$\int_{w_1} \chi(u)du = -\int_{w_2} \chi(u)du - \int_{w_3} \chi(u)du. \tag{3.110}$$

To solve each integral in (3.109), we have to solve the two integrals on the right-hand side of (3.110) in the limit $D \to +\infty$

$$\Phi_{cor} = -\lim_{D\to+\infty}\left(\int_{w_2} \chi(u)du + \int_{w_3} \chi(u)du\right).$$

The first integral is written as

$$\lim_{D\to+\infty}\int_{w_2} \chi(u)du = \lim_{D\to+\infty}\int_0^{\pi/2} jD\chi\left(De^{j\eta}\right)d\eta.$$

For large values of η,

$$e^{jku} = e^{jkDe^{j\eta}} = e^{jkD(\cos\eta + j\sin\eta)} = e^{-kD\sin\eta}e^{jkD\cos\eta},$$

therefore, the exponential term is approximately

$$e^{jk\rho_{0sL}\cosh u} \approx e^{-0.5k\rho_{0sL}e^D\sin\eta}e^{j0.5k\rho_{0sL}e^D\cos\eta}.$$

In the limit $D \to +\infty$, and for $0 < \eta < \pi/2$, $\chi\left(De^{j\eta}\right) \to 0$, so that $\lim_{D\to+\infty}\int_{w_2} \chi(u)du = 0$ and

$$\Phi_{cor} = -\lim_{D\to+\infty}\left(\int_{w_3} \chi(u)du\right) = \lim_{D\to+\infty}\left(\int_{-w_3} \chi(u)du\right) = \int_0^{+\infty} \chi(u(\eta))\frac{du}{d\eta}d\eta.$$

Hence, the correction term is finally given by:

$$\Phi_{cor} = -\frac{\alpha\varepsilon\gamma^2}{2\pi}e^{-j(\omega t_L - kR_{0sL})}\left\{\frac{Mx_L}{R_{0sL}}\frac{1}{\varepsilon(\alpha R_{0sL} + \nu)}\right.$$
$$\left. + \int_0^{\infty} \frac{ke^{-k\eta}d\eta}{\sqrt{\left(\varepsilon^2(j\eta + R_{0sL}) + \alpha\nu\right)^2 + \left(\alpha^2 - \varepsilon^2\right)\left(\varepsilon^2 R_{0sL}^2 - \nu^2\right)}}\right.$$

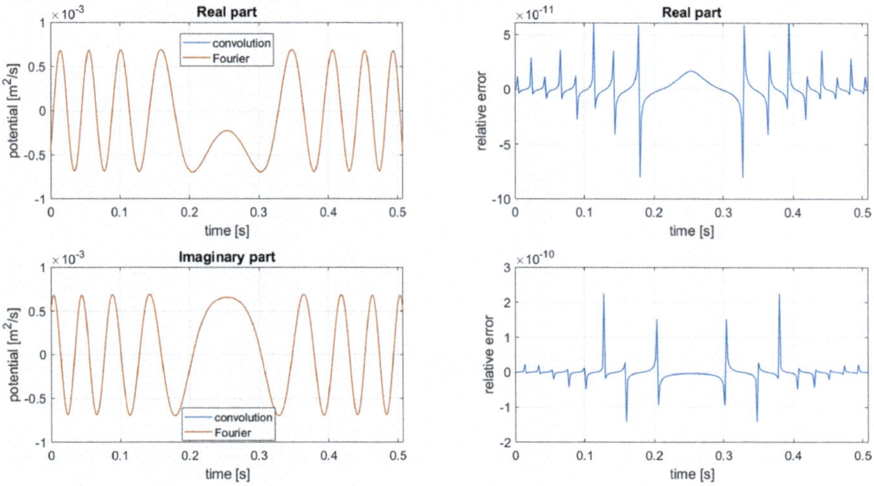

Fig. 3.22 Comparison of the correction term for the mass character impedance

$$+ jM\varepsilon^2 \int_0^\infty \frac{e^{-k\eta}\left[(\alpha(j\eta + R_{0sL}) + \nu)\alpha M + \left(\alpha^2 - \varepsilon^2\right)x_L\right]}{\left[\left(\varepsilon^2(j\eta + R_{0sL}) + \alpha\nu\right)^2 + \left(\alpha^2 - \varepsilon^2\right)\left(\varepsilon^2 R_{0sL}^2 - \nu^2\right)\right]^{3/2}} d\eta \Bigg\}. \tag{3.111}$$

Now, both integrals in (3.111) converge due to the exponentially decaying term $e^{-k\eta}$. Additionally, the integrals in (3.111) are less time-consuming as the integral (3.106) since the latter is a double integral.

Figures 3.22 and 3.23 show a comparison of the correction term computed with each formulation. The present approach is labeled as "convolution" and the solution from Wang et al. as "Fourier". The agreement between the results is excellent in both cases, with a relative error below 10^{-10}. Hence, the formulas developed here are validated.

3.4.4 Example: Source Emitting a "Sin^2-Pulse"

As an example, the sound radiation of a pulse of width $\Delta T = \pi/\Omega$ emitted at the time $t = \mu$ is computed. The source function is defined as

$$q(t) = \begin{cases} \sin^2 \Omega(t-\mu), & 0 < t - \mu < \Delta T \\ 0, & else \end{cases}.$$

Figure 3.24 shows the sound pressure at two different points P_1 and P_2 for three different velocities and the two types of ground impedance. The mass-like impedance

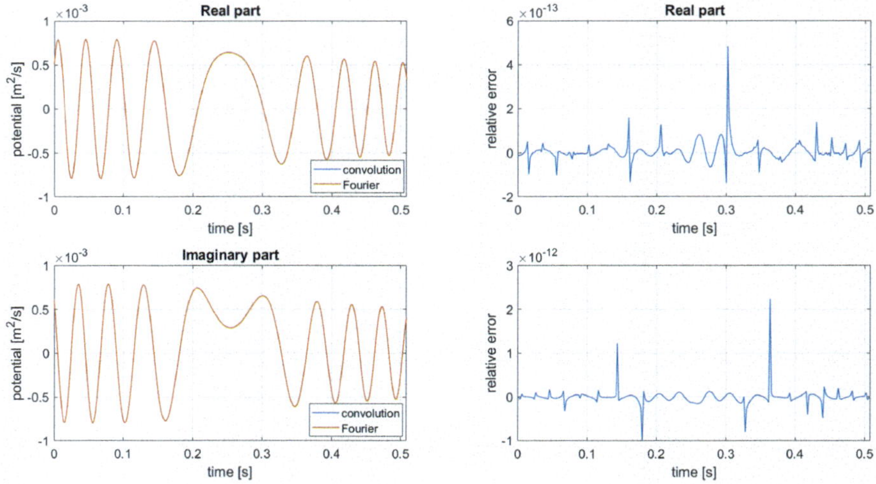

Fig. 3.23 Comparison of the correction term for the absorbing character impedance

has a mass density $m\prime\prime = 1$ kg/m^2, and the absorbing impedance a normalized impedance $Z_0 = 0.5$. When the pulse is emitted, the source is moving away from P_1 and moving toward P_2.The effect of the velocity is small but can be observed in the plots. When the source is moving away from the observer, the amplitude of the pulse decreases and its width broadens by increasing velocity, while the contrary takes place when the source is coming toward the observer. The effect is more evident in the direct wave than in the reflected one.

Figure 3.25 shows the pressure distribution above the ground at a certain time for the two types of impedances. Part of the direct wavefront has reached the ground, and a reflected wave is originated.

3.4.5 *Example: Sequence of "Sin^2-Pulses"*

Now, the monopole emits a sequence of sin^2-pulses with a constant time separation between pulses. Figure 3.26 shows the sound pressure at a fixed point above the ground for a velocity $M = 0.2$. The first four pulses are emitted when the source is approaching the listener, and the last three, when the source has passed the listener and is moving away from him. This can be observed in the plot since the first four pulses are coming closer while the last three are getting apart.

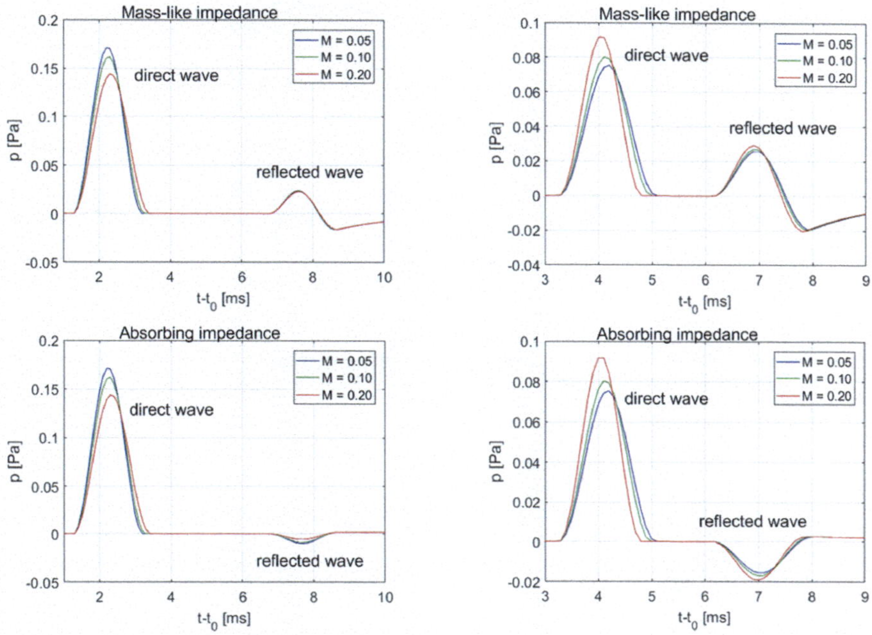

Fig. 3.24 Velocity potential at points P_1 (left) and P_2 (right)

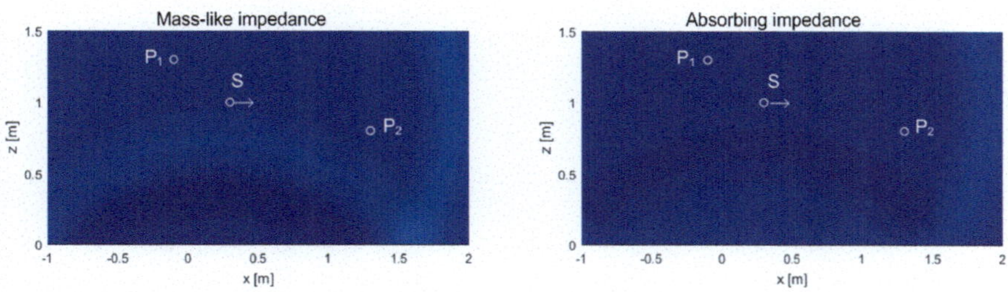

Fig. 3.25 Direct and reflected wavefronts for mass-like and absorbing grounds

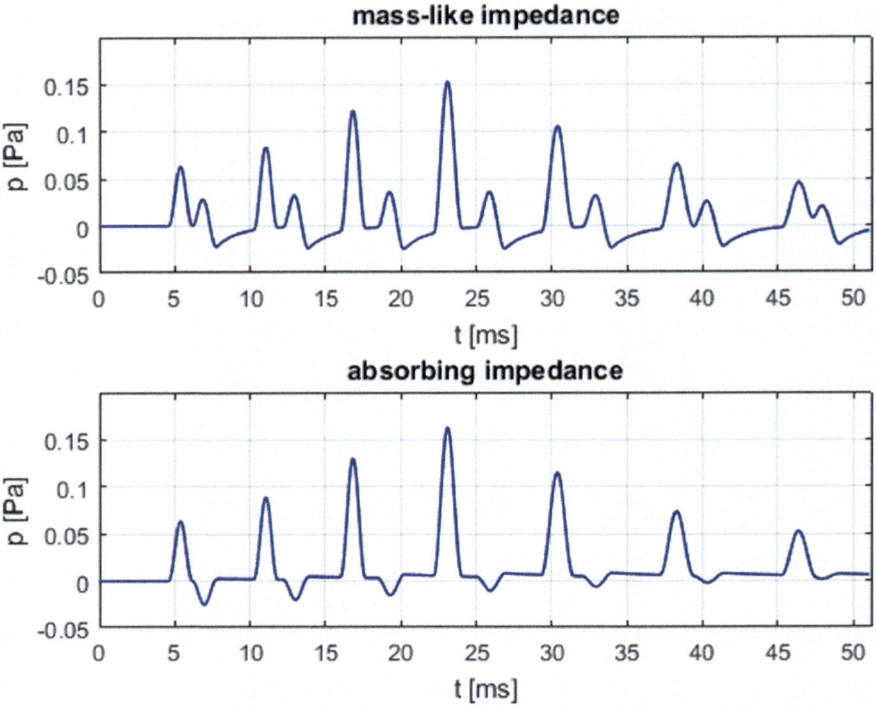

Fig. 3.26 Sound pressure due to a sequence of pulses at a fixed listener position

References

1. K. Aki, P.G. Richards, *Quantitative Seismology*, 2nd edn, Chap. 6, (University Science Books, California, 2009)
2. K. Attenborough, Sound propagation close to the ground. Annu. Rev. Fluid Mech. **34**, 51–82 (2002)
3. J. Delfs, *Basics of Aeroacoustics*, Lecture notes, WS 2021/2022 at Technische Universität Braunschweig, 2021: https://dlr.de/as/Portaldata/5/Resources/dokumente/abteilungen/abt_ta/Notes_Basics_of_Aeroacoustics_Delfs.pdf
4. D. Dragna, P. Blanc-Benon, Sound radiation by a moving line source above an impedance plane with frequency-dependent properties. J. Sound Vib. **349**, 259–275 (2015)
5. M. Ochmann, Closed form solutions for the acoustical impulse response over a masslike or an absorbing plane. J. Acoust. Soc. Am. **129**(6) (2011)
6. M. Ochmann, Exact solution for the acoustical impulse response of a line source above an absorbing plane. J. Acoust. Soc. Am. **144**(3), 1539–1549 (2018)
7. M. Ochmann, R. Piscoya, *Theory and Application of Acoustic Sources Using Complex Analysis*, 1st edn. (Springer Nature, Singapore, 2021)
8. C. Schwartz, Numerical calculation of Bessel functions. Int. J. Mod. Phys. C **23**(12), 1250084 (8 pages) (2012)
9. Y. Wang, B. Tong, K.M. Li, Sound fields generated by a monopole pointsource moving above a locally reacting surface, in *INTER-NOISE and NOISE-CON Congress, Conference Proceedings*, vol. 252. Institute of Noise Control Engineering, 2016, pp. 683–694

10. Y. Wang, K.M. Li, D. Dragna, P. Blanc-Benon, On the sound field from a source moving above non-locally reacting grounds. J. Sound Vib. **464**, Article 114975 (2020)
11. G.N. Watson, *A Treatise on the Theory of Bessel Functions* (Cambridge University Press, 1922)

Chapter 4
The Equivalent Source Method in Time Domain

The equivalent source method (ESM) is a simulation tool for estimating the sound radiation or scattering from complex-shaped structures into the three-dimensional space. The basic idea of the method consists of replacing the structure with a system of acoustical sources placed in the interior of the structure, whose strengths are determined by satisfying certain boundary conditions. For stationary structures and harmonic sound sources, the ESM can be implemented in the frequency domain, but for moving bodies or transient sound sources, it is more appropriate to work in the time domain. General features of the method are first explained in the frequency domain approach, and then the time domain approach is presented.

4.1 ESM in Frequency Domain

The ESM in the frequency domain is easier to implement than in the time domain since we can factorize the oscillating harmonic part in any acoustic quantity, and a time differentiation can be replaced by a multiplication, $\partial/\partial t \rightarrow j\omega t$. Additionally, most acoustic properties are frequency-dependent and can be better modeled in the frequency domain. A computation is performed at each frequency separately.

We consider a body at rest with the surface S and the normal vector $\hat{n}$ pointing toward the outside of the body. The acoustic domain where the sound is propagated is the volume B_e outside S. On the surface of the body, an acoustic quantity, pressure p_s, normal particle velocity v_{ns} or normalized impedance Z_0 has to be known.

In the scattering case, an incident wave is present, while there is no incident wave in the radiation case. Figure 4.1 illustrates the parameters of the method.

The sound pressure at a frequency ω is modeled as the contribution of each individual equivalent source

M. Ochmann and R. Piscoya, *Moving Acoustic Sources*,
Springer-AAS Acoustics Series, https://doi.org/10.1007/978-981-96-0871-3_4

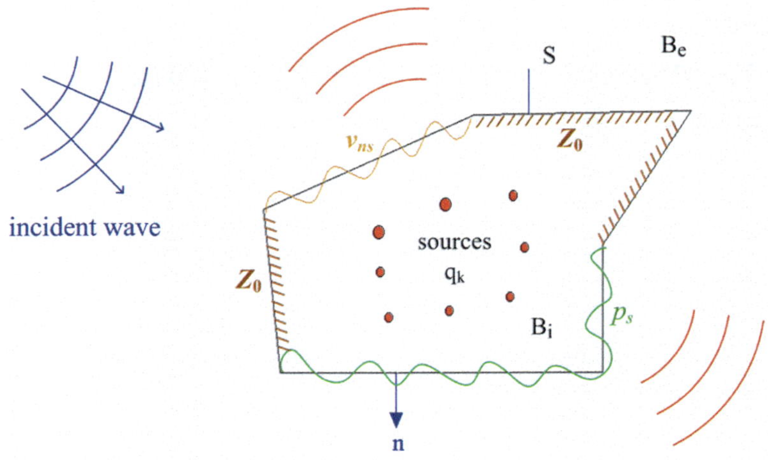

Fig. 4.1 Radiating/scattering body and the equivalent sources

$$p(\vec{r}) = \sum_{i=1}^{N} c_i \Psi\left(\vec{r}, \vec{r}_i'\right), \tag{4.1}$$

where c_i are the strengths of the sources, Ψ the radiating source functions, $\vec{r}_{i'}$ the position of the i-th source and $\vec{r}$ an arbitrary field point. The particle velocity can be obtained from the pressure using the momentum equation:

$$\rho \frac{\partial \left(\vec{v} e^{j\omega t}\right)}{\partial t} = -\nabla \left(p e^{j\omega t}\right) \rightarrow \vec{v} = \frac{j}{\rho\omega} \nabla p.$$

Hence, the velocity is given by

$$\vec{v}(\vec{r}) = \frac{j}{\rho\omega} \sum_{i=1}^{N} c_i \nabla_{\vec{r}} \Psi\left(\vec{r}, \vec{r}_i'\right). \tag{4.2}$$

The general form of the boundary condition that needs to be satisfied can be written as

$$R(p) = a\frac{\partial p}{\partial n} + bp = f. \tag{4.3}$$

When the pressure on the surface, p_s, is known, $a = 0$, $b = 1$ and $f = p_s$. When the normal velocity on the surface, v_{ns}, is given, $a = \frac{j}{\rho\omega}$, $b = 0$ and $f = v_{ns}$. When the normalized impedance Z_0 of the surface is provided, $a = \frac{j}{k}$, $b = \frac{1}{Z_0}$ and $f = 0$.

The strengths c_i can be found in different ways, but the simplest one is fulfilling the boundary condition (4.3) on a set of points on the surface $\vec{r}_\ell$ denoted as "collocation points". Inserting (4.1) into (4.3), we obtain

$$\sum_{i=1}^{N} c_i\left(a\frac{\partial \Psi_{\ell i}}{\partial n} + b\Psi_{\ell i}\right) = f_\ell, \quad \ell = 1, 2, \ldots, M, \tag{4.4}$$

with $\Psi_{\ell i} = \Psi(\vec{r}_\ell, \vec{r}_i')$ and $f_\ell = f(\vec{r}_\ell)$. Equation (4.4) is a set of M linear equations with N unknowns. We can write (4.4) in matrix form as $Ax = b$, where A is the $M \times N$ system matrix, b the $M \times 1$ right-hand side vector and x the $N \times 1$ vector of source strengths.

One possible type of radiating source function is the set of spherical functions, i.e., $\Psi \epsilon \psi_{nm}^{c,s}$.

$$\Psi(\vec{r}, \vec{r}')\epsilon \left\{ \begin{array}{l} \psi_{nm}^{c}(\vec{r}, \vec{r}') = h_n^{(2)}(kR)P_n^m(\cos\gamma)\cos m\delta \\ \psi_{nm}^{s}(\vec{r}, \vec{r}') = h_n^{(2)}(kR)P_n^m(\cos\gamma)\sin m\delta \end{array} \right\}, \quad n = 0, 1, \ldots, m = 0, 1, \ldots, n, \tag{4.5}$$

where $R = |\vec{r} - \vec{r}'|$, $\cos\gamma = \frac{z-z'}{R}$, $\tan\delta = \frac{y-y'}{x-x'}$, $h_n^{(2)}$ is the spherical Hankel functions and P_n^m is the associated Legendre functions.

The spherical functions represent an alternative form of higher order multipoles to the usual form defined as $\phi_{\alpha\beta l} = \frac{\partial^\alpha}{\partial x^\alpha}\frac{\partial^\beta}{\partial y^\beta}\frac{\partial^l}{\partial z^l}\left(\frac{e^{-jkR}}{R}\right)$.

For $n = 0$, there is only one spherical function,

$$\psi_{00}^{c} = h_n^{(2)}(kR),$$

that corresponds to the usual monopole.

For $n = 1$, there are three spherical functions:

$$\psi_{10}^{c} = h_1^{(2)}(kR)\cos\gamma, \quad \psi_{11}^{c} = -h_1^{(2)}(kR)\sin\gamma\cos\delta \quad \psi_{11}^{s} = -h_1^{(2)}(kR)\sin\gamma\sin\delta$$

that correspond to three dipoles in perpendicular directions.

For $n = 2$, there are five spherical functions that correspond to quadrupoles.

For an arbitrary n, the number of spherical functions is $2n + 1$.

Since, in most cases, the number of collocation points is larger than the number of equivalent sources ($M > N$), there is no unique solution and the strengths of the sources are obtained by solving the least squares problem:

$$\min_x \|Ax - b\|_2. \tag{4.6}$$

The solution of (4.6) can be found by applying the singular value decomposition (SVD) to the system matrix A. Through the SVD, any $M \times N$ matrix A can be written as the product of three matrices U, Σ and V:

$$A = U\Sigma V^T, \tag{4.7}$$

where $U = [u_1 u_2 \dots u_M]$ is a $M \times M$ matrix, $\Sigma = \text{diag}(\sigma_1 \sigma_2 \dots \sigma_N)$ is a $M \times N$ diagonal matrix and $V = [v_1 v_2 \dots v_N]$ is a $N \times N$ matrix. U and V have orthonormal columns (u_i and v_i), i.e., $U^T U = I$ and $V^T V = I$, and Σ has non-negative elements appearing in non-increasing order. The numbers σ_i are the singular values of A and the vectors u_i and v_i are the left and right singular vectors of A, respectively. The solution vector x can be written as

$$x_{lsq} = V_c \Sigma_c^{-1} \left(U_c^T b \right), \tag{4.8}$$

where $U_c = \left[u_1 u_2 \dots u_Q \right]$, $V_c = \left[v_1 v_2 \dots v_Q \right]$ and $\Sigma_c^{-1} = diag\left(1/\sigma_1 1/\sigma_2 \dots 1/\sigma_Q \right)$, with $Q = \min(M, N)$. The condition number of A is equal to the ratio $\kappa = \sigma_1 / \sigma_Q$.

4.1.1 Example: Radiating Multipoles of 4th Order

We consider a spherical surface of radius 1 m and center at $C = (0, 0, 0)$ vibrating with the normal velocity equal to the normal velocity generated by a set of nine multipoles of 4th order ($n = 4$) as given in (4.5) and all located at $\vec{r}' = (-0.2, -0.2, -0.2)$ m.

We simulate the sound field using multipoles of different orders sitting at nine positions inside the sphere. One equivalent source position in each octant of the sphere and one in the center are chosen (see Fig. 4.2f). Five calculations are performed, the first one with sources of order zero and each following simulation increases the order of sources by one, i.e., at each source position, there are $(n + 1)^2$ sources. Since the sound field of the real source is given by (4.5), the exact pressure and power can be determined exactly. The accuracy of the simulated results is obtained by comparison with the exact values. Figure 4.2 presents a comparison of the results. Figure 4.2a shows the curves of sound power level in comparison to the exact curve, while Fig. 4.2c shows the difference in dB with respect to the exact values. The differences decrease with increasing order of the multipoles. They remain below 3 dB when $n \geq 2$ and below 1 dB for $n \geq 3$. The condition number of the system matrix is plotted in Fig. 4.2b, and it increases as the order of the equivalent sources increases. In all cases, by a fixed source order, the condition number decreases as the frequency increases. The maximum condition number $\kappa_{\max}$ reaches 10^6 at 100 Hz. Despite the large values of κ, the results are reliable, since the inverse of κ is very much larger than the floating-point relative error eps, $\kappa^{-1} \gg eps$. Figure 4.2d shows the relative error of the normal velocity on the spherical surface. The surface error decreases with increasing order of multipoles. The surface error remains below 20% for $n \geq 3$. Figure 4.2e displays the radiation pattern of the source at 350 Hz, computed on a circle of radius 3 m in the XZ-plane. The results of the simulations improve by increasing the order of multipoles and show a good agreement already for $n \geq 2$. These results indicate that a few multipoles of higher order distributed on the interior of the body can simulate a source with a complicated radiation pattern with reasonable accuracy. Increasing the order of the multipoles would improve the accuracy of the results but also increase the condition number of the matrix.

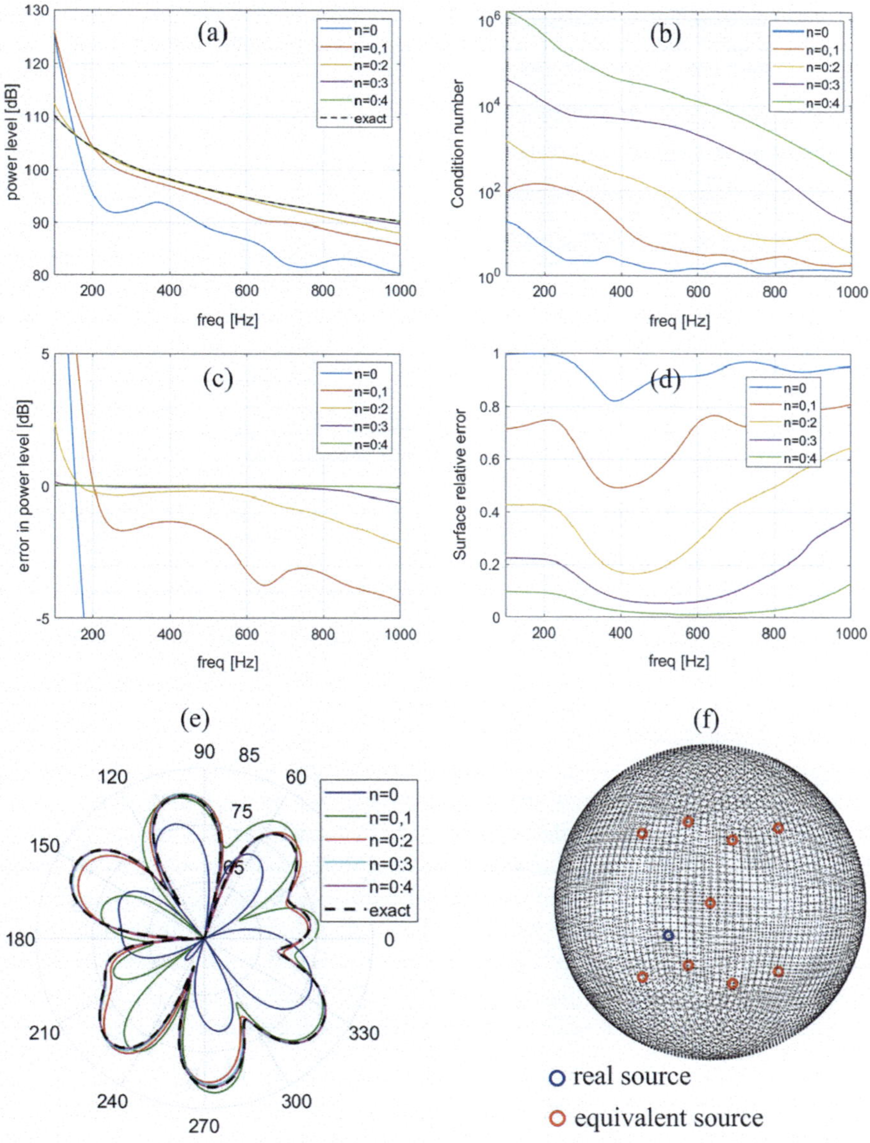

Fig. 4.2 Comparison of the solutions for different multipole orders

4.1.2 Example: Scattering from a Rigid Sphere in Front of a Point Source

This is a benchmark problem since there is an analytical solution. We consider a rigid sphere of radius $R = 1$ m centered at the origin and a point source located at the z-axis with coordinates $\vec{r}_q = \left(0, 0, z_q\right)$ with $z_q = 2m$. The total pressure and particle velocity will be given as the sum of the incident and scattered wave.

$$p_{tot} = p_{inc} + p_{sc}, \quad \vec{v}_{tot} = \vec{v}_{inc} + \vec{v}_{sc}. \tag{4.9}$$

The incident wave is given by

$$p_{inc} = p_0 h_0^{(2)}\left(k\left|\vec{r} - \vec{r}_q\right|\right); \quad \vec{v}_{inc} = \frac{jp_0}{\rho c} h_0^{\prime(2)}\left(k\left|\vec{r} - \vec{r}_q\right|\right) \frac{\left(\vec{r} - \vec{r}_q\right)}{\left|\vec{r} - \vec{r}_q\right|}, \tag{4.10}$$

and the scattered pressure is expressed as

$$p_{sc} = -p_0 \sum_{m=0}^{\infty} (2m+1) \frac{j_m^{\prime}(kR)}{h_m^{\prime(2)}(kR)} h_m^{(2)}\left(kz_q\right) h_m^{(2)}(kr) P_m(\cos\theta). \tag{4.11}$$

Since the sphere is rigid, the boundary condition on the surface of the sphere is

$$v_{ninc} + v_{nsc} = 0 \rightarrow v_{nsc} = -v_{ninc}. \tag{4.12}$$

We repeat the study made in the previous example, i.e., we consider the same nine source positions and vary the order of the multipoles from 0 to 4. The condition number will then be the same as before since only the right-hand side vector changes. The results are presented in Fig. 4.3.

Similar to the radiation case, the surface error decreases with increasing order of the multipoles, but for a fixed n, the surface error increases with the frequency (see Fig. 4.3a). Figure 4.3b–d show the comparison of the scattered pressure at field points on a circle of radius 3 m around the sphere for three different frequencies. Reasonable results are obtained again for the simulations with $n \geq 3$. The better agreement is obtained at the lower frequency (250 Hz). The results indicate that for the scattering case, the chosen distribution of equivalent sources is good for frequencies up to 500 Hz but less accurate for higher frequencies. For these frequencies, an increase in the number of equivalent sources should provide better results.

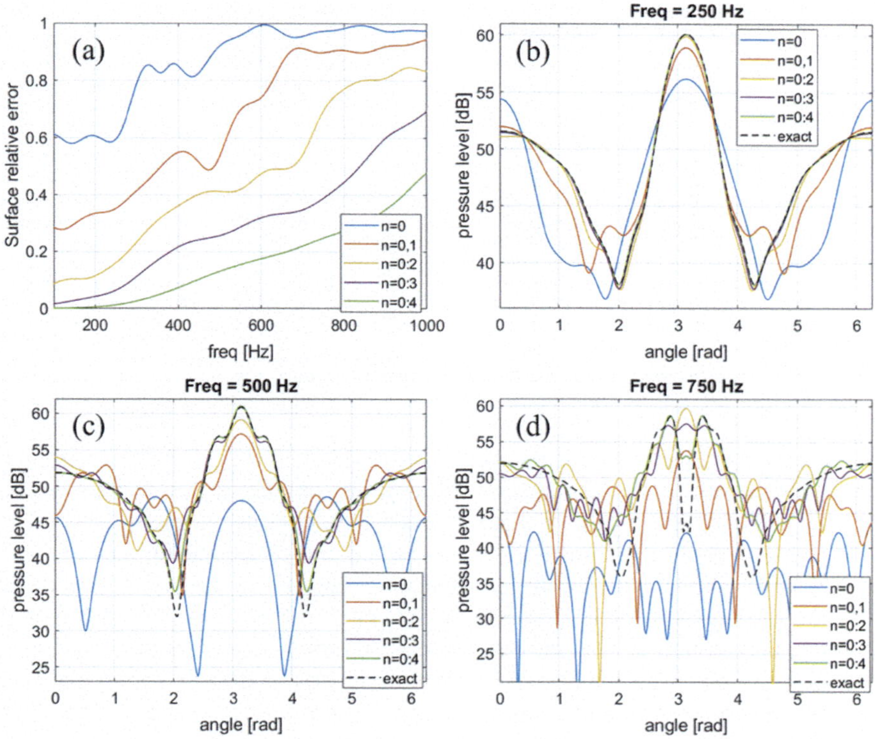

Fig. 4.3 Comparison of the simulated and analytical solutions

4.1.3 Source Distribution on an Interior Surface Conformal to the Body Surface

Monopoles are clearly the simplest radiating functions, but the studies in Sects. 4.1.1 and 4.1.2 show that only a few monopoles will not produce accurate results if the sound radiation or scattering presents a complex spatial distribution. However, only monopoles can still be used if many monopoles are uniformly distributed inside the body. An extended practice is to distribute the monopoles over a surface S_e located inside the body that is conformal to the body surface S. The closer S_e is to S, the singular the system matrix would become. On the other hand, if S_e is moved away from S, it will reduce its size, the monopoles will be closer to each other and their individual effect will merge. Hence, a commitment can be reached by taking S_e with a size around $0.65S$ to $0.85S$.

We consider the same scattering problem as in Sect. 4.1.2, but this time using four different numbers of equivalent monopoles ($N_e = 152, 294, 602, 1238$) located on the surface of a sphere with radius $R_e = 0.8$ m. Figure 4.4 shows the variation of the condition numbers (4.4a) and the surface error (4.4b) with the frequency.

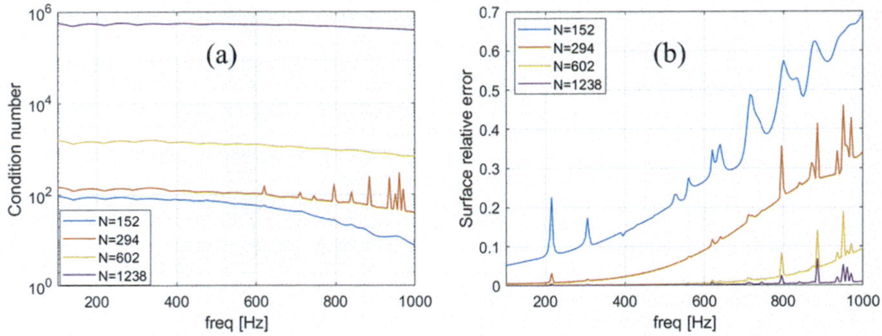

Fig. 4.4 Condition number and surface error for conformal surface

For a fixed N_e, the condition number is almost constant over all frequencies, and the values increase with the number of monopoles. In contrast, the surface errors increase with the frequency for a fixed N_e and decrease for increasing N_e.

Figure 4.5 compares the scattered pressures at points on a circle of radius 3 m in the XZ-plane. The agreement between simulations and exact values is very good for all four distributions at 250 Hz. For 500 Hz, the first distribution with 152 monopoles shows bigger differences with the exact values. For 750 Hz, only the last two distributions agree well with the analytical solution. Finally, at 1000 Hz, only the last distribution with 1236 monopoles provides accurate results. The results indicate that the proper number of monopoles should increase with the frequency.

4.1.4 Sound Radiation of a Loudspeaker Box

In this last example, a more realistic object was considered. It is a circular membrane of radius 0.04 m embedded in a rectangular rigid box with dimensions 0.2 m × 0.2 m × 0.3 m. The membrane oscillates harmonically with a normal velocity of $v_n = 1$ m/s. The surface mesh is made of 2422 quad elements. As in the previous example, we choose a set of monopoles on an interior conformal surface. Figure 4.6 shows the surface mesh, the normal velocity distribution and the position of the equivalent monopoles. We have performed three simulations with 247, 529 and 1059 monopoles.

To evaluate the accuracy of the results of the ESM, a BEM simulation was performed, and its results are taken as reference. The comparison is presented in Fig. 4.7. The surface errors due to the three monopole distributions are high at all frequencies (Fig. 4.7a), with a minimum of 60% and a maximum of near 85% at 1000 Hz. It seems that the jump of the normal velocity from zero at almost the whole surface to 1 m/s at the small circular membrane is difficult to reproduce with monopoles that produce a continuous spatial velocity distribution. The condition numbers increase with the number of monopoles as expected, reaching 10^{10}, but it

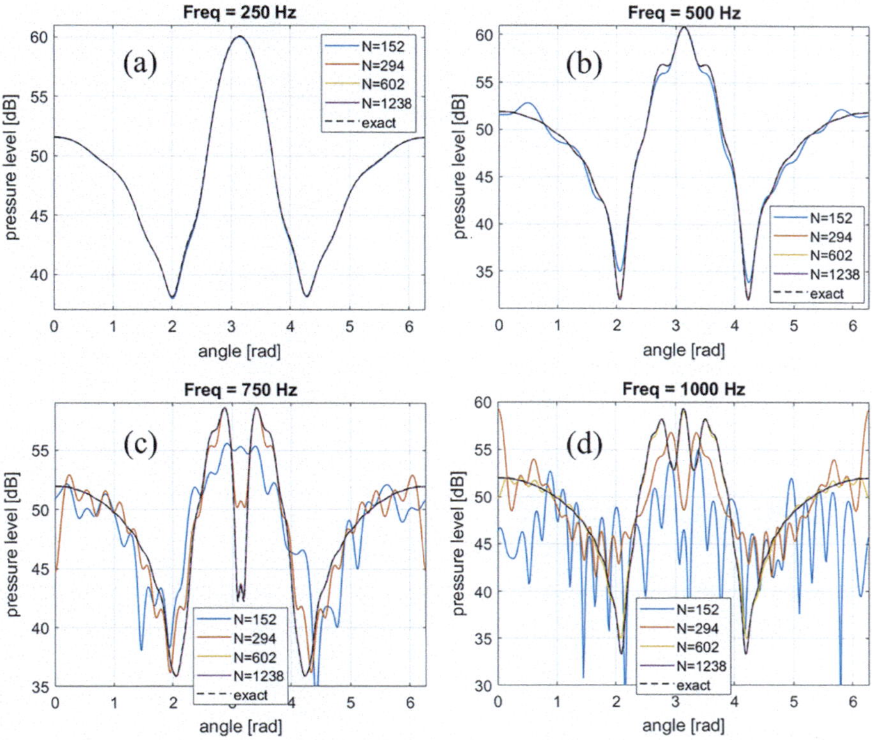

Fig. 4.5 Comparison of the scattered pressure around the sphere

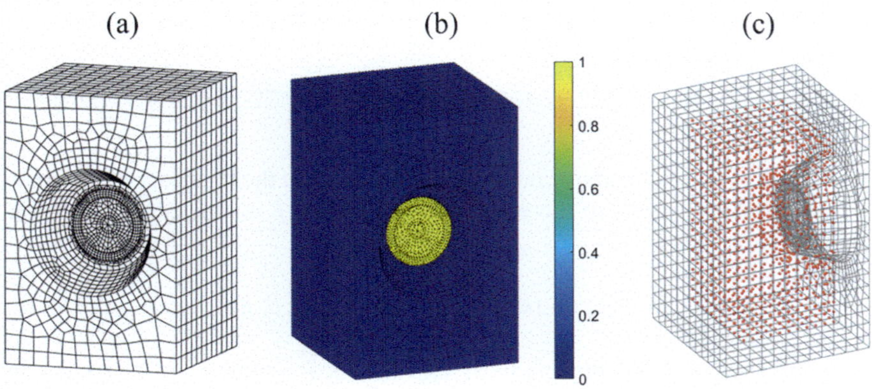

Fig. 4.6 Loudspeaker box: **a** surface mesh; **b** velocity distribution and **c** position of equivalent monopoles

is still manageable (Fig. 4.7b). The ESM underestimates the sound power at low frequencies, especially the apparent resonance around 1000 Hz. The difference is not greater than 5 dB (Fig. 4.7c). There is not much difference between the results from the ESM, which indicates that the increase in the number of equivalent sources does not improve the accuracy much. The sound pressure was computed at 360 field points located on a circle of radius 3 m placed on the xz-plane. At these points, the pressure reasonably matches the maximum value, but there are noticeable differences away from the main lobe. The increase in number of monopoles does not improve the accuracy either (Fig. 4.7d, e). In the curve of the sound power computed by the BEM, we can observe one big fall at around 1500 Hz, which shows that for exterior problems, the BEM cannot provide good results at certain frequencies known as "critical frequencies" that correspond to resonances of the interior volume delimited by S. Such problems can be avoided with the ESM.

4.2 ESM in Time Domain

In the same way as in the frequency domain approach, the sound pressure is modeled as the sum of the contribution of all equivalent sources

$$p(\vec{r},t) = \sum_{i=1}^{N} c_i(\tau)\Psi\left(\vec{r},\vec{r}_i',t,\tau\right), \tag{4.13}$$

where t is the reception time, τ the emission time, c_i is the time-dependent strength and Ψ is the source function. In the time domain, the source function is given by

$$\Psi\left(\vec{r},\vec{r}_i',t,\tau\right) = \frac{1}{4\pi R\left(\vec{r},\vec{r}_i',t,\tau\right)}.$$

A time-dependent boundary condition $f(t)$ is known at the boundary of the body. To compute the strengths of the equivalent sources c_i, f is defined at a set of K equidistant time steps.

$$t^k = t_{\min} + (k-1)\Delta t, \quad k = 1, 2, \ldots, K. \tag{4.14}$$

Usually, the reception times are fixed (e.g., in a measurement), but we are free to choose the emission times. Therefore, we choose a set of J equidistant time steps

$$\tau^j = \tau_{\min} + (j-1)\Delta\tau, \quad j = 1, 2, \ldots, J, \tag{4.15}$$

where the strengths of the equivalent sources will be calculated. Usually, the same time resolution is chosen $\Delta\tau = \Delta t$. The present method will only consider monopoles as equivalent sources and two different cases: (a) stationary bodies and (b) moving bodies.

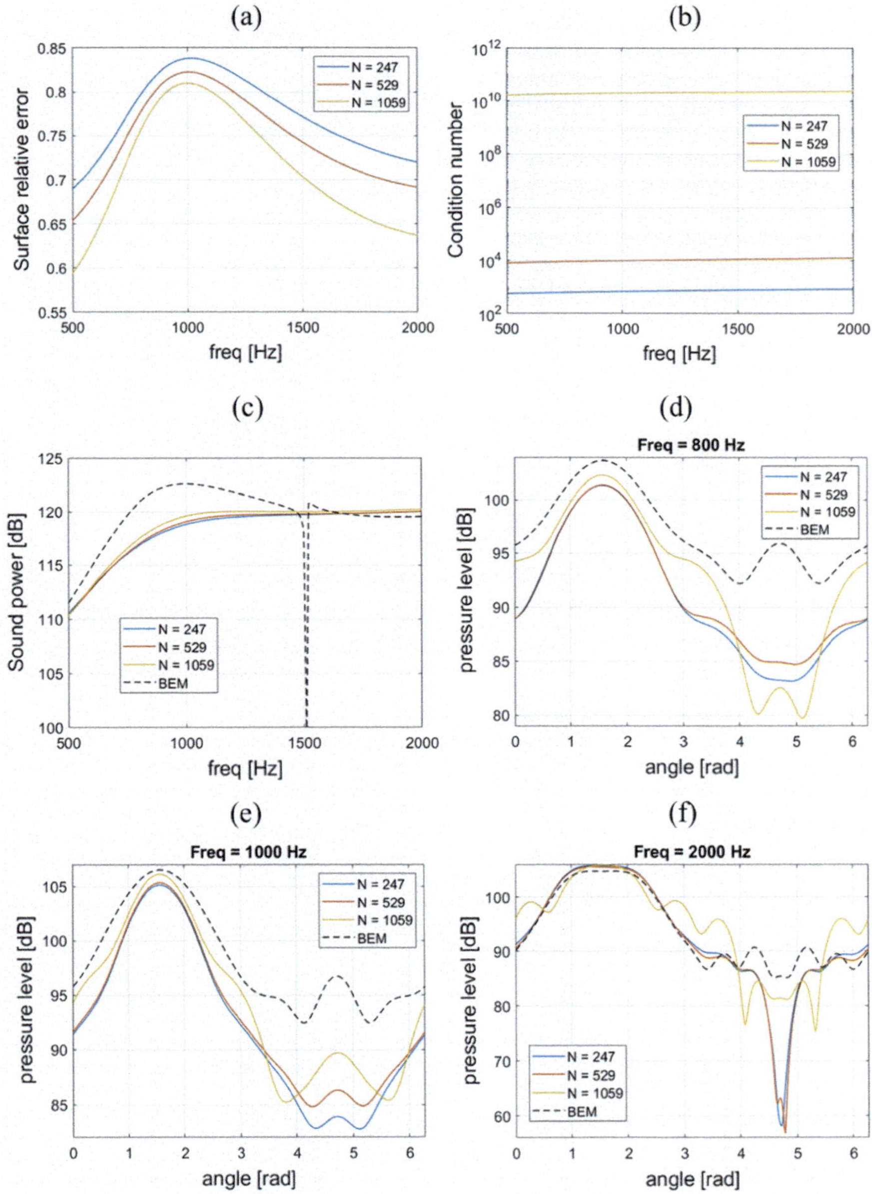

Fig. 4.7 Results of the ESM simulation of the loudspeaker box

4.2.1 Stationary Bodies

Considering again M collocation points and N equivalent sources, there are $N \times M$ constant distances $R_{\ell i}$ from all collocation points to all equivalent sources, i.e., for each reception time t^k, there are $N \times M$ emission times $\tau_{\ell i}^k = t^k - R_{\ell i}/c$. Hence, for the reception time t^k, we can express the sound pressure as

$$p\left(\vec{r}_\ell, t^k\right) = \sum_{i=1}^{N} \frac{c_i\left(\tau_{\ell i}^k\right)}{4\pi R_{\ell i}}, \tag{4.16}$$

and its normal derivative as

$$\frac{\partial p}{\partial n_\ell}\left(\vec{r}_\ell, t^k\right) = -\frac{1}{4\pi} \sum_{i=1}^{N} \left(\frac{\dot{c}_i\left(\tau_{\ell i}^k\right)}{c R_{\ell i}} + \frac{c_i\left(\tau_{\ell i}^k\right)}{R_{\ell i}^2} \right) \frac{\vec{R}_{\ell i} \cdot \hat{n}_\ell}{R_{\ell i}}, \tag{4.17}$$

with $\dot{c}_i(\tau) = \frac{\partial c_i(\tau)}{\partial \tau}$ and $\hat{n}_\ell$ the normal unit vector.

Since the emission times $\tau_{\ell i}^k$ do not usually match the emission times τ^j, the strengths of the sources need to be interpolated at those discrete time steps. The simplest interpolation is the linear one (Fig. 4.8). So, $c_i\left(\tau_{\ell i}^k\right)$ is approximated as

$$c_i\left(\tau_{\ell i}^k\right) = \sum_{n=1}^{J} \phi_n\left(\tau_{\ell i}^k\right) \hat{c}_i^n, \; \hat{c}_i^n = c_i\left(\tau^n\right), \tag{4.18}$$

and its derivative as

$$\dot{c}_i\left(\tau_{\ell i}^k\right) = \sum_{n=1}^{J} \dot{\phi}_n\left(\tau_{\ell i}^k\right) \hat{c}_i^n, \tag{4.19}$$

with

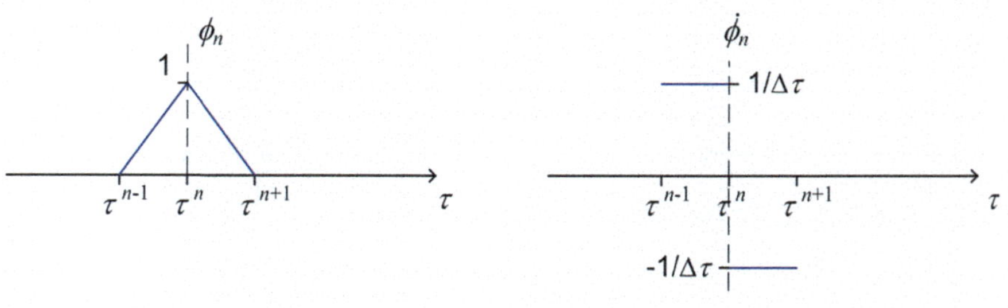

Fig. 4.8 Interpolation function and its derivative

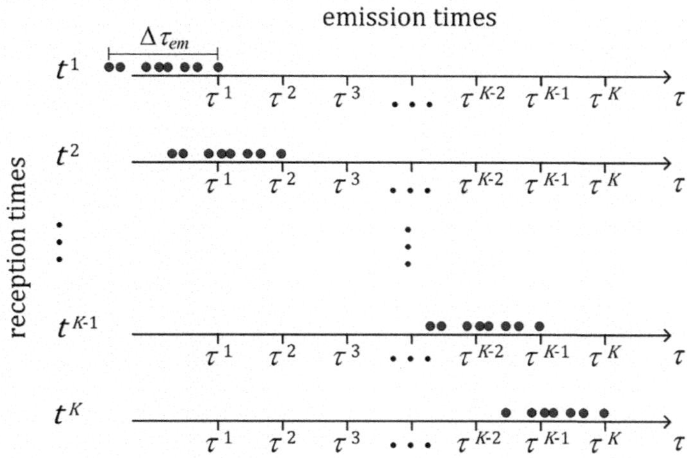

Fig. 4.9 Relation between emission and reception times

$$\phi_n(\tau) = \begin{cases} \frac{\tau-\tau^{n-1}}{\tau^n-\tau^{n-1}}, & \tau^{n-1} < \tau < \tau^n \\ \frac{\tau^{n+1}-\tau}{\tau^{n+1}-\tau^n}, & \tau^n < \tau < \tau^{n+1} \end{cases}, \quad \dot{\phi}_n(\tau) = \begin{cases} \frac{1}{\Delta\tau}, & \tau^{n-1} < \tau < \tau^n \\ \frac{-1}{\Delta\tau}, & \tau^n < \tau < \tau^{n+1} \end{cases}.$$

Inserting (4.18) and (4.19) into (4.16) and (4.17), we obtain

$$p(\vec{r}_\ell, t^k) = \frac{1}{4\pi}\sum_{i=1}^{N}\sum_{n=1}^{J}\frac{\phi_n(\tau_{\ell i}^k)}{R_{\ell i}}\hat{c}_i^n, \tag{4.20}$$

$$\frac{\partial p}{\partial n_\ell}(\vec{r}_\ell, t^k) = -\frac{1}{4\pi}\sum_{i=1}^{N}\sum_{n=1}^{J}\left(\frac{\dot{\phi}_n(\tau_{\ell i}^k)}{cR_{\ell i}} + \frac{\phi_n(\tau_{\ell i}^k)}{R_{\ell i}^2}\right)\frac{\vec{R}_{\ell i}\cdot\hat{n}_\ell}{R_{\ell i}}\hat{c}_i^n. \tag{4.21}$$

The emission times $\tau_{\ell i}^k$ for the reception time t^k are distributed within a time interval $\Delta\tau_{em}$ that covers a certain number of emission times τ^j. Since the distances between collocation points and equivalent sources are constant, by increasing the reception time by Δt, we shift the interval $\Delta\tau_{em}$ also by Δt. So, if we choose $\tau_{\min} = t_{\min} - R_{\ell i}/c$, we can ensure that all emission times $\tau_{\ell i}^k$ will lie before τ^k $\left(\tau_{\ell i}^k \leq \tau^k\right)$ and the sum in (4.18) will be limited to $n = k$.

$$c_i(\tau_{\ell i}^k) = \sum_{n=1}^{k}\phi_n(\tau_{\ell i}^k)\hat{c}_i^n. \tag{4.22}$$

Equations (4.20) or (4.21) must satisfy the boundary condition at all collocation points. In most cases, the normal derivative of the pressure is known, $f = \frac{\partial p_s}{\partial n}$. Equation (4.21) becomes

$$-\frac{1}{4\pi}\sum_{i=1}^{N}\sum_{n=1}^{k}\left(\frac{\dot{\phi}_n\left(\tau_{\ell i}^k\right)}{cR_{\ell i}}+\frac{\phi_n\left(\tau_{\ell i}^k\right)}{R_{\ell i}^2}\right)\frac{\vec{R}_{\ell i}\cdot\hat{n}_\ell}{R_{\ell i}}\hat{c}_i^n=f\left(\vec{r}_\ell,t^k\right),\quad \ell=1,2,\ldots,M. \tag{4.23}$$

The sum over n is divided into two terms, the first with the sum from $n = 1, \ldots, k-1$, and the second for $n = k$. The first term is then moved to the right side

$$\frac{1}{4\pi}\sum_{i=1}^{N}\left(\frac{\dot{\phi}_k\left(\tau_{\ell i}^k\right)}{cR_{\ell i}}+\frac{\phi_k\left(\tau_{\ell i}^k\right)}{R_{\ell i}^2}\right)\frac{\vec{R}_{\ell i}\cdot\hat{n}_\ell}{R_{\ell i}}\hat{c}_i^k=-f\left(\vec{r}_\ell,t^k\right)$$
$$-\frac{1}{4\pi}\sum_{i=1}^{N}\sum_{n=1}^{k-1}\left(\frac{\dot{\phi}_n\left(\tau_{\ell i}^k\right)}{cR_{\ell i}}+\frac{\phi_n\left(\tau_{\ell i}^k\right)}{R_{\ell i}^2}\right)\frac{\vec{R}_{\ell i}\cdot\hat{n}_\ell}{R_{\ell i}}\hat{c}_i^n \tag{4.24}$$

Equation (4.24) indicates that the strengths of the equivalent sources at emission time τ^k can be obtained from the strengths of the sources at the previous $k-1$ time steps and the boundary condition at reception time t^k.

Denoting the source functions $\psi_{\ell i}^{nk}$ and the right-hand side functions f_ℓ^k as

$$\psi_{\ell i}^{kn}=\left(\frac{\dot{\phi}_n\left(\tau_{\ell i}^k\right)}{cR_{\ell i}}+\frac{\phi_n\left(\tau_{\ell i}^k\right)}{R_{\ell i}^2}\right)\frac{\vec{R}_{\ell i}\cdot\hat{n}_\ell}{R_{\ell i}},\quad f_\ell^k=f\left(\vec{r}_\ell,t^k\right). \tag{4.25}$$

we can build the matrices Ψ^{kn} with elements $\psi_{\ell i}^{kn}$, the vectors F^k with elements f_ℓ^k and the vector of strengths C^k with elements $\hat{c}_i^k$.

$$\Psi^{kn}=\begin{pmatrix}\psi_{11}^{kn} & \psi_{12}^{kn} & \cdots & \psi_{1N}^{kn}\\ \psi_{21}^{kn} & \psi_{22}^{kn} & \cdots & \psi_{2N}^{kn}\\ \vdots & \vdots & \ddots & \vdots\\ \psi_{M1}^{kn} & \psi_{M2}^{kn} & \cdots & \psi_{MN}^{kn}\end{pmatrix},\quad F^k=\begin{pmatrix}f_1^k\\ f_2^k\\ \vdots\\ f_M^k\end{pmatrix},\quad C^k=\begin{pmatrix}\hat{c}_1^k\\ \hat{c}_2^k\\ \vdots\\ \hat{c}_M^k\end{pmatrix} \tag{4.26}$$

The iterative procedure will run as follows:

$$\begin{aligned}\Psi^{11}C^1&=F^1,\\ \Psi^{22}C^2&=F^2-\Psi^{21}C^1,\\ \Psi^{33}C^3&=F^3-\Psi^{32}C^2-\Psi^{31}C^1,\\ \Psi^{44}C^4&=F^4-\Psi^{43}C^3-\Psi^{42}C^2-\Psi^{41}C^1,\\ &\vdots\\ \Psi^{JJ}C^J&=F^J-\Psi^{JJ-1}C^{J-1}-\Psi^{JJ-2}C^{J-2}-\cdots-\Psi^{J2}C^2-\Psi^{J1}C^1,\end{aligned}$$

which can be summarized as

$$\Psi^{kk}C^k = F^k - \sum_{n=1}^{k-1} \Psi^{kn}C^n. \tag{4.27}$$

As has been seen previously, the number of time steps needed to compute the next time step is limited and determined by Δt_{em} and is the same for all receiving time steps. This can be approximated by

$$n_{\Delta\tau} = ceil\left(\frac{\max(R_{\ell i}) - \min(R_{\ell i})}{c\Delta\tau}\right). \tag{4.28}$$

So, the number of terms needed in the sum can be reduced, and (4.27) becomes

$$\Psi^{kk}C^k = F^k - \sum_{n=n_+}^{k-1} \Psi^{kn}C^n, \quad n_+ = \max(k - n_{\Delta\tau}, 1). \tag{4.29}$$

Since the distance between sources and receivers is the same at all times, it can be shown that for an integer U, such that $U > j$ and $U > k$.

$$\Psi^{jk} = \Psi^{(j-U)(k-U)}.$$

By taking $U = k - 1 - n_{\Delta\tau}$, (4.29) reads

$$\Psi^{(n_{\Delta\tau}+1)(n_{\Delta\tau}+1)}C^k = F^k - \sum_{n=n_+}^{k-1} \Psi^{(n_{\Delta\tau}+1)(n_{\Delta\tau}+1-(k-n))}C^n.$$

If we make the change of variable $\mu = k - n$, (4.29) can be finally rewritten as

$$\Psi^{(n_{\Delta\tau}+1)(n_{\Delta\tau}+1)}C^k = F^k - \sum_{\mu=1}^{n_-} \Psi^{(n_{\Delta\tau}+1)(n_{\Delta\tau}+1-\mu)}C^{k-\mu}, \tag{4.30}$$

with $n_- = \min(n_{\Delta\tau}, k - 1)$.

We can see that it is not necessary to calculate all $J \times K$ matrices Ψ, but only $n_{\Delta\tau} + 1$ and only $\Psi^{(n_{\Delta\tau}+1)(n_{\Delta\tau}+1)}$ needs to be inverted.

The number of collocation points M is usually larger than the number of equivalent sources N; therefore, the series of Eq. (4.30) has no unique solution, and they have to be solved in the least-square sense.

In the frequency domain, one matrix has to be inverted at each frequency and the results at that particular frequency are independent of the results at other frequencies. So despite the condition number may be large, reasonable results can still be obtained, as has been seen in Sect. 4.1. But in the time domain, the results of one-time step affect the results of later times. Hence, the method is more sensitive to the ill-conditioning of the matrices since errors at early times can be transferred to later times and amplified.

Therefore, the time domain ESM most definitely requires the application of some type of regularization.

4.2.2 Regularization Techniques

The least-square solution without any regularization of the linear system $Ax = b$ is obtained by solving the minimization problem

$$\min_x \|Ax - b\|_2.$$

Applying the singular value decomposition (SVD) to the matrix A, the solution vector is given in (4.8).

$$x_{lsq} = V_c S_c^{-1} \left(U_c^T b \right) = \sum_{m=1}^{Q} \frac{\left(u_m^T b \right)}{\sigma_m} v_m. \tag{4.31}$$

The problems arise when there are many eigenvalues of A near zero, which is common in inverse problems like the ESM. The errors in the directions of the small eigenvalues are magnified and surpass the information provided by the larger eigenvalues. Hence, the effect of the regularization will be to reduce or eliminate the contribution of the small eigenvalues. Following this idea, the regularized solution should be expressed as

$$x_{reg} = \sum_{m=1}^{Q} f_m \frac{\left(u_m^T b \right)}{\sigma_m} v_m, \tag{4.32}$$

where f_m are the so-called "filter factors" with values $0 \leq f_m \leq 1$. The filter factors are expected to have values near one for the larger eigenvalues and close to zero for the small eigenvalues.

Here, we will present three techniques: (a) Tikhonov regularization; (b) truncated SVD and (c) conjugated gradient algorithm.

4.2.2.1 Tikhonov Regularization

The regularized solution x_λ is the minimizer of the weighted combination of the residual norm and the solution norm:

$$x_\lambda = \min_x \left(\|Ax - b\|_2^2 + \lambda^2 \|x\|_2^2 \right),$$

where λ is the regularization parameter that controls how much weight is given to the minimization of $\|x\|_2$ in relation to the minimization of the residual norm $\|Ax - b\|_2$. A large value of λ favors a small solution norm at the cost of a large residual norm, while a small value of λ prioritizes the residual norm over the solution norm. The regularized solution is given by

$$x_\lambda = \sum_{m=1}^{Q} \frac{\sigma_m \left(u_m^T b\right)}{\sigma_m^2 + \lambda^2} v_m. \tag{4.33}$$

The corresponding filter factors are then $f_m = \frac{\sigma_m^2}{\sigma_m^2 + \lambda^2}$.

One can easily see that for $\lambda = 0$, the least-square solution is recovered ($f_m = 1$), and for $\sigma_m \ll \lambda$, the contribution of the corresponding terms in the summation to the final solution is negligible ($f_m = 0$).

The norms of the residual and the solution can be expressed in terms of the filter factors as

$$\|Ax_\lambda - b\|_2^2 = \sum_{m=1}^{Q} \left((1 - f_m) u_m^T b\right)^2, \tag{4.34}$$

$$\|x_\lambda\|_2^2 = \sum_{m=1}^{Q} \left(f_m \frac{u_m^T b}{\sigma_m}\right)^2. \tag{4.35}$$

4.2.2.2 Truncated SVD

The technique consists of simply truncating the summation in (4.31) at an upper limit $k < Q$, before the small eigenvalues start to dominate

$$x_{tr} = \sum_{m=1}^{k} \frac{\left(u_m^T b\right)}{\sigma_m} v_m. \tag{4.36}$$

The corresponding filter factors are then $f_m = \begin{cases} 1 & m \le k \\ 0 & m > k \end{cases}$.

The norms of residual and the solution are given by

$$\|Ax_k - b\|_2^2 = \sum_{m=k+1}^{Q} \left(u_m^T b\right)^2, \tag{4.37}$$

$$\|x_k\|_2^2 = \sum_{m=1}^{k} \left(\frac{u_m^T b}{\sigma_m}\right)^2. \tag{4.38}$$

4.2.2.3 Conjugate Gradient Method CG

The CG algorithm is an iterative procedure used to solve a sparse system of equations with a symmetric positive definite coefficients matrix. Interestingly, when applied to the modified system $A^T Ax = A^T b$, the low-frequency components of the solution tend to converge faster than the high-frequency components. Hence, the number of iterations can act as a regularization parameter in a similar way as the upper limit k in the truncated SVD since a small number of iterations can filter out the contribution of the higher frequency components. The explicit form of the filter factors is not as straightforward as in the previous two cases, and its deduction exceeds the scope of this study.

A detailed description of the three regularization methods mentioned here can be found in the manual for the Matlab package of regularization tools from Hansen [2].

4.2.2.4 Choice of Optimum Regularization Parameter

The accuracy of the regularized solution depends on a proper choice of the regularization parameter. A proper value of the regularization parameter normally provides a fair balance in keeping the residual norm and the solution norm small. We present two methods proposed to determine the optimum parameter: (a) the L-curve criterion and (b) the generalized cross validation (GCV).

(a) The L-Curve Criterion

Since the norms of the residual and the solution are sought to be both small, it is convenient to look for a plot that displays both at the same time. By denoting $\rho = \|Ax - b\|_2$ and $\eta = \|x\|_2$, the L-curve is a log–log plot of ρ versus η. Its name comes from the fact that for ill-posed problems like the ESM, the L-curve usually has the shape of an L with a distinct corner that separates the horizontal from the vertical part of the curve. The vertical part shows the region that is more sensitive to changes in the solution norm (regularization error), while the horizontal part shows the region that is more sensitive to changes in the residual norm (perturbation error). Therefore, the corner of the curve corresponds to a balance between the minimization of both norms, and the associated regularization parameter can be considered an optimum one.

To find the corner of the curve, Hansen and O'Leary [1] suggest finding the point of maximum curvature. The L-curves for the Tikhonov and the truncated SVD methods can be easily constructed since ρ and η are explicitly given in (4.34) and (4.38). In the former method, a smooth parametric curve in λ can be generated, while in the latter method, the curve is discrete and has only as many points as eigenvalues. In the CG method, we can still construct an L-curve by computing ρ and η for each iteration, which will be discrete with a number of points equal to the number of iterations.

For the Tikhonov regularization, the curvature κ of the L-curve can be obtained by computing the formula

$$\kappa(\lambda) = \frac{\hat{\rho}\prime\hat{\eta}'' - \hat{\eta}\prime\hat{\rho}''}{\left((\hat{\rho}\prime)^2 + (\hat{\eta}\prime)^2\right)^{3/2}}, \tag{4.39}$$

where $'$ denotes differentiation with respect to λ, $\hat{\rho} = \log \rho$ and $\hat{\eta} = \log \eta$.

For the other two methods, the L-curve is not a parametric function but a set of discrete points. Therefore, Hansen and O'Leary [1] suggest first to perform a smoothing of the original L-curve and then fit the smoothed curve to cubic splines which are twice differentiable, so that the curvature can be determined using (4.39).

(b) The Generalized Cross Validation

The idea of the cross validation is that if one arbitrary data point b_i is left out, and the reduced problem is solved, the estimation of b_i computed with the regularized solution should be a good one. Additionally, the value of the optimum regularization parameter should be independent of an orthogonal transformation of b. The regularization parameter that balances the regularization and perturbation errors is the one that minimizes the function [6].

$$G = \frac{\left\|Ax_{reg} - b\right\|_2^2}{\left(M - N - \sum_{m=1}^{Q}(1 - f_m)\right)^2}. \tag{4.40}$$

In the present study, (4.40) will be applied to the Tikhonov and the truncated SVD methods. Figure 4.10 shows the location of the optimum regularization parameter on the L-curve and the GCV curve.

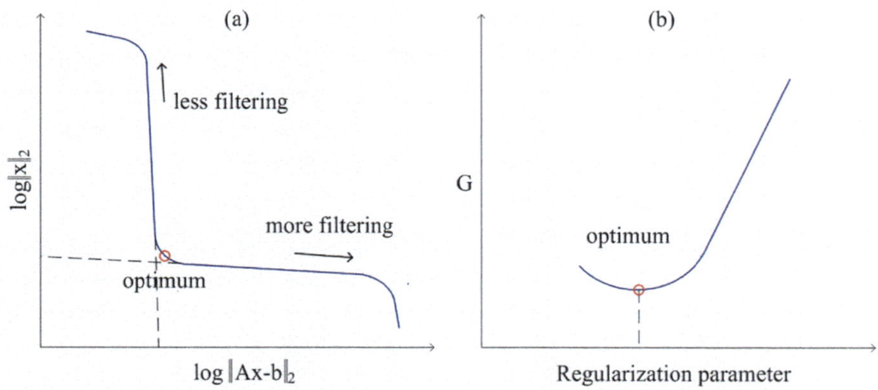

Fig. 4.10 Illustration of the curves used for the determination of the optimum regularization parameter, **a** L-curve; **b** GCV

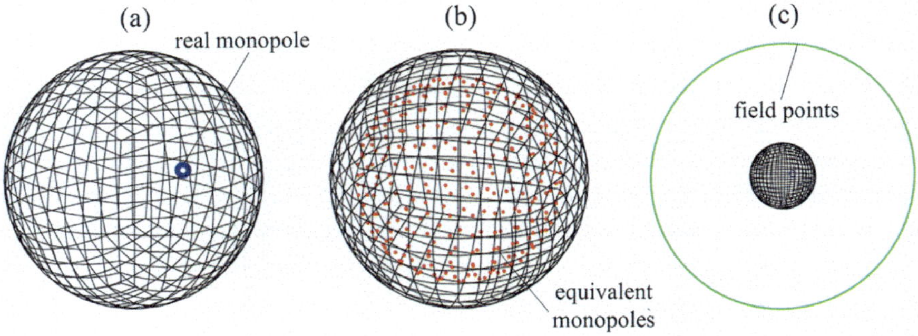

Fig. 4.11 Position of real and equivalent sources and field points

4.2.3 *Example: Radiation of a Monopole*

This example is equivalent to the monopole test in the frequency domain. Here, we consider a sphere of radius $R = 1$ m and center $(0, 0, 0)$ m, whose surface is vibrating with a normal velocity equal to the particle velocity of a monopole. The real monopole is located inside the sphere at the position $\vec{r}_q = (0.25, -0.3, 0.1)$ m.

The surface is discretized with 600 quad elements. The maximum size of the elements ensures that the rule of thumb of at least six elements per wavelength is fulfilled up to 300 Hz. The monopole radiates with N_f frequencies f_i within the interval [100, 300] Hz. The pressure and the normal velocity generated by the monopole are given by

$$p(\vec{r}, t) = \frac{1}{4\pi} \sum_{i=1}^{N_f} \frac{e^{j(\Omega_i t - k_i R)}}{R}, \quad R = \left|\vec{r} - \vec{r}_q\right|, \quad k_i = \frac{\Omega_i}{c} = \frac{2\pi f_i}{c},$$

$$\frac{\partial p}{\partial n}(\vec{r}, t) = -\frac{1}{4\pi} \frac{\vec{R} \cdot \hat{n}}{R} \sum_{i=1}^{N_f} \left(jk_i + \frac{1}{R}\right) \frac{e^{j(\Omega_i t - k_i R)}}{R}.$$

For the present computation, 240 reception time steps were considered with a resolution $\Delta t = 4.2 \times 10^{-4}$ s. We apply the ESM to simulate the sound radiation from the vibrating sphere. The equivalent monopoles are located inside the sphere over a conformal spherical surface of radius 0.8 m. To find the solution, we test the different regularization procedures explained in the previous section and evaluate their performance. In total, we conduct seven calculations:

(i) TSVD with a fixed truncation term for all time steps.
(ii) Tikhonov regularization with a fixed parameter λ for all time steps.
(iii) Tikhonov regularization with the parameter λ obtained at each time step from the corner of the L-curve.
(iv) Tikhonov regularization with the parameter λ obtained at each time step with the GCV technique.

(v) TSVD with a truncation term obtained at each time step from L-curve technique.
(vi) TSVD with a truncation term obtained at each time step using the GCV technique.
(vii) CG method applied to the least-square problem. The solution is obtained at each time step by finding the corner of the L-curve.

From the seven computations, only four provided a solution. In tests (iv), (v) and (vi), the errors in the strengths were amplified at each time step and jumped to infinity. Figure 4.12 shows the values of the optimum λ computed at each time step in tests (ii), (iii) and (iv). One can see that with the GCV method (green line), the values of λ are small at the first few time steps but then start growing, and after some steps oscillate between a large and a small value. At that stage, the procedure becomes unstable and the strengths of the equivalent sources increase and increase so that the iteration is stopped. The optimum values of λ obtained with the L-curve (red line) vary more at the beginning and end of the calculation, but remain relatively uniform during the rest of the time steps, at a value smaller than the fixed value in test (ii).

Figure 4.13 shows the variation of the optimum truncation terms for the TSVD for tests (i), (v) and (vi) and the CG method for test (vii). Tests (v) and (vi) (red and green lines) are stopped before even reaching half of the time steps since the strengths just increase at each time step. By the CG method, 50 iterations are performed at each time step, and the optimum number of iterations obtained from the L-curve remains low, less than 20 in the whole time interval.

Figure 4.14 shows the residual norm for the four tests that provide results. The method that provides the smaller residual is the TSVD with a constant truncation number for all time steps (test (ii)). This number results from setting the ratio between

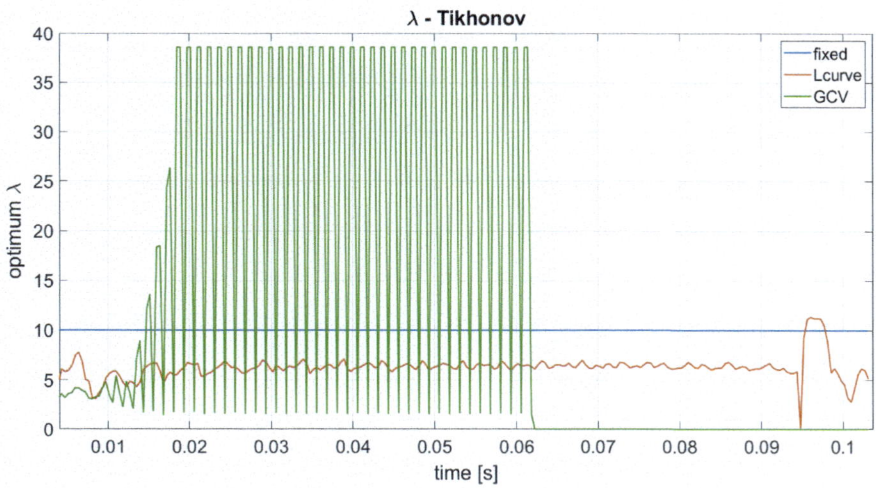

Fig. 4.12 Optimum value of the regularization parameter for the Tikhonov method

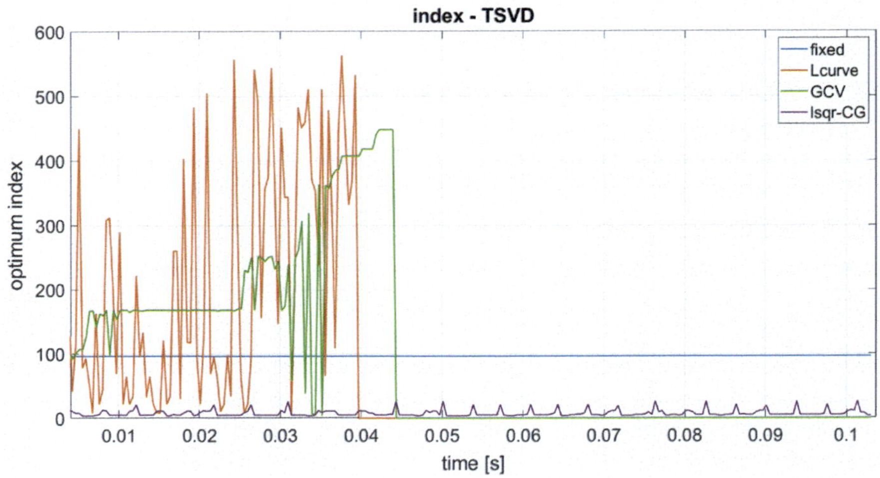

Fig. 4.13 Optimum truncation term for the TSVD and the CG method

the maximum eigenvalue to the smaller eigenvalue included $w = \frac{\sigma_{\max}}{\sigma_{\min,included}} = 0.3$. The second best results come from the Tikhonov method with constant λ (test i)).

In Fig. 4.15, we can see the solution norm for the four successful tests. The four methods provide similar norms. The method that provides the smaller solution norm is the Tikhonov method with fixed λ. The second best is the Tikhonov method with optimized λ. Finally, we compare in Fig. 4.16 the sound pressures at one field point computed with the different approaches with the analytical solution. The agreement is very good with all four approaches, but the approaches (i) and (ii) are slightly

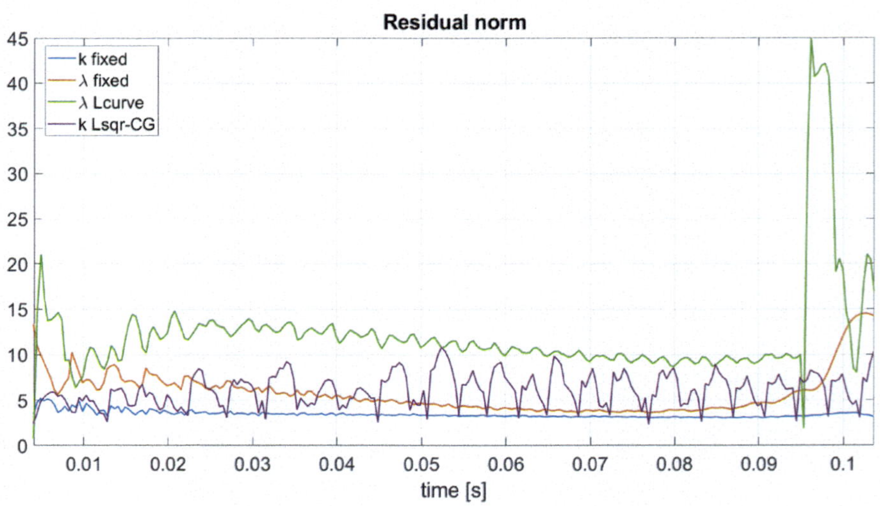

Fig. 4.14 Residual norm for the four successful tests

better. Since the iterative procedure uses previous time steps to calculate the next one, it seems more suitable to use one regularization parameter for all time steps than look for the optimum one at each step. In the next examples, we will use the approaches (i) or (ii).

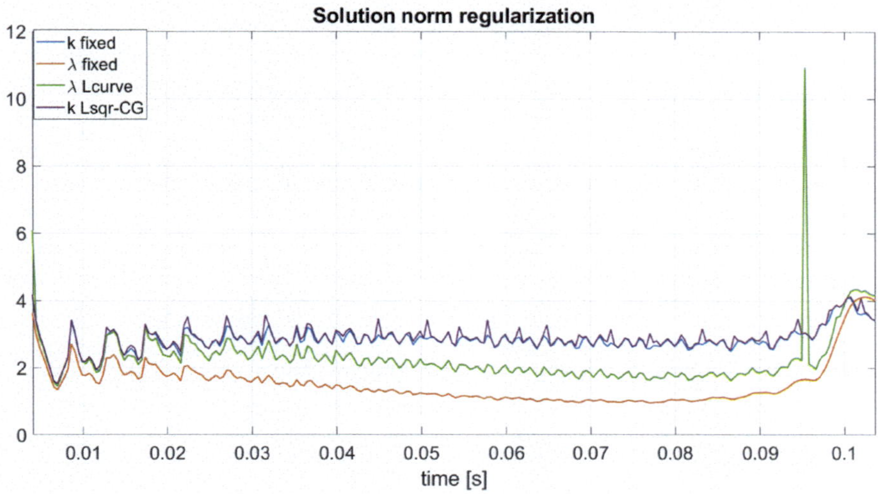

Fig. 4.15 Solution norm for the four successful tests

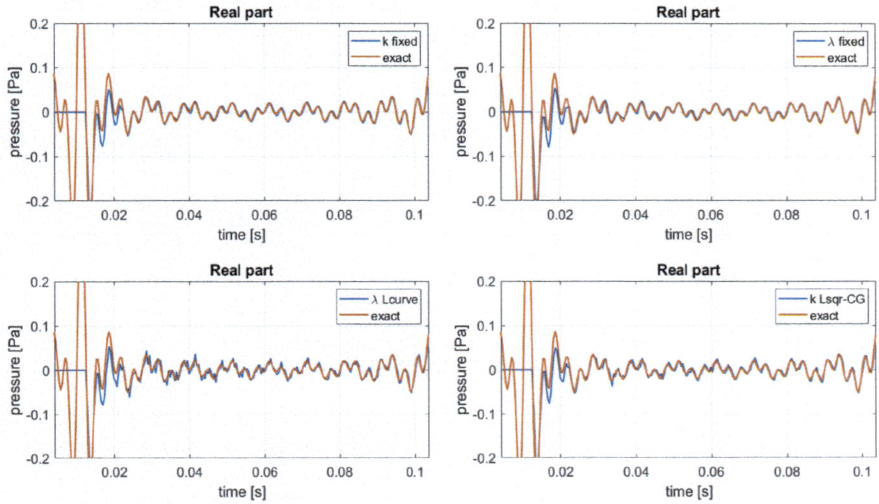

Fig. 4.16 Comparison of pressure at one field point obtained with the ESM with the analytical solution

4.2.4 Example: Scattering of a Rigid Sphere in Front of a Point Source

In Sect. 4.1.2, this case was simulated in the frequency domain and was chosen because it has an analytical solution. The analytical formulas for the incident and scattered pressures in the frequency domain are given in expressions (4.10) and (4.11). The expressions in the time domain are obtained just by multiplying the frequency expressions by a factor $e^{j\Omega t}$, and Ω is the circular frequency radiated by the point source. The boundary condition for a rigid sphere is already given in (4.12) and imposes that the normal velocity vanishes. Since, for harmonic sound waves, the velocity is proportional to the pressure gradient, we set the normal derivative of the pressure to zero on the surface, $\frac{\partial p}{\partial n} = 0 \rightarrow \frac{\partial p_{sc}}{\partial n} = -\frac{\partial p_{inc}}{\partial n}$.

For the simulation, we consider the same time resolution, number of time steps and number of equivalent monopoles as the example in Sect. 4.2.3. The radiated frequency is f = 275 Hz. For the regularization, the TSVD with a constant ratio $w = 0.3$ is chosen.

Figure 4.17 shows the strengths of all 600 equivalent sources in the emission time interval. Most equivalent sources contribute less to the scattered pressure. The periods of the oscillation are clearly seen for the few sources that contribute more to the scattered field.

A comparison of the simulated and exact scattered pressure at a point field is presented in Fig. 4.18. The field point is located on the z-axis, at a distance of 4 m from the center of the sphere. The agreement is excellent, the simulated pressure shows the maximum difference at the first time steps, and after that, the relative errors remain below 10% at the rest of the time steps (Fig. 4.19).

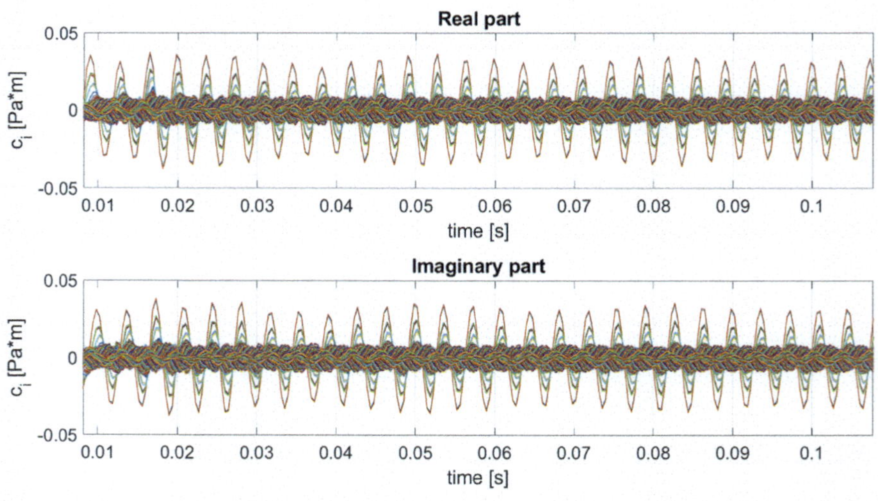

Fig. 4.17 Strengths c_i of all equivalent sources versus time

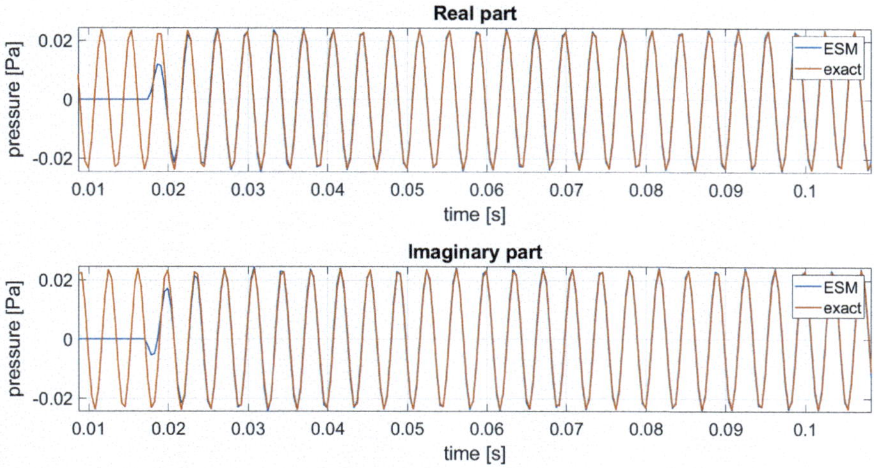

Fig. 4.18 Simulated and analytical scattered pressure at one field point

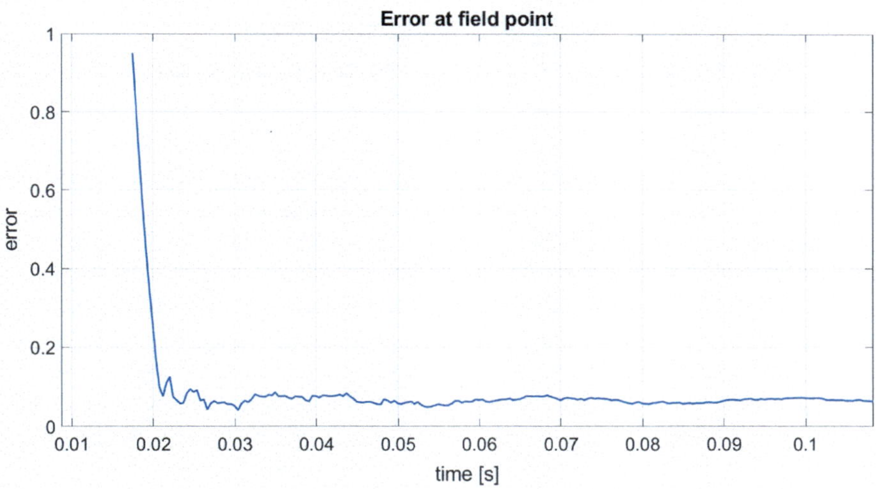

Fig. 4.19 Relative error of the scattered pressure at the field point

The directivity of the scattered sound is obtained by calculating the pressure on a circle of radius 4 m with its center in the center of the sphere and computing the RMS value over the time steps.

Figure 4.20 plots a comparison of the RMS values of the scattered pressure levels. The ESM is able to reproduce the main lobe of the radiation very well. The differences in the radiation away from the main lobe are larger but do not exceed the mark of 1 dB.

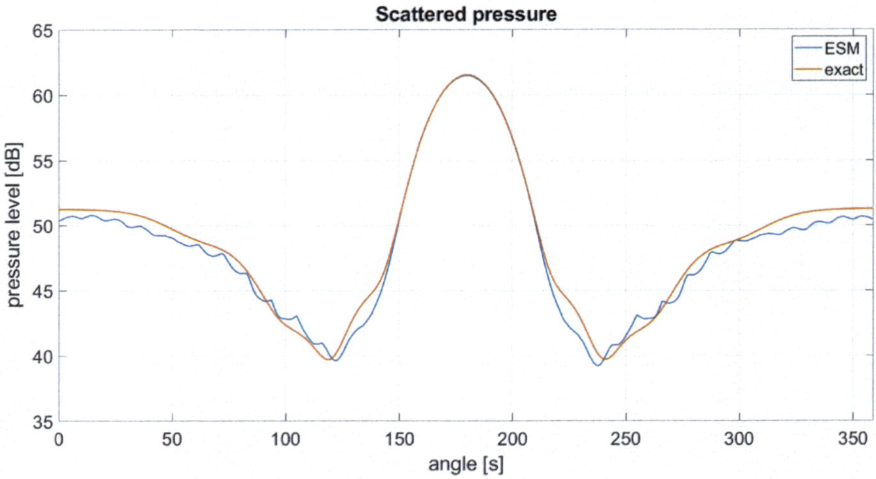

Fig. 4.20 Directivity of the scattered pressure

4.2.5 Example: Scattering of a Rigid Sphere Due to a Harmonically Radiating Moving Point Source

The true advantage of the time domain ESM is that it is able to deal with problems where sources emit transient signals or the sources are moving. Hence, in the third example, we take the monopole of the previous example and make it move with a constant velocity Mc parallel to the x-axis and pass by the rigid sphere. The difference between this case and the previous one is the incident wave. Hence, the previous computation can be repeated, changing only the normal derivative of the incident pressure on the surface of the sphere. This case has no analytical solution to evaluate the simulation, but the results obtained are very similar to those of Lee et al. [4].

To determine the incident wave, we consider that the point source is located at $\vec{r}_{q0}$ at $t = 0$. Hence, the position of the source at an arbitrary time t is given by $\vec{r}_q(t) = \vec{r}_{q0} + \vec{M}\,ct$. In Chap. 3, the expression for the velocity potential of a moving point source in the Lorentz space is given in Eq. (3.84). It was assumed that the volume flow $q(t)$ in m^3/s was known. In the present case, it is assumed that the point source has a known pressure strength $q(t) = Q_0 e^{j\omega t}$, Q_0 in $Pa\cdot m$. The sound pressure and its gradient in the real space are then written as

$$p(\vec{r}, t) = \frac{Q_0 e^{j\omega(t-R_e/c)}}{4\pi R_e(1 - M\cos\theta_e)}. \tag{4.41}$$

$$\nabla p(\vec{r}, t) = -\frac{Q_0 e^{j\omega(t-R_e/c)}}{4\pi R_e(1 - M\cos\theta_e)^2}\left(jk\vec{R}_e + \frac{(1-M^2)\vec{R}_e}{R_e(1 - M\cos\theta_e)} - \vec{M}\right), \tag{4.42}$$

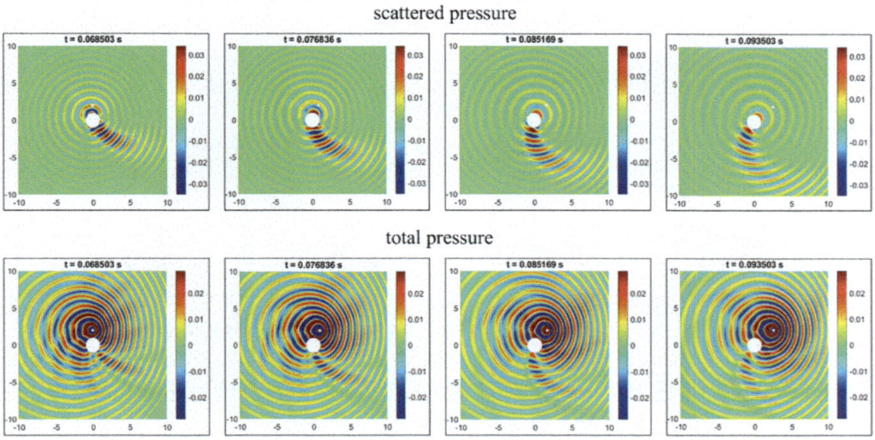

Fig. 4.21 Four snapshots of the sound of a moving monopole scattered by a rigid sphere

where $\vec{R}_e = \vec{r} - \vec{r}_q(\tau_e)$, $R_e = \left|\vec{R}_e\right|$, $\cos\theta_e = \frac{\vec{M}\cdot\vec{R}_e}{MR_e}$, and τ_e is the emission time. The emission time is the solution of the equation: $c(t-\tau_e) = \left|\vec{r} - \vec{r}_q(\tau_e)\right|$ and is given by

$$\tau_e = \frac{t - \vec{M}\cdot(\vec{r}-\vec{r}_{q0})/c - \sqrt{\left(t - \vec{M}\cdot(\vec{r}-\vec{r}_{q0})/c\right)^2 + (1-M^2)\left(\left|\vec{r}-\vec{r}_{q0}\right|^2/c^2 - t^2\right)}}{1-M^2}. \tag{4.43}$$

Figure 4.21 shows four snapshots of the sound pressure. The monopole (tiny white circle) moves from left to right with $M = 0.3$. The upper plots show only the scattered pressure, and the lower plots the total pressure. The main lobe of the scattered pressure moves in the opposite direction of the moving monopole and bends at some distance behind the sphere (big white circle) due to the motion of the source.

In Fig. 4.22 we can see the strengths of all 600 equivalent sources in the time and in the frequency domain. The strengths of the more relevant sources increase up to about 0.05 s and then decrease again. The maximum strength must occur when the monopole is at its closest distance to the sphere. Due to the motion of the monopole, the sound arriving to the sphere contains not only the frequency of the source but all frequencies within the interval $\left[\frac{f}{1+M}, \frac{f}{1-M}\right]$ because of the Doppler effect; therefore, the spectra of the equivalent sources have also those frequency components.

We also look at the spectra of three field points. The first point (Pt. 1) is located at a position where the monopole is always going away during the measurement time. The second point (Pt. 2) is located at a position on the opposite side of the source trajectory, i.e., the source is always getting closer. The third point (Pt. 3) is located near the middle of the trajectory so that the source gets closer half of the time, passes by, and then goes away the other half of the time. The spectra of the three points are plotted in Fig. 4.23.

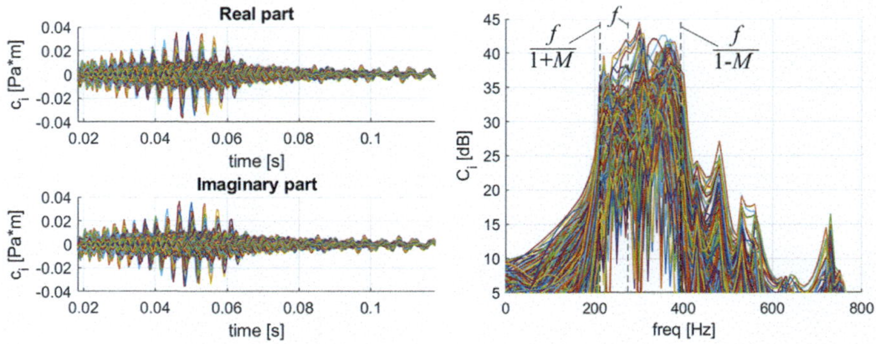

Fig. 4.22 Strengths of the equivalent sources in time domain (left) and in frequency domain (right)

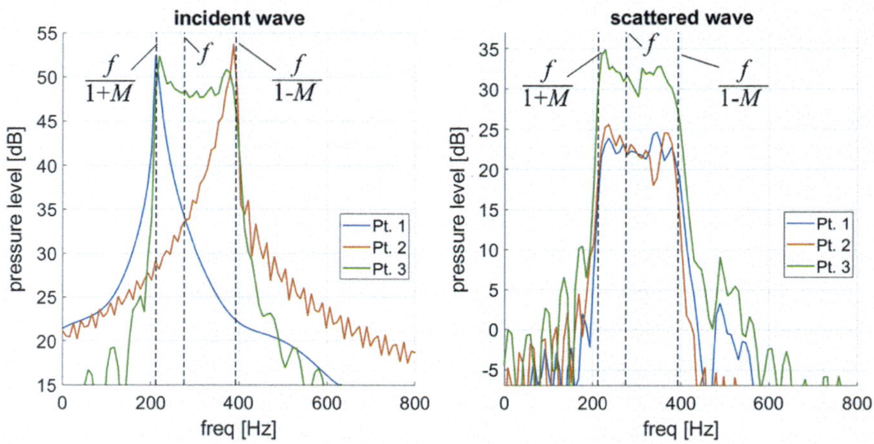

Fig. 4.23 Spectra of the sound pressure at three field points for the moving point source and stationary rigid sphere

The plot on the left shows the incident pressure. One can observe the differences between points 1 and 2. When the source is going away (Pt. 1), there is a frequency shift toward the lower limit frequency $\frac{f}{1+M}$, and when the source is always getting closer to the receiver (Pt. 2), the shift is toward the upper limit frequency $\frac{f}{1-M}$. For the receiver that sees the source pass by, the change in the frequency is to all frequencies within the whole Doppler interval.

The plot on the right shows the spectra of the scattered wave. There is practically no difference in the form of the spectra for all three points. Only the amplitude changes. The amplitude at the third point is higher than at the other two points that are equal. One can expect no difference between the three points since there is no relative motion between the equivalent sources and the field points.

4.2.6 *Moving Bodies*

We assume now that the body is not at rest but moves with Mach number $\vec{M}$. For simplicity, we consider a constant Mach number, i.e., the body carries out a rectilinear movement. The equivalent monopoles must, therefore, travel with the same velocity as the body. Since the collocation points are located on the surface, they also move with the same velocity. The sound pressure of the body is modeled as the contribution of the sound from all moving equivalent sources, Eq. (4.13), where the source functions are the functions of moving monopoles (Eq. 4.41). The pressure and its gradient at the collocation point $\vec{r}_\ell$ and receiving time t^k are written as

$$p(\vec{r}_\ell, t^k) = \sum_{i=1}^{N} \frac{c_i(\tau_{\ell i}^k)}{4\pi R_{\ell i}^k (1 - M \cos\theta_{\ell i}^k)}, \tag{4.44}$$

$$\begin{aligned} \nabla p(\vec{r}_\ell, t^k) = \frac{-1}{4\pi} \sum_{i=1}^{N} & \frac{\dot{c}_i(\tau_{\ell i}^k)}{cR_{\ell i}^k (1 - M \cos\theta_{\ell i}^k)^2} \frac{\vec{R}_{\ell i}^k}{R_{\ell i}^k} \\ & + \frac{c_i(\tau_{\ell i}^k)}{(R_{\ell i}^k)^2 (1 - M \cos\theta_{\ell i}^k)^2} \left(\frac{1 - M^2}{1 - M \cos\theta_{\ell i}^k} \frac{\vec{R}_{\ell i}^k}{R_{\ell i}^k} - \vec{M} \right), \end{aligned} \tag{4.45}$$

where $\vec{R}_{\ell i}^k = \vec{r}_\ell^s(t^k) - \vec{r}_i^q(\tau_{\ell i}^k)$, $\vec{r}_\ell^s(t) = \vec{r}_{\ell 0}^s + \vec{M}ct$, $\vec{r}_i^q(t) = \vec{r}_{i0}^q + \vec{M}ct$, $R_{\ell i}^k = \left|\vec{R}_{\ell i}^k\right|$, $\cos\theta_{\ell i}^k = \frac{\vec{M} \cdot \vec{R}_{\ell i}^k}{MR_{\ell i}^k}$, $M = \left|\vec{M}\right|$ and $\tau_{\ell i}^k$ is the emission time. The emission time is the solution of the equation: $c(t^k - \tau_{\ell i}^k) = \left|\vec{r}_\ell^s(t^k) - \vec{r}_i^q(\tau_{\ell i}^k)\right|$ and is given by

$$\tau_{\ell i}^k = t^k - \frac{1}{c} \left(\frac{\vec{M} \cdot (\vec{r}_{\ell 0}^s - \vec{r}_{i0}^q) + \sqrt{\left(\vec{M} \cdot (\vec{r}_{\ell 0}^s - \vec{r}_{i0}^q)\right)^2 + (1 - M^2)\left|\vec{r}_{\ell 0}^s - \vec{r}_{i0}^q\right|^2}}{1 - M^2} \right). \tag{4.46}$$

The expression in brackets in (4.46) is independent of the time. This was expected since equivalent sources and collocation points move with the same velocity. That means that an increase of Δt in the receiving time leads to an increase of Δt in the emitting time, and the similar iterative approach applied to stationary bodies can be used with moving bodies. All the considerations about the emission time discretization and the interpolation scheme for the strengths of equivalent sources explained for the stationary case are valid for the moving case. The strengths of the equivalent sources are obtained by matching a condition on the boundary. For a given normal pressure derivative on the boundary, f, the boundary condition for collocation point $\vec{r}_\ell$ and reception time t^k reads

$$-\frac{1}{4\pi}\sum_{i=1}^{N}\sum_{n=1}^{k}\left(\frac{\phi_n\left(\tau_{\ell i}^k\right)}{\left(R_{\ell i}^k\right)^2\left(1-M\cos\theta_{\ell i}^k\right)^2}\left(\frac{1-M^2}{1-M\cos\theta_{\ell i}^k}\frac{\vec{R}_{\ell i}^k\cdot\hat{n}_\ell}{R_{\ell i}^k}-\vec{M}\right)\right.$$
$$\left.+\frac{\dot{\phi}_n\left(\tau_{\ell i}^k\right)}{cR_{\ell i}^k\left(1-M\cos\theta_{\ell i}^k\right)^2}\frac{\vec{R}_{\ell i}^k\cdot\hat{n}_\ell}{R_{\ell i}^k}\right)\hat{c}_i^n=f\left(\vec{r}_\ell,t^k\right),\quad \ell=1,2,\ldots,M, \tag{4.47}$$

and after rearrangement of terms into the iteration form, it is expressed as

$$\frac{1}{4\pi}\sum_{i=1}^{N}\left(\frac{\phi_k\left(\tau_{\ell i}^k\right)}{\left(R_{\ell i}^k\right)^2\left(1-M\cos\theta_{\ell i}^k\right)^2}\left(\frac{1-M^2}{1-M\cos\theta_{\ell i}^k}\frac{\vec{R}_{\ell i}^k\cdot\hat{n}_\ell}{R_{\ell i}^k}-\vec{M}\right)\right.$$
$$\left.+\frac{\dot{\phi}_k\left(\tau_{\ell i}^k\right)}{cR_{\ell i}^k\left(1-M\cos\theta_{\ell i}^k\right)^2}\frac{\vec{R}_{\ell i}^k\cdot\hat{n}_\ell}{R_{\ell i}^k}\right)\hat{c}_i^k$$
$$=-\frac{1}{4\pi}\sum_{i=1}^{N}\sum_{n=1}^{k-1}\left(\frac{\phi_n\left(\tau_{\ell i}^k\right)}{\left(R_{\ell i}^k\right)^2\left(1-M\cos\theta_{\ell i}^k\right)^2}\left(\frac{1-M^2}{1-M\cos\theta_{\ell i}^k}\frac{\vec{R}_{\ell i}^k\cdot\hat{n}_\ell}{R_{\ell i}^k}-\vec{M}\right)\right.$$
$$\left.+\frac{\dot{\phi}_n\left(\tau_{\ell i}^k\right)}{cR_{\ell i}^k\left(1-M\cos\theta_{\ell i}^k\right)^2}\frac{\vec{R}_{\ell i}^k\cdot\hat{n}_\ell}{R_{\ell i}^k}\right)\hat{c}_i^n-f\left(\vec{r}_\ell,t^k\right). \tag{4.48}$$

From (4.48), one can deduce an iteration scheme similar to (4.30)

$$\Upsilon^{(n_{\Delta\tau}+1)(n_{\Delta\tau}+1)}C^k=F^k-\sum_{\mu=1}^{n_-}\Upsilon^{(n_{\Delta\tau}+1)(n_{\Delta\tau}+1-\mu)}C^{k-\mu}, \tag{4.49}$$

for $=1,2,\ldots,K$, with $\Upsilon^{kn}=\begin{pmatrix}\chi_{11}^{kn} & \chi_{12}^{kn} & \cdots & \chi_{1N}^{kn}\\ \chi_{21}^{kn} & \chi_{22}^{kn} & \cdots & \chi_{2N}^{kn}\\ \vdots & \vdots & \ddots & \vdots\\ \chi_{M1}^{kn} & \chi_{M2}^{kn} & \cdots & \chi_{MN}^{kn}\end{pmatrix}$ and

$$\chi_{\ell i}^{kn}=\frac{\phi_n\left(\tau_{\ell i}^k\right)}{\left(R_{\ell i}^k\right)^2\left(1-M\cos\theta_{\ell i}^k\right)^2}\left(\frac{1-M^2}{1-M\cos\theta_{\ell i}^k}\frac{\vec{R}_{\ell i}^k\cdot\hat{n}_\ell}{R_{\ell i}^k}-\vec{M}\right)$$
$$+\frac{\dot{\phi}_n\left(\tau_{\ell i}^k\right)}{cR_{\ell i}^k\left(1-M\cos\theta_{\ell i}^k\right)^2}\frac{\vec{R}_{\ell i}^k\cdot\hat{n}_\ell}{R_{\ell i}^k}.$$

To test the effectiveness of the method, we give two examples. In those examples, the medium is at rest.

4.2.6.1 Example: Monopole Test

Once again, we use this benchmark test to evaluate the accuracy of the method since there is an analytical solution. We consider a so-called “permeable surface”, i.e., a fictitious surface that encloses all acoustic sources and allows the passage of fluid through it. Hence, the motion of the surface does not disturb the medium, and the variations in the fluid motion outside the surface are of acoustic nature alone. In our case, the surface is a sphere of radius $R = 1m$ moving parallel to the x-axis with $M = 0.3$. The normal derivative of the pressure on the surface of the sphere is equal to the normal derivative of the pressure of a moving monopole (4.42).

For the simulation, the same sphere with 600 quad elements from 4.2.3 is used. At $t = 0s$, the real monopole is located at $\vec{r}_{Q0} = (5.3644, -0.2, -0.2)$ m and radiates at $f = 275$ Hz. The center of the sphere is located at $O_s = (5.5644, 0, 0)m$. The number of equivalent sources is $N = 600$, and they are located on an interior spherical surface with radius $R_{se} = 0.9$.

The strengths of the equivalent sources are obtained by applying a truncated SVD regularization with a truncation ratio $w = 0.5$ to solve the equation system. To evaluate the accuracy of the simulation, the pressure at three field points is compared with the analytical values. The points are located at: $P1 = (-6.0863, 0, 3)m$, $P2 = (6.5644, 0, 3)m$ and $P3 = (0, 0, 3)m$, which correspond to the start, end and middle of the body’s trajectory.

Figure 4.24 shows the comparison of the sound pressure at the three points. The agreement is very good, and the bigger differences are seen at the first time steps, but they decrease at advanced time steps. The spectra corresponding to the time signals in Fig. 4.24 are shown in Fig. 4.25. They are very similar to the spectra of the incident sound in example 4.2.5, produced by the moving monopole. Obviously, this result is to be expected since the permeable sphere just emulates the monopole.

Figure 4.26 shows the calculated strengths of the equivalent sources in the time domain (left) and in the frequency domain (right). A distinctive and narrow peak at the emitted frequency Ω is clearly noticeable, but also other much smaller peaks are observed. Those peaks are fictitious frequencies introduced by the ill-conditioning of the system matrix. So, there is no Doppler shift, since there is no relative motion between equivalent sources and collocation points.

4.2.6.2 Example: Scattering from a Moving Rigid Sphere

In example 4.2.5, we had a moving monopole and a stationary rigid scattering sphere. We invert the situation here and we have now a stationary monopole and a moving rigid scattering sphere.

For this case, the fluid velocity $\vec{u}$ can be split into a uniform component and a fluctuating component $\vec{u} = \vec{u}_0 + \vec{u'}$. The boundary condition for a rigid moving body in a fluid at rest ($\vec{u}_0 = 0$) is then

$$\vec{u'} \cdot \hat{n} = c\vec{M} \cdot \hat{n}, \tag{4.50}$$

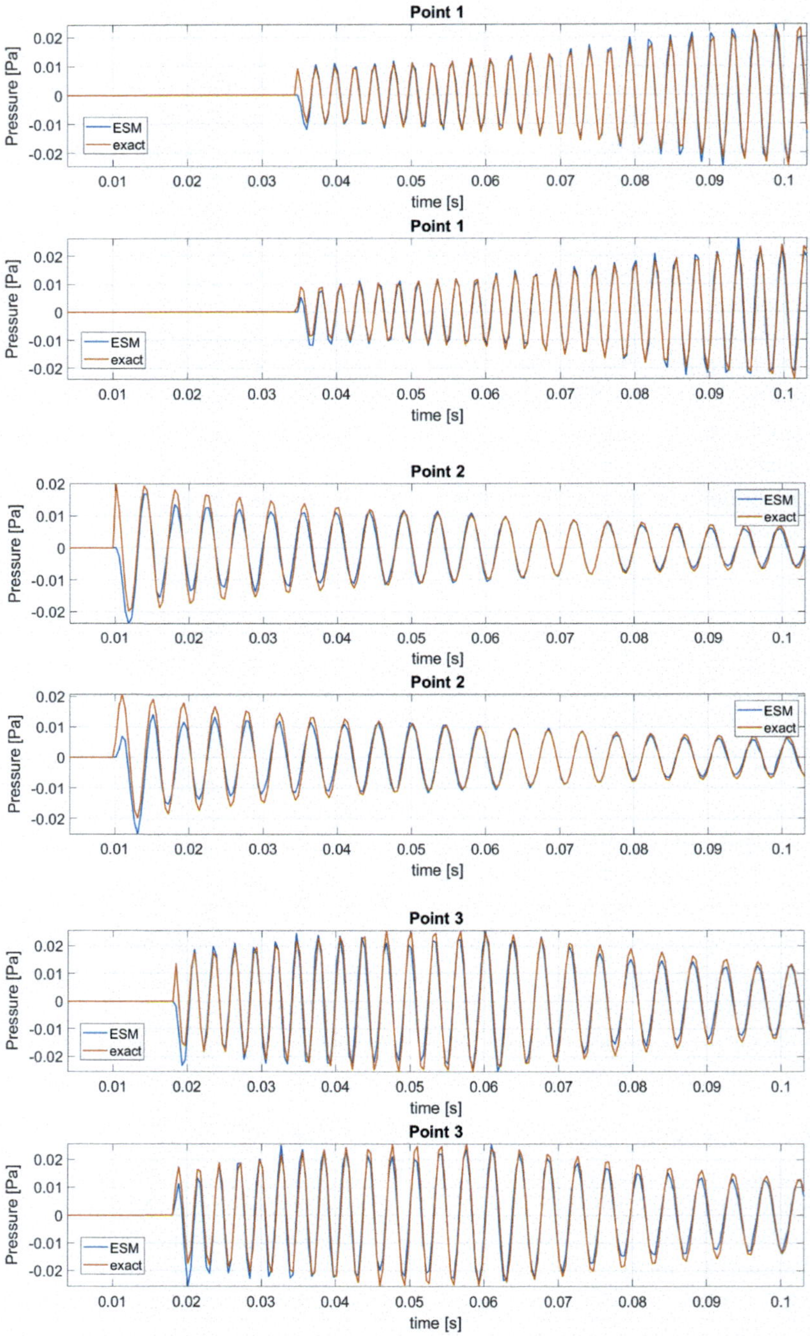

Fig. 4.24 Comparison of sound pressure from ESM and analytical solution

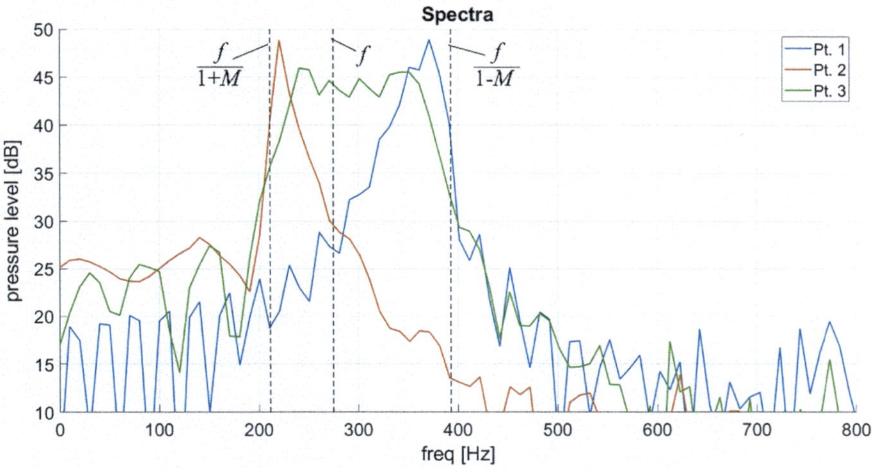

Fig. 4.25 Spectra of the sound pressure at three field points for the monopole test

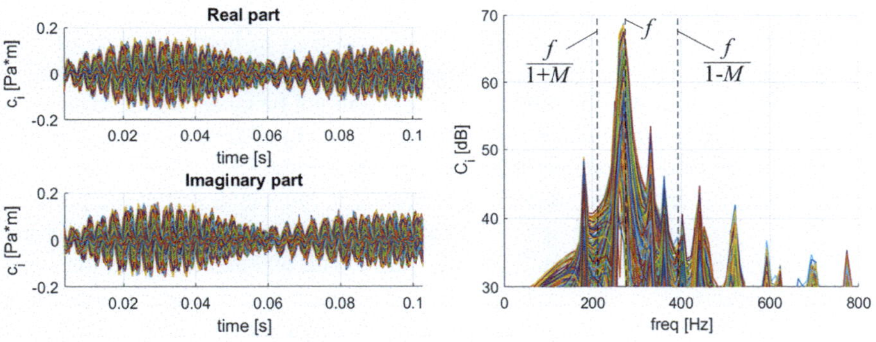

Fig. 4.26 Strengths of the equivalent sources in time and frequency domains for the monopole test

where $c\vec{M}$ is the velocity of the body, and $\hat{n}$ is its unit normal vector. When moving, a solid body (impermeable surface) displaces the surrounding fluid, creating a transient non-uniform flow (aerodynamic perturbation), so we can write $\vec{u}\prime = \vec{u}_{aer} + \vec{u}_{ac}$. If the body is thin, $\vec{u}_{aer}$ is small, and there is no interaction between aerodynamic and acoustic components. Then, the condition (4.50) can be separated into two uncoupled problems [5]

$$\vec{u}_{aer} \cdot \hat{n} = c\vec{M} \cdot \hat{n}, \vec{u}_{ac} \cdot \hat{n} = 0. \tag{4.51}$$

For the present study, we are only concerned about the second problem. The linearized momentum equation is given by

$$\rho_0 \frac{\partial \vec{u}\prime}{\partial t} = -\nabla p\prime. \tag{4.52}$$

By splitting also the pressure fluctuations p' into aerodynamic and acoustic components, $p' = p_{aer} + p_{ac}$. we obtain two separate equations

$$\rho_0 \frac{\partial \vec{u}_{aer}}{\partial t} = -\nabla p_{aer}, \rho_0 \frac{\partial \vec{u}_{ac}}{\partial t} = -\nabla p_{ac}. \tag{4.53}$$

If we multiply the second equation with the normal unit vector $\hat{n}$, we obtain [3]

$$\rho_0 \frac{\partial \vec{u}_{ac}}{\partial t} \cdot \hat{n} = -\nabla p_{ac} \cdot \hat{n} \to \rho_0 \left(\frac{\partial (\vec{u}_{ac} \cdot \hat{n})}{\partial t} - \vec{u}_{ac} \cdot \frac{\partial \hat{n}}{\partial t} \right) = -\frac{\partial p_{ac}}{\partial n}. \tag{4.54}$$

Using (4.51) and the condition that the motion is rectilinear $\left(\frac{\partial \hat{n}}{\partial t} = 0\right)$, we find that the normal derivative of the acoustic pressure vanishes on the surface. Finally, separating the acoustic pressure into incident and scattered components, $p_{ac} = p_{inc} + p_{sc}$, we find that on the surface

$$\frac{\partial p_{sc}}{\partial n} = -\frac{\partial p_{inc}}{\partial n}. \tag{4.55}$$

Therefore, (4.55) applies to bodies at rest as well as to bodies in a rectilinear motion.

In the present example, the refraction effects will not be taken into account even though the sphere is not a slender body. However, for a low Mach number and a low frequency, it can be a reasonable first order approximation.

The incident sound is the sound of a harmonic monopole at rest at the position $\vec{r}_q = (0, 0, 2)m$ m while the sphere of radius 1 m moves with $M = 0.3$ in the $-x$ direction. For the simulation, the same sphere with 600 elements and the same time discretization are considered. At $t = 0$, the center of the sphere is located at $O_s = (7.0753, 0, 0)m$. To obtain the strengths of the equivalent sources, a Tikhonov regularization with the same regularization parameter $\lambda = 15$ for all time steps, was applied.

In Fig. 4.27, we see on the left the variation of the strengths of the equivalent sources with time and on the right the corresponding spectra. The frequency content of the sources covers almost the entire Doppler interval $\left[\frac{f}{1+M}, \frac{f}{1-M}\right]$ since the collocation points move relative to the stationary source.

Figure 4.28 shows four snapshots of the sound field. In the upper figures, only the scattered pressure is illustrated. In the lower plots, we see the total pressure. The sphere moves from right to left so that the relative motion between the sphere and the source is the same as in example 4.2.5. The main lobe of the scattered pressure has a very similar evolution in time in both cases, showing a sort of spiral. In this example, we can observe the compression of the scattered wavefronts in front of the sphere and the separation of the wavefronts behind it due to the motion, while in the other case, the wavefronts are equally separated in front and behind the sphere.

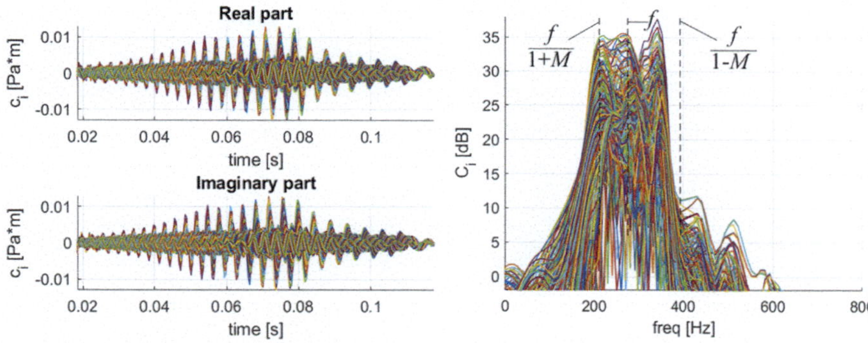

Fig. 4.27 Strengths of the equivalent sources in time and frequency domain for the moving rigid sphere

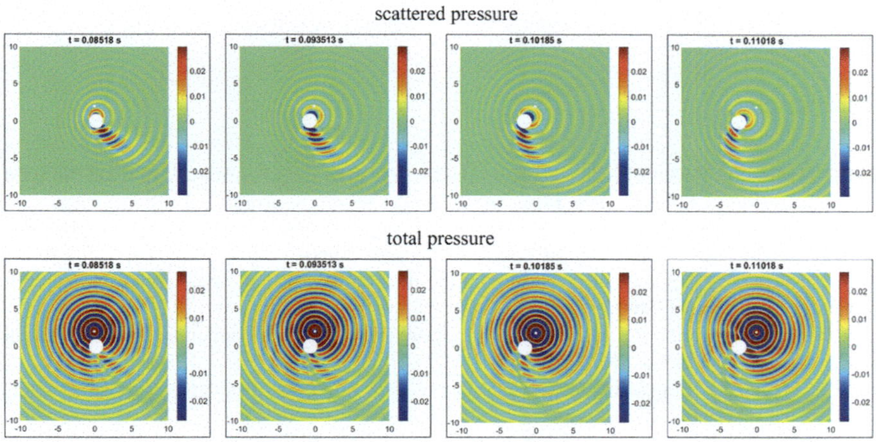

Fig. 4.28 Four snapshots of the scattering from a moving rigid sphere

In the plots of the total pressure, compression and separation of the wavefronts, characteristic of the Doppler effect is no longer noticeable because the incident pressure is much larger than the scattered pressure. That can be better seen in Fig. 4.29, where the spectra of incident and scattered waves at three field points are plotted. The amplitude of the incident pressure is about to 10 to 15 dB larger than the scattered pressure. The scattered wave shows a broadening of the spectra around the radiating frequency and differs from the spectrum of a moving monopole.

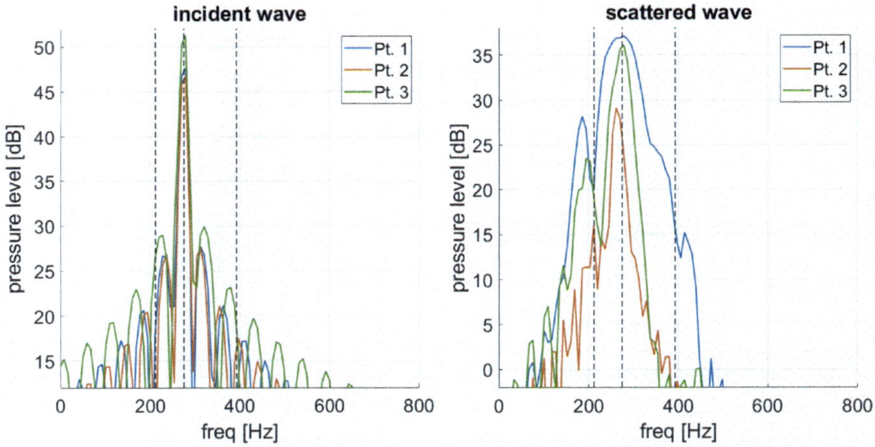

Fig. 4.29 Spectra of the sound pressure at three field points for te moving rigid sphere

References

1. P.C. Hansen, D. Prost O'Leary, The use of the L-curve in the regularization of discrete ill-posed problems, SIAM J. Sci. Compt. **14**(6), 1487–1503 (1993)
2. P.C. Hansen, *Regularization Tools A Matlab Package for Analysis and Solution of Discrete Ill-Posed Problems* (Technical University of Denmark, 2008)
3. S. Lee, Prediction of acoustic scattering in the time domain and its applications to rotorcraft noise, Ph.D. Thesis (2009)
4. S. Lee, K.S. Brentner, P.J. Morris, Acoustic scattering in the time domain using an equivalent source method. AIAA J. **48**(12) (2010)
5. M.K. Myers, J.S. Hausmann, Computation of acoustic scattering from a moving rigid surface. JASA **91**(5), 2594–2605 (1992)
6. G. Wahba, Spline models for observational data, in *CBMS-NSF Regional Conference Series in Applied Mathematics*, vol. 59 (Society for Industrial and Applied Mathematics, Philadelphia, PA, 1990)

Chapter 5
The Equivalent Source Method for Localization of Moving Sources

In the previous chapter, we have used the ESM to simulate the sound field from a radiating or scattering body. The equivalent sources are placed inside the body, and their location is not relevant as long as the exterior sound field is reproduced accurately. Different source distributions can approximate the boundary condition with a similar degree of accuracy.

In this chapter, however, we are interested in finding the true position of the sound sources since once they are located, measures for their attenuation can be carried out. The moving body is replaced by a set of equivalent sources placed on the surface of the body or on an exterior surface close to it that is assumed to be a projection of the real sources (source surface). Their strengths are determined by matching the sound pressure due to the sum of their contributions with the pressure measured by a set of microphones positioned at some distance from the moving body. The refraction effects due to the non-uniform flow produced by the motion are not taken into account.

5.1 Localization Procedure Through an Iterative Approach

In the present approach, the source surface is a rectangular plane in front of the body, and the equivalent sources are uniformly distributed on the plane. The center of the plane moves along the trajectory of the moving body with the same velocity $\vec{M}c$. The trajectory needs to be well- known during the measurement time. For the actual method, we consider a linear trajectory. For simplicity, the microphones are assumed to be mounted on a rigid ground at positions $\vec{r}^A$, i.e., the sound pressure at the microphones is doubled. Figure 5.1 illustrates the parameters of the procedure.

As in the previous chapter, we need to discretize reception and emission times as

$$t^k = t_{\min} + (k-1)\Delta t, \quad k = 1, 2, \ldots, K, \tag{5.1}$$

M. Ochmann and R. Piscoya, *Moving Acoustic Sources*,
Springer-AAS Acoustics Series, https://doi.org/10.1007/978-981-96-0871-3_5

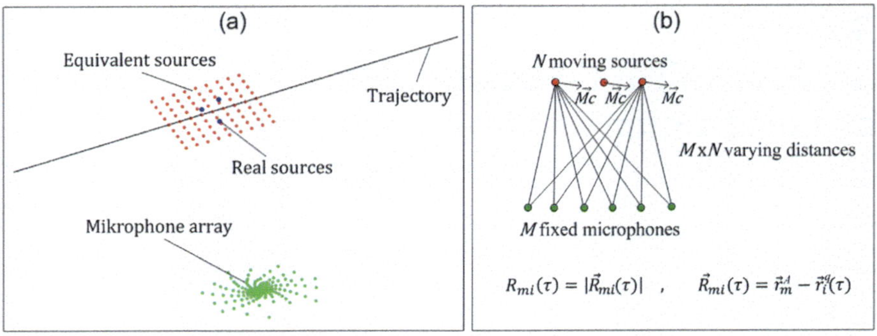

Fig. 5.1 Setup of the localization procedure

$$\tau^j = \tau_{\min} + (j - 1)\Delta\tau, \quad j = 1, 2, \ldots, J. \tag{5.2}$$

The sound pressure at the m-th reception point $\vec{r}_m^A$ and reception time t^k was given in (4.4) and is written once again here

$$p\left(\vec{r}_m^A, t^k\right) = \sum_{i=1}^{N} \frac{c_i\left(\tau_{mi}^k\right)}{4\pi R_{mi}^k\left(1 - M\cos\theta_{mi}^k\right)}, \tag{5.3}$$

where $\vec{R}_{mi}^k = \vec{r}_m^A - \vec{r}_i^q\left(\tau_{mi}^k\right)$, $\vec{r}_i^q(t) = \vec{r}_{i0}^q + \vec{M}ct$, $R_{mi}^k = \left|\vec{R}_{mi}^k\right|$, $\cos\theta_{mi}^k = \frac{\vec{M}\cdot\vec{R}_{mi}^k}{MR_{mi}^k}$, $M = \left|\vec{M}\right|$ and τ_{mi}^k is the emission time. The emission time is the solution of the equation: $c\left(t^k - \tau_{mi}^k\right) = \left|\vec{r}_m^A - \vec{r}_i^q\left(\tau_{mi}^k\right)\right|$ and is given by

$$\tau_{mi}^k = t^k - \frac{1}{c}\left(\frac{\vec{M}\cdot\vec{R}_{mi}^k + \sqrt{\left(\vec{M}\cdot\vec{R}_{mi}^k\right)^2 + \left(1 - M^2\right)\left(R_{mi}^k\right)^2}}{1 - M^2}\right). \tag{5.4}$$

Since the microphone positions are fixed, the distances R_{mi}^k vary with time, i.e., an increase in Δt in the reception time t will not imply an increase of the emission time τ in Δt. That is a problem because now it is not possible to choose an initial emission time $\tau_{\min}$ that ensures that $\tau_{mi}^k \leq \tau^k$ for all k, which is the necessary condition to apply the iterative method described in Chap. 4. Nevertheless, we can fix this problem by defining a new set of reception times $\hat{t}^k$ as follows:

- For $k = 1$, we set $\hat{t}^1 = t^1$, calculate all emission times τ^k_{mi} using (5.2) and define $\tau_{\min} = \max(\tau^k_{mi})$
- For $k = 2, \ldots, \min(J, K)$, we compute all distances $R^k_{mi} = |\vec{r}^A_m - \vec{r}^q_i(\tau^k)|$ and set $\hat{t}^k = \min\left(\frac{R^k_{mi}}{c}\right)$

The elements of the new set $\hat{t}^k$ are not equally spaced in time, and since the pressure is known at the set t^k, the pressure at the new reception times ($\hat{p}(\hat{t})$) needs to be interpolated. For simplicity, a linear interpolation is performed, and the pressure at the times $\hat{t}^k$ that are not within the limits of the set t^k is set to zero. The application of the interpolation can introduce some high-frequency noise in the microphone data.

The strengths of the equivalent sources, c_i, are determined by matching the modeled pressure with the interpolated pressure data

$$\sum_{i=1}^{N} \frac{c_i(\hat{\tau}^k_{mi})}{4\pi \hat{R}^k_{mi}(1 - M \cos\hat{\theta}^k_{mi})} = \hat{p}(\vec{r}^A_m, \hat{t}^k), \tag{5.5}$$

where $\vec{\hat{R}}^k_{mi} = \vec{r}^A_m - \vec{r}^q_i(\hat{\tau}^k_{mi})$, $\hat{R}^k_{mi} = \left|\vec{\hat{R}}^k_{mi}\right|$, $\cos\hat{\theta}^k_{mi} = \frac{\vec{M}\cdot\vec{\hat{R}}^k_{mi}}{M\hat{R}^k_{mi}}$, $M = \left|\vec{M}\right|$ and $\hat{\tau}^k_{mi}$ is the emission time. The emission time is the solution of the equation: $c(\hat{t}^k - \hat{\tau}^k_{mi}) = |\vec{r}^A_m - \vec{r}^q_i(\hat{\tau}^k_{mi})|$ and is given by

$$\hat{\tau}^k_{mi} = \hat{t}^k - \frac{1}{c}\left(\frac{\vec{M}\cdot\vec{\hat{R}}^k_{mi} + \sqrt{\left(\vec{M}\cdot\vec{\hat{R}}^k_{mi}\right)^2 + (1 - M^2)(\hat{R}^k_{mi})^2}}{1 - M^2}\right). \tag{5.6}$$

With the new set $\hat{t}^k$, we now can ensure that $\hat{\tau}^k_{mi} \leq \tau^k$ for all k. A linear interpolation of the source strengths at the equidistant emission times τ^j (4.18) is once again necessary. Hence (5.5) becomes

$$\frac{1}{4\pi}\sum_{i=1}^{N}\sum_{n=1}^{k} \frac{\phi_n(\hat{\tau}^k_{mi})\hat{c}^n_i}{\hat{R}^k_{mi}(1 - M\cos\hat{\theta}^k_{mi})} = \hat{p}(\vec{r}^A_m, \hat{t}^k). \tag{5.7}$$

Separating the sum over n into two terms and rearranging terms, we obtain

$$\frac{1}{4\pi}\sum_{i=1}^{N} \frac{\phi_k(\hat{\tau}^k_{mi})\hat{c}^k_i}{\hat{R}^k_{mi}(1 - M\cos\hat{\theta}^k_{mi})} = \hat{p}(\vec{r}^A_m, \hat{t}^k) - \frac{1}{4\pi}\sum_{i=1}^{N}\sum_{n=1}^{k-1} \frac{\phi_n(\hat{\tau}^k_{mi})\hat{c}^n_i}{\hat{R}^k_{mi}(1 - M\cos\hat{\theta}^k_{mi})}. \tag{5.8}$$

Denoting the source functions $\hat{\gamma}_{mi}^{nk}$ and the right-hand side functions $\hat{p}_m^k$ as

$$\hat{\gamma}_{mi}^{kn} = \frac{\phi_n\left(\hat{\tau}_{mi}^k\right)}{4\pi \hat{R}_{mi}^k\left(1 - M\cos\hat{\theta}_{mi}^k\right)} \quad \hat{p}_m^k = \hat{p}\left(\vec{r}_m^A, \hat{t}^k\right), \tag{5.9}$$

we can build the matrices $\hat{\Gamma}^{kn}$ with elements $\hat{\gamma}_{\ell i}^{kn}$, the vectors $\hat{P}^k$ with elements $\hat{p}_m^k$ and the vector of strengths C^k with elements $\hat{c}_i^k$

$$\hat{\Gamma}^{nk} = \begin{pmatrix} \hat{\gamma}_{11}^{kn} & \hat{\gamma}_{12}^{kn} & \cdots & \hat{\gamma}_{1N}^{kn} \\ \hat{\gamma}_{21}^{kn} & \hat{\gamma}_{22}^{kn} & \cdots & \hat{\gamma}_{2N}^{kn} \\ \vdots & \vdots & \ddots & \vdots \\ \hat{\gamma}_{M1}^{kn} & \hat{\gamma}_{M2}^{kn} & \cdots & \hat{\gamma}_{MN}^{kn} \end{pmatrix} \quad \hat{P}^k = \begin{pmatrix} f_1^k \\ f_2^k \\ \vdots \\ f_M^k \end{pmatrix} \quad C^k = \begin{pmatrix} \hat{c}_1^k \\ \hat{c}_2^k \\ \vdots \\ \hat{c}_M^k \end{pmatrix}. \tag{5.10}$$

An iterative procedure similar to that of Chap. 4 is obtained and written as

$$\hat{\Gamma}^{kk} C^k = \hat{P}^k - \sum_{n=1}^{k-1} \hat{\Gamma}^{kn} C^n. \tag{5.11}$$

We evaluate the performance of this approach with the next examples.

5.1.1 *Example: Moving Monopole Radiating at a Discrete Series of Frequencies*

We consider a monopole moving parallel to the x-axis with a Mach number $M = 0.25$ at a height $h = 10$ m passing over a spiral array of radius $R_A = 1$ m and center (0, 0, 0). At $t = 0$, the monopole is located at $\vec{r}_{q0} = (-2.5, -0.3, 10)$ and radiates at a set of N_f frequencies f_μ within the interval [100, 500] Hz. The sound pressure at any receiving point is given by

$$p(\vec{r}, t) = \frac{\sum_{\mu=1}^{N_f} A_\mu e^{j\Omega_\mu (t - R_e/c)}}{4\pi R_e(1 - M\cos\theta_e)}, \tag{5.12}$$

where A_μ is the amplitude of the oscillation at the frequency f_μ ($\Omega_\mu = 2\pi f_\mu$) and R_e and θ_e are given after (4.42).

For the simulation, a sampling frequency of 2000 Hz, i.e., a time step of $\Delta t = 1.25 \times 10^{-4}$ s and $J = 128$ emission time steps is considered. We use a microphone array with $M = 100$ microphones displayed in a spiral arrangement. The positions of the microphones, $\vec{r}_m^A$, are defined according to Sarradj (2015)

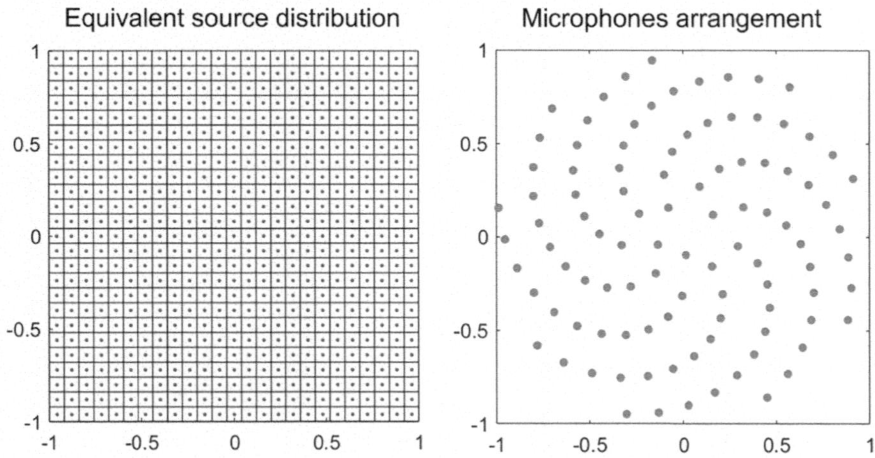

Fig. 5.2 Illustration of distribution of equivalent sources and microphones

$$\vec{r}_m^A = (D_m \cos\theta_m, D_m \sin\theta_m, 0) \quad D_m = R_A\sqrt{\frac{m}{M}} \quad \theta_m = 2\pi m\left(\frac{1+\sqrt{V}}{2}\right), \tag{5.13}$$

where V is a parameter that controls the distribution of points over the circle. In the present study, we chose $V = 6.5$. The arrangement of the 100 microphones is shown in Fig. 5.2. The reception times are defined from the emission times and the delay from source to the microphones

For the source localization, a total of 625 equivalent sources are uniformly distributed on a square of side $L_e = 2\,\text{m}$ that moves with velocity $(Mc, 0, 0)$. To obtain a uniform distribution, we draw a rectangular grid on the source plane and place the equivalent sources at the center of each element of the grid (Fig. 5.2). At time $t = 0\,\text{s}$, the center of the square is located at $(-2.72, 0, 10)\,\text{m}$.

The strength of the monopole in the time domain and its corresponding spectrum is shown in the upper plots of Fig. 5.3, and the pressure captured by one microphone and its spectrum are depicted in the lower plots of Fig. 5.3.

When we look closer at (5.10) and (5.11), we note that an equivalent source i will be taken into account at the k-th iteration step, only if at least one of the M emission times from that source to the receivers lies within the interval of emission times $[\tau^{k-1}, \tau^k]$; otherwise, the method will ignore it (invisible), since all coefficients of the i-th column in the matrix Γ^{kk} will be zero.

Figure 5.4 shows the position of the plane of equivalent sources at six different times during its motion above the microphone array. The "visible" sources are marked in red. For the specific size of the plane, its height, the position of the microphones and the time step of this example, many equivalent sources are ignored at the start and (due to symmetry) at the end of the emission time.

We performed a simulation using the iterative approach (5.11), considering only the emission times that take into account all equivalent sources. A regularization is

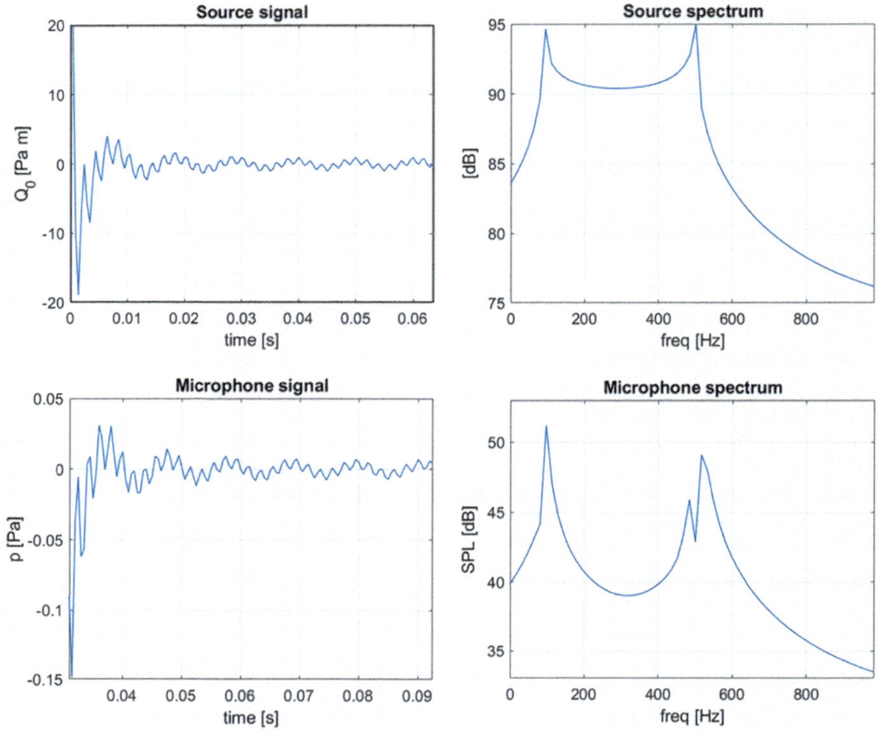

Fig. 5.3 Source and microphone signals in time and frequency domains

required, and for the present case, it was found that the application of the Tikhonov regularization with parameter $\lambda_k = 0.05\, w^k_{\max}$, where $w^k_{\max}$ is the maximum singular value of $\hat{\Gamma}^{kk}$, provides reasonable results.

The computed strengths of the 625 equivalent sources and their corresponding spectra are presented in the upper plot of Fig. 5.5. The lower plots of Fig. 5.5 show a comparison of time signal and spectrum between the ESM simulation and the analytical solution at certain field point between the source plane and the microphone array. Since not all emission times are considered, also not all reception times are included in the simulation. The spectra were obtained by performing the FFT of the time signals. The comparison of the pressure at the field point in the time domain shows a very good agreement. The differences in the spectra are approximately constant at around 3 dB between 200 and 600 Hz and increase above 600 Hz.

The previous results show that the iterative ESM approach is capable of reproducing the radiated sound field of a moving source with the chosen distribution of equivalent sources. Now, it remains to see if the source can also be properly localized. Figure 5.6 shows color maps of the strengths of the equivalent sources at six different 1/3-octave bands. At each band, the strengths have been normalized to the maximum value and are presented in dB scale, i.e., in each map, the maximum value is 0 dB.

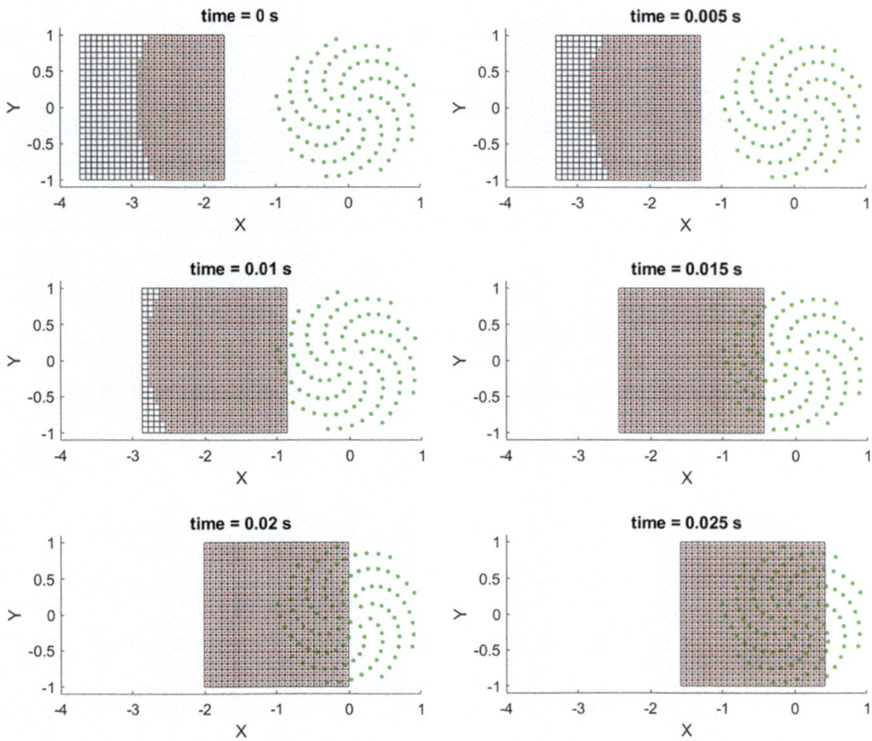

Fig. 5.4 "Visible" equivalent sources (red points) at different emission times when passing over the microphone array

The color scale is limited to 16 dB which is a reasonable range for the scaling the strengths. For reference, the white circle indicates the position of the real source.

Unfortunately, the source was not properly localized. Instead, we observe that the equivalent sources near the border of the plane of equivalent sources ($x = 1$) obtain the larger strengths, while the equivalent sources around the real source have strengths that are 8 dB smaller or more.

The reason for the failure of the method for source localization may rest in the sparsity character of the expected solution since only the equivalent sources close to the real source should obtain significant strengths.

The Tikhonov regularization falls under the category of a Ridge regression. By this type of regression, the coefficients of the solution vector can become very small but not zero. That is a disadvantage because the unimportant unknowns will always be included in the system and there are too little observations to correctly determine all the unknowns.

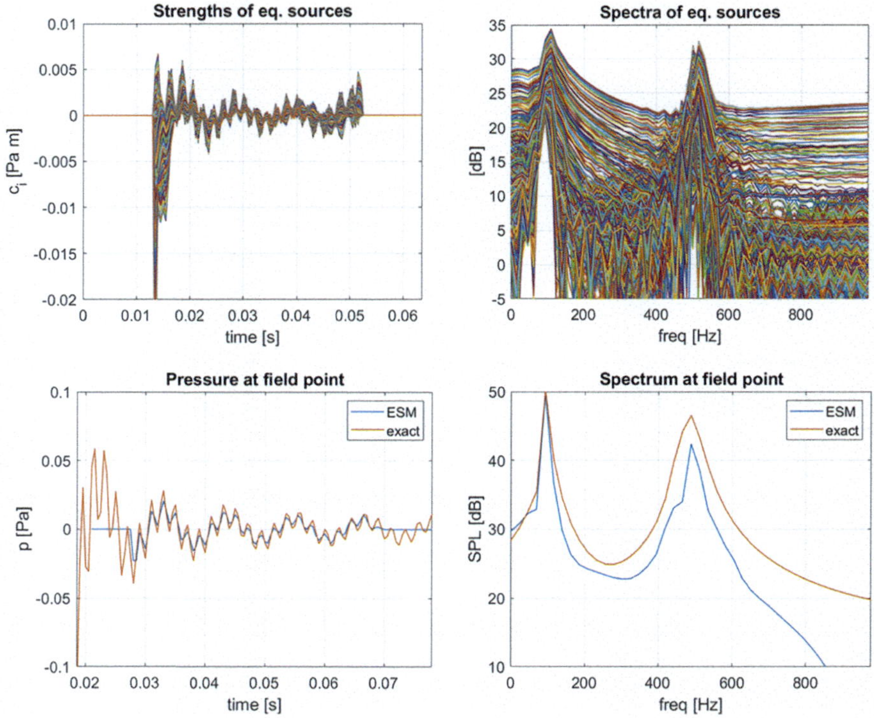

Fig. 5.5 Results of the simulation using an iterative approach and Tikhonov regularization

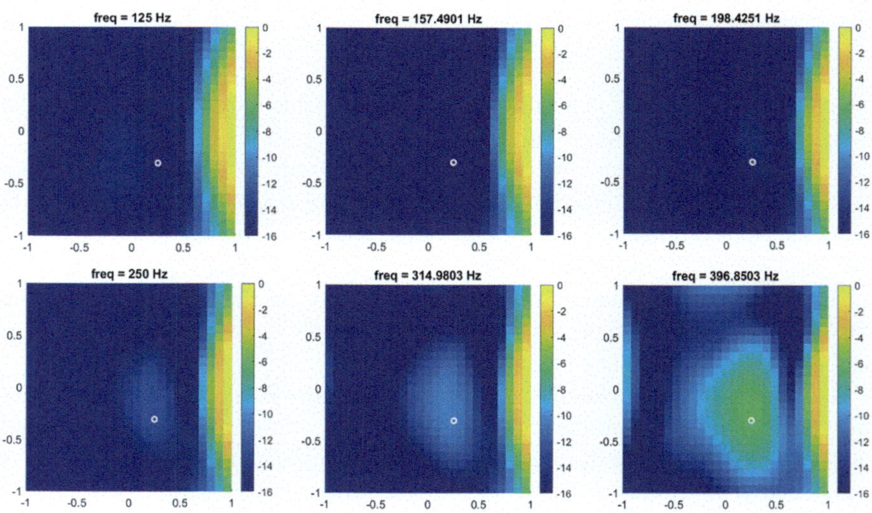

Fig. 5.6 Strengths of the equivalent sources at different 1/3-octave bands obtained with least squares and Tikhonov regularization

5.1.2 *Least Absolute Shrinkage and Selection Operator (LASSO)*

An alternative to the Ridge regression is the LASSO regression. This method minimizes the sum of the squared residual and an additional penalty term over the solution vector.

$$\min_{C}\left(\|\Gamma C - P\|_2^2 + \lambda\|C\|_1\right). \tag{5.14}$$

In contrast to the Ridge regression, LASSO does set coefficients of the solution to zero with increasing λ. Hence, LASSO is a useful tool to eliminate irrelevant variables that are not related to the response variable. If λ is large enough, the method will give the null vector as the result. It is possible to approximate the minimum value of λ (λ_m) that yields the null vector as a solution, namely

$$\lambda_m = 2\left\|\Gamma^T P\right\|_{\infty}. \tag{5.15}$$

Therefore, the optimum parameter lies within the interval$]0, \lambda_m[$.

The strengths of the equivalent sources of the previous case were now computed with the iterative approach, replacing the Tikhonov regularization with the LASSO regression at each time step. It was found that the parameter $\lambda_{opt} = 0.25\lambda_m$ provided good results.

The upper plots of Fig. 5.7 show the strengths of the equivalent sources and their corresponding spectra.

We observe that the spectra of the equivalent sources differ from the source spectrum in Fig. 5.5. However, the comparison of the pressure in the time domain at field points provides a good agreement between simulation and exact values, although not as good as with the Ridge regression. The spectrum obtained with the LASSO regression shows the same general form as the exact one but also large differences at particular frequencies.

Figure 5.8 shows the color maps of the strengths for the 1/3-octave bands between 125 and 400 Hz. In contrast to the ridge regression, with the LASSO regression, one obtains much fewer equivalent sources whose strengths lie within the interval [−16, 0] dB and are placed near to the real source. However, the LASSO regression applied to the iterative approach is also not able to localize the real source with enough accuracy. A possible reason for the lack of accuracy lies in the propagation of errors from one time step to the next in the iterative approach.

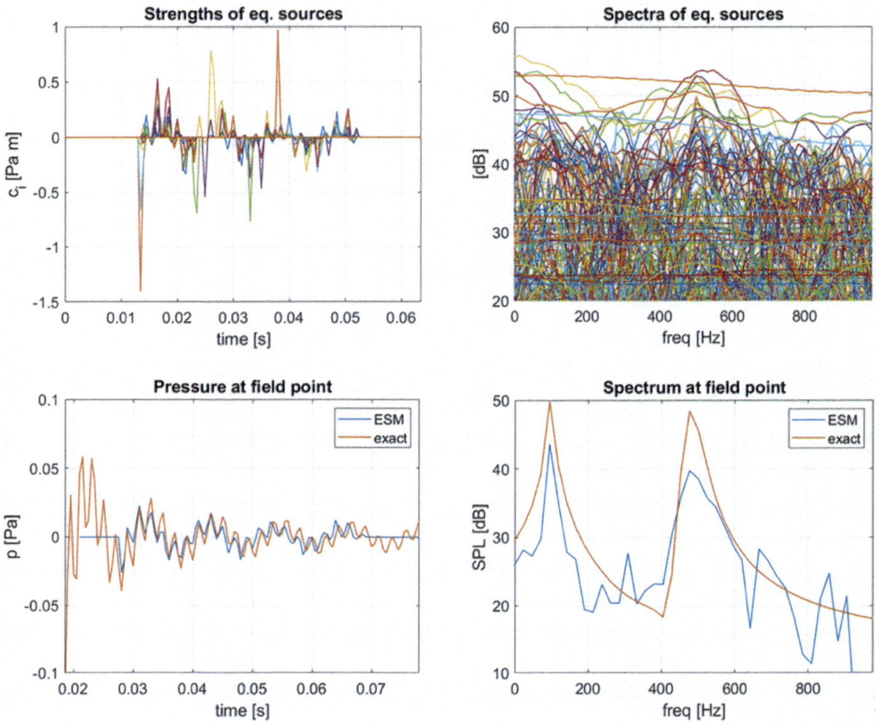

Fig. 5.7 Results of the simulation using iterative approach and LASSO regression

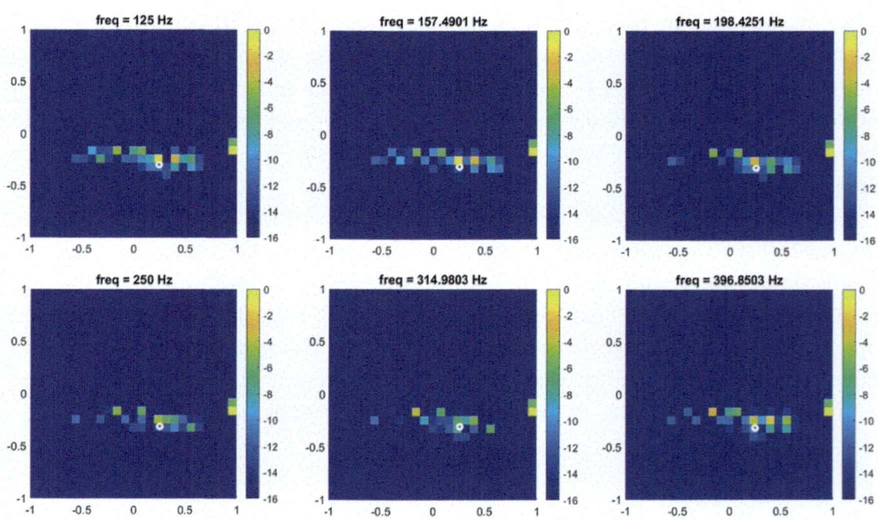

Fig. 5.8 Strengths of the equivalent sources at different 1/3-octave bands obtained with the LASSO regression

5.2 Localization Procedure Through the Solution of a Global System

Since the iterative method can ignore certain equivalent sources and errors can propagate during the iteration, the alternative is to solve the whole system of equations in one step. Instead of separating the strengths and pressures for each time step, we build global vectors for strengths and pressures and a global matrix that connects both vectors. For this approach, it is not necessary to resample the receiving times, i.e., they are given by (5.1).

If we choose $\tau_{\min} = t_{\min} - \max\left(\frac{R^1_{mi}}{c}\right)$, we ensure that $\tau^1_{mi} \geq \tau^1$.

The sound pressure at the m-th receiving point $\vec{r}^A_m$ at reception time t^k is given by (5.3). Once again, the strengths of the equivalent sources need to be interpolated at the discretized emission times (5.2) using the linear interpolation scheme (4.18). Equation (5.3) becomes

$$p(\vec{r}^A_m, t^k) = \sum_{i=1}^{N}\sum_{n=1}^{J} \frac{\phi_n(\tau^k_{mi})\hat{c}^n_i}{4\pi R^k_{mi}(1 - M\cos\theta^k_{mi})}. \tag{5.16}$$

Similar to (5.9), we denote the source functions Υ^{kn}_{mi} and the right-hand side functions p^k_m as

$$\gamma^{kn}_{mi} = \frac{\phi_n(\tau^k_{mi})}{4\pi R^k_{mi}(1 - M\cos\theta^k_{mi})} \qquad p^k_m = p(\vec{r}^A_m, t^k), \tag{5.17}$$

and if we write the sum over n in (5.16) explicitly, we have

$$p^k_m = \sum_{i=1}^{N}\gamma^{k1}_{mi}\hat{c}^1_i + \sum_{i=1}^{N}\gamma^{k2}_{mi}\hat{c}^2_i + \cdots + \sum_{i=1}^{N}\gamma^{kJ-1}_{mi}\hat{c}^{J-1}_i + \sum_{i=1}^{N}\gamma^{kJ}_{mi}\hat{c}^J_i, \tag{5.18}$$

and in matrix form

$$\begin{aligned} p^k_m = {} & \left[\gamma^{k1}_{m1}\ \gamma^{k1}_{m2} \cdots \gamma^{k1}_{mN}\right]\begin{bmatrix}\hat{c}^1_1\\ \hat{c}^1_2\\ \vdots\\ \hat{c}^1_N\end{bmatrix} + \left[\gamma^{k2}_{m1}\ \gamma^{k2}_{m2} \cdots \gamma^{k2}_{mN}\right]\begin{bmatrix}\hat{c}^2_1\\ \hat{c}^2_2\\ \vdots\\ \hat{c}^2_N\end{bmatrix} + \cdots + \\ & \left[\gamma^{kJ-1}_{m1}\ \gamma^{kJ-1}_{m2} \cdots \gamma^{kJ-1}_{mN}\right]\begin{bmatrix}\hat{c}^{J-1}_1\\ \hat{c}^{J-1}_2\\ \vdots\\ \hat{c}^{J-1}_N\end{bmatrix} + \left[\gamma^{kJ}_{m1}\ \gamma^{kJ}_{m2} \cdots \gamma^{kJ}_{mN}\right]\begin{bmatrix}\hat{c}^J_1\\ \hat{c}^J_2\\ \vdots\\ \hat{c}^J_N\end{bmatrix}. \end{aligned} \tag{5.19}$$

Extending (5.19) to all microphones and building matrices and vectors as in (5.10)

$$\Gamma^{kn} = \begin{pmatrix} \gamma_{11}^{kn} & \gamma_{12}^{kn} & \cdots & \gamma_{1N}^{kn} \\ \gamma_{21}^{kn} & \gamma_{22}^{kn} & \cdots & \gamma_{2N}^{kn} \\ \vdots & \vdots & \ddots & \vdots \\ \gamma_{M1}^{kn} & \gamma_{M2}^{kn} & \cdots & \gamma_{MN}^{kn} \end{pmatrix} \quad P^k = \begin{pmatrix} p_1^k \\ p_2^k \\ \vdots \\ p_M^k \end{pmatrix}, \tag{5.20}$$

we can write the system of equations for the k-th receiving time

$$P^k = \left[\Gamma^{k1} \; \Gamma^{k2} \ldots \Gamma^{kJ} \right] \begin{bmatrix} C^1 \\ C^2 \\ \vdots \\ C^J \end{bmatrix}.$$

Finally, considering all receiving times, we can construct a system of $M \cdot K$ equations with $N \cdot J$ unknowns

$$\begin{bmatrix} \Gamma^{11} & \Gamma^{12} & \ldots & \Gamma^{1J} \\ \Gamma^{21} & \Gamma^{22} & \ldots & \Gamma^{2J} \\ \vdots & \vdots & \ddots & \vdots \\ \Gamma^{K1} & \Gamma^{K2} & \ldots & \Gamma^{KJ} \end{bmatrix} \begin{bmatrix} C^1 \\ C^2 \\ \vdots \\ C^J \end{bmatrix} = \begin{bmatrix} P^1 \\ P^2 \\ \vdots \\ P^K \end{bmatrix} \rightarrow \Gamma C = P. \tag{5.21}$$

While the number of emission and reception times is similar, $K \cong J$, the number of equivalent sources is usually much larger than the number of microphones $M \ll N$. Therefore, our system is highly underdetermined. On the other hand, since each reception time is related to a limited number of emission times, the system matrix is sparse.

Since a sparse solution vector is expected, the global matrix will be solved again with the LASSO regression. For the following examples, the regularization parameter is chosen as $\lambda_{opt} = \lambda_m/10^4$.

The matrix Γ has the size $(M \cdot K) \times (N \cdot J)$. For large values of those parameters, the matrix would be too large to be solved with Matlab's function "lasso". Therefore, the size of the problem that can be solved is limited.

5.2.1 *Test Case: Harmonic Sources Radiating at a Single Frequency*

We performed a calculation considering real sources distributed over a square plane of side L_e, moving with a Mach number $M = 0.25$ parallel to the x-axis. The sources move at a height h and pass over 200 microphones disposed in a spiral

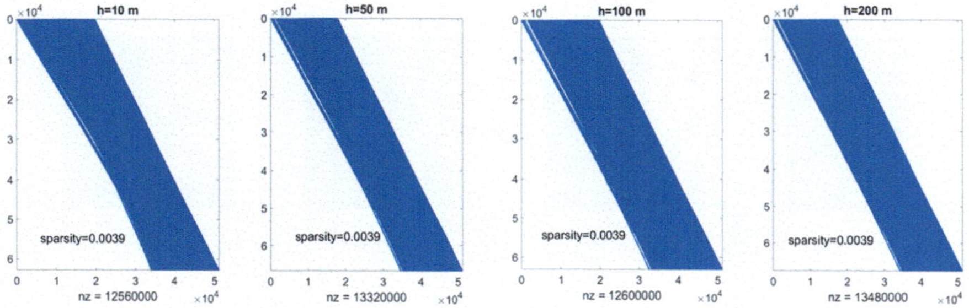

Fig. 5.9 Density of the system matrices for the four different calculations

array of radius R_a. The sources radiate at a frequency $f = 250\,\text{Hz}$. We considered only 100 equivalent sources located at the center of each element of a 10×10 grid and 0.064 s emission time. The calculation was performed for four different heights $h = 10, 50, 100, 200$ m. We found that in order to obtain accurate results, the size of the plane and the radius of the array had to increase with increasing height. The radius of the array was chosen to be $R_a = \sqrt{2}L_e/2$. The strength of the equivalent sources in the frequency domain is determined by performing the FFT numerically. The sparsity of the matrices (the number of zero-valued elements divided by the total number of elements) is shown in Fig. 5.9. In the four cases, the density is extremely low and equal to 0.0039.

According to the position of the real sources with respect to the equivalent sources, we considered the following cases:

(a) Best case: It takes place when the position of the real sources coincides with the positions of equivalent sources. We consider here five sources, hence, we expect that the method provides a much larger strength for the corresponding five equivalent sources than for their neighboring sources. The upper plots in Fig. 5.10 present the strengths of the equivalent sources at the radiated frequency in the form of a color map with a color range of 16 dB. The positions of the real sources are marked with black small circles. We see that the positions of the real sources are well localized since the strengths of the five corresponding equivalent sources are clearly larger than those of the remaining 95. The lower plots in Fig. 5.10 show the spectra of all 100 equivalent sources. The spectra show the peak at 250 Hz for the five sources and the lower strengths of the other sources. At $h = 50\,\text{m}$, one equivalent source has constant spectra over all frequencies, implying a very narrow pulse, which is an error in the solution. Fortunately, its amplitude is much smaller than those of the relevant sources.

(b) Worst case: It takes place when the position of the real sources lies exactly between four equivalent sources (corner of an element in the grid). We expect that the method will provide a similar strength for all four equivalent sources around the real source. For this case, we consider three real sources. The results are presented in Fig. 5.11 in a similar way as for the best case. The method highlights the equivalent sources around the real ones and gives them similar

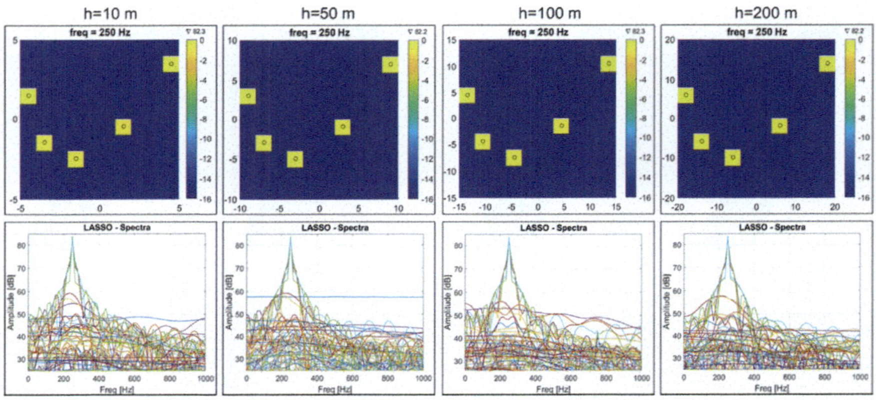

Fig. 5.10 Best case study; up: strength of the equivalent sources at 250 Hz; down: spectra of equivalent sources

strengths as expected. At $h = 10\,\text{m}$, one equivalent source obtains a constant spectrum with the maximum amplitude, which indicates an unrealistic strong pulse.

(c) General case: The real sources are located inside an element of the plane but neither in the center nor in a corner. We consider again five sources and four values of h. Figure 5.12 shows the strengths at 250 Hz on the source plane (upper plots) and the spectra (lower plots) of the equivalent sources. The method is able to identify the five real sources, but the neighboring equivalent sources are also visible in the color map although with a lower strength. The spectra at $h = 10\,\text{m}$ and $h = 200\,\text{m}$ show one equivalent source with a constant spectrum which as mentioned before, implies a fictitious pulse.

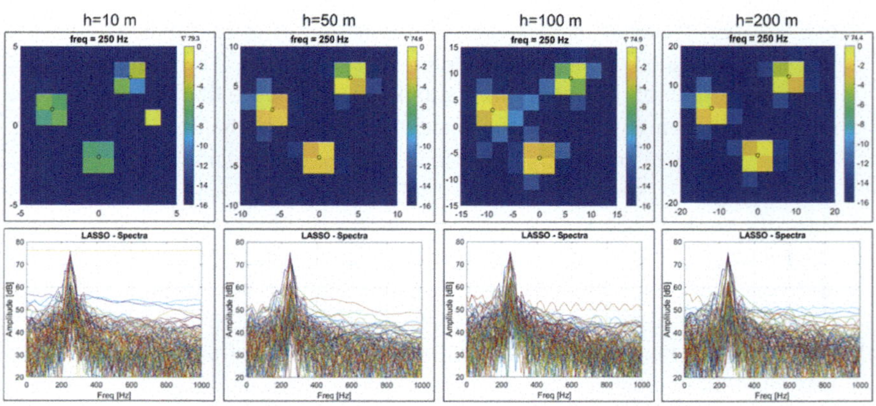

Fig. 5.11 Worst case, up: strength of the equivalent sources; down: spectra of equivalent sources

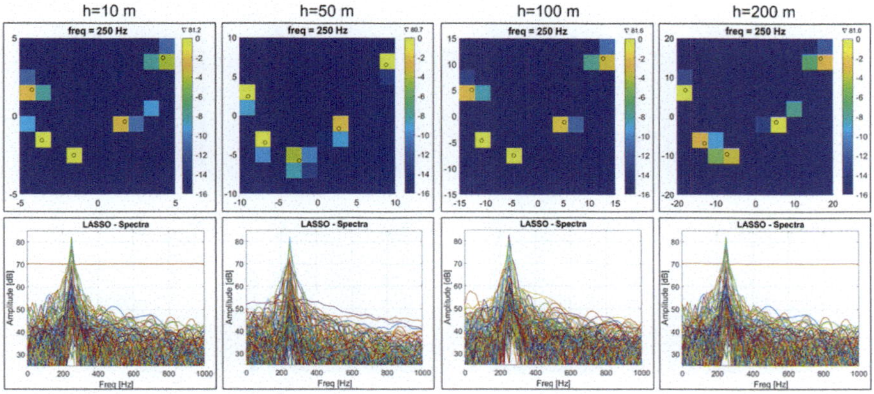

Fig. 5.12 General case, up: strength of the equivalent sources; down: spectra of equivalent sources

5.2.2 *Test Case: Harmonic Sources Radiating at Different Frequencies*

As a next example, we consider again the five sources at $h = 100$ m, but this time, each source radiates with a different frequency, namely 125, 250, 500, 1000 and 2000 Hz. We performed two calculations, one where the five real sources meet the best case and the other where they meet the general case. Figure 5.13 shows the spectra of the equivalent sources in both cases.

Both spectra present peaks at the five radiated frequencies indicating that the method could estimate them with accuracy. Figure 5.14 shows the strengths of the equivalent sources on the source plane for the four different frequencies as color maps, the best case in the upper plots and the general case in the lower plots. In the best case, the method is able to clearly identify the equivalent sources that match the real sources. In the general case, the three sources radiating the lowest frequencies

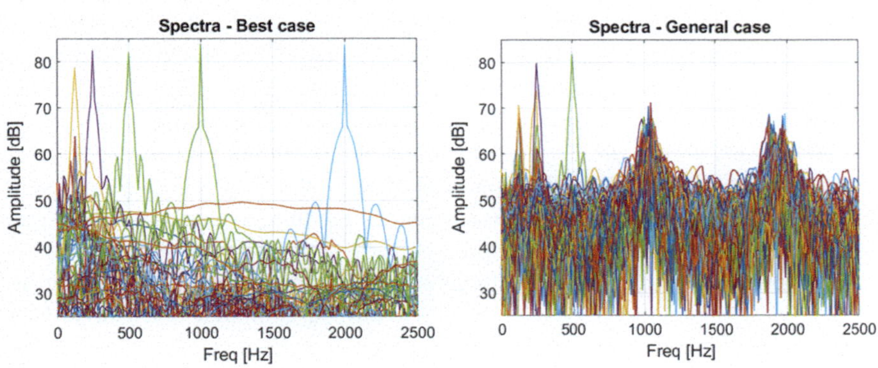

Fig. 5.13 Comparison of the spectra of the equivalent sources between best and general cases

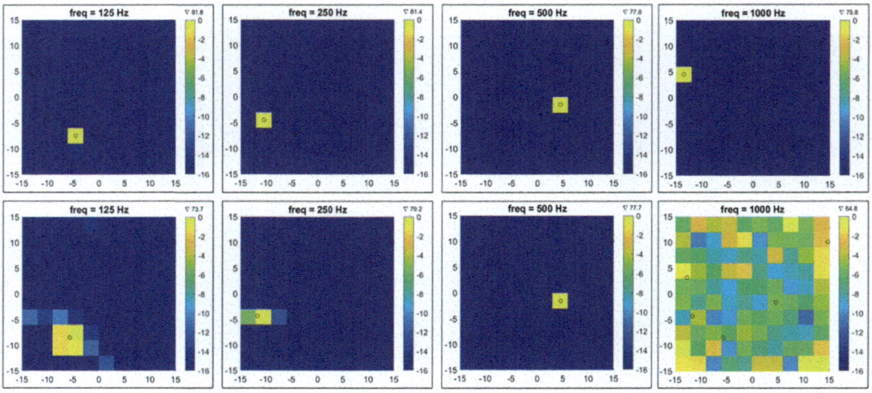

Fig. 5.14 Comparison of the strengths of the equivalent sources between up: best case and down: general case

are well identified, but the two sources radiating the higher frequencies could not be properly localized.

These results indicate that the procedure is highly sensitive to the position of the equivalent sources and provides the most accurate results when the real sources coincide with equivalent sources.

5.2.3 Test Case: Broadband Noise

Machines do not generate only pure tones; they usually they also produce broadband noise. The present ESM approach should be able to localize such sources of sound as well. To evaluate the performance of the present approach, we should generate noise. We simulate white noise with Matlab, generating a series of random numbers using the function “randn”. For this example, we consider a frequency sample of 16 kHz. Hence, to avoid the aliasing effect, the white noise is passed through a low-pass filter with cut-off frequency around 6 kHz. For the filtering, we use the Matlab function “filtfilt”. Once again, we consider five sources located within the 10×10 grid on the plane of equivalent sources and the 200 microphones. The plane moves at a height $h = 200$ m with $M = 0.25$. Figure 5.15 shows on the left the strengths of the real sources in time and frequency domains. The effect of the filter is clearly observed. On the right of Fig. 5.15, the total pressure due to the five sources captured by one microphone and its spectrum are plotted.

The spectra of the calculated strengths of the equivalent sources are plotted in Fig. 5.16. A small number of equivalent sources have an amplitude much larger than the rest (around 20 dB) and a frequency response that agrees well with those of the sources.

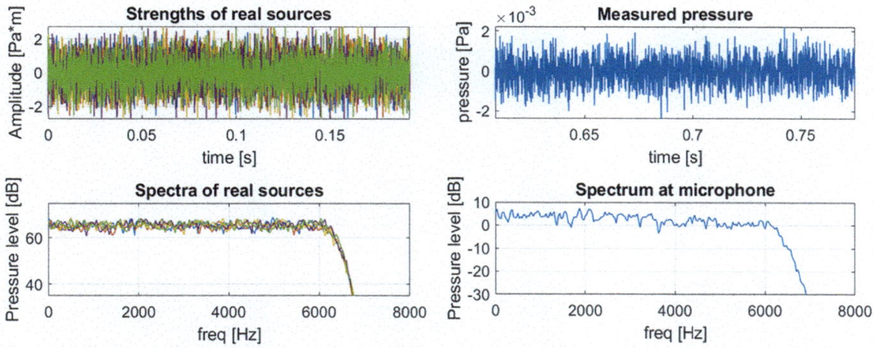

Fig. 5.15 Noise from source and microphone in time and frequency domains

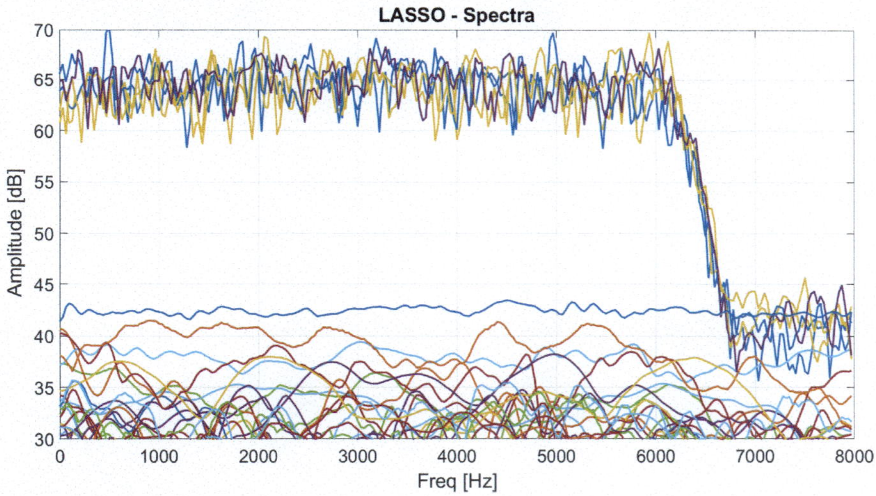

Fig. 5.16 Spectra of the equivalent sources due to noise sources (best case)

Figure 5.17 shows the amplitudes of the equivalent sources at different frequencies. The five real sources are very well identified in the 16 dB range.

We repeated the computation with the real sources not coinciding with the positions of equivalent sources (general case). The spectra of the simulated equivalent sources are depicted in Fig. 5.18. There is no set of equivalent sources clearly distinguishing themselves from the rest as in the best case. All equivalent sources appear to have similar amplitudes, and their frequency responses have the same form.

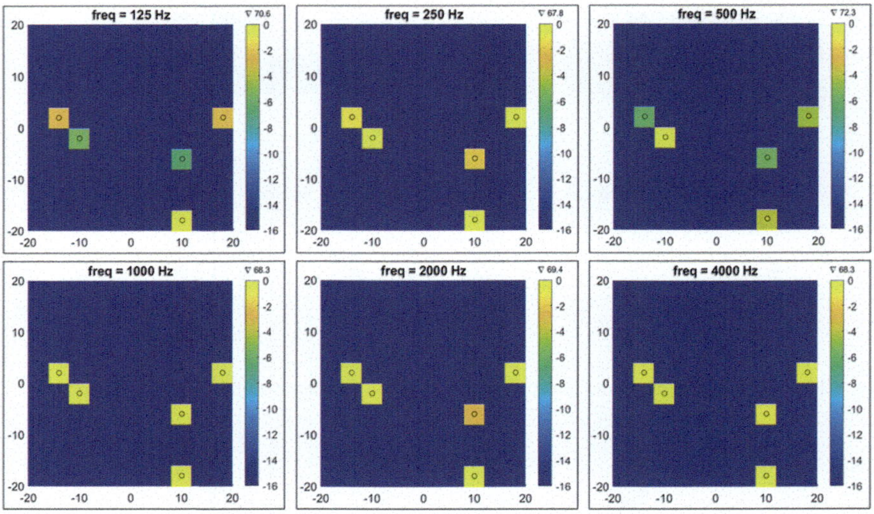

Fig. 5.17 Localization of sources radiating white noise (best case)

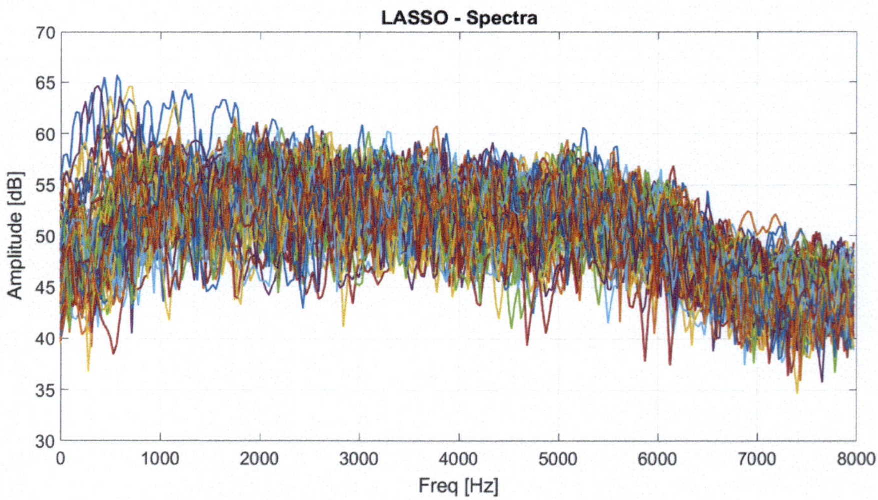

Fig. 5.18 Spectra of the equivalent sources due to noise sources (general case)

In Fig. 5.19, the amplitudes of the equivalent sources on the source plane for different frequencies are presented. The source identification is not clear up to 1000 Hz and is not good at all beyond 2000 Hz. This study confirms the sensitivity of the method to the position of the equivalent sources, and it is higher for broadband noise than for harmonic sources.

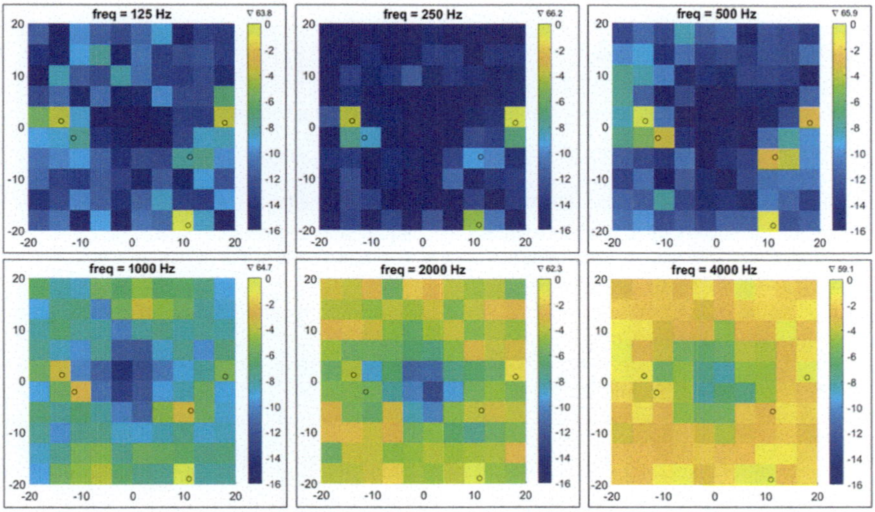

Fig. 5.19 Localization of sources radiating white noise (general case)

5.3 Multigrid Approach

The number of equivalent sources cannot be very large because the size of the matrix would make it impossible to solve using Matlab's function LASSO, since it does not accept the matrix in sparse format. As we have seen in the previous studies, the method is very sensitive to the positions of the equivalent sources; therefore, a high source density on the source plane may usually be required. Additionally, the moving body may contain several compact sources distributed on its surface. So, if the plane of equivalent sources is big, it will not be possible to have a high source density. However, it is still possible to have a high source density only in certain zones of the plane. We introduce a multigrid approach, which starts with a plane with low uniform source density, and in successive refinement steps, the source density increases only in the zones where the real sources are most likely to be. The key point of this approach is how to find or select the zones to be refined. We considered as the selection criterion that the strength of the equivalent sources is not less than 16 dB of the maximum value.

We can express the approach as the following algorithm:

Algorithm: Multigrid technique
Given are
$\vec{r}_i^{(0)}$: Position of equivalent sources (center of elements in original coarse grid)
N: Number of refinement steps
for $k = 0, 1, \ldots, N$ **do**

- Determine the strength of the equivalent sources in time domain $c_i(t)$ with LASSO
- Compute the spectra of equivalent sources $\hat{c}_i(f)$ and the maximum value $\hat{c}_{max}(f)$
- Elements with sources whose amplitudes are not less than 16 dB of the maximum value ($\hat{q}_{max}(f) - \hat{q}_i(f) < 16$), are selected
- **if** $0 < k < N$ **then**
  - Refine selected elements
  - Determine the position of the new equivalent sources $\vec{r}_i^{(k)}$

end if
end for

We evaluate the performance of the present approach with the next examples.

5.3.1 *Test Case: Harmonic Sources*

For the following three examples, we consider a square source plane of side $L_e = 3$ m with an initial 5×5 coarse grid moving parallel to the x-axis at a height $h = 10$ m with Mach number $M = 0.25$. The spiral microphone array has a radius $R_a = \sqrt{2}L_e/2$ and 200 microphones. We want to keep the number of equivalent sources around 100, so we consider only two refinement steps.

(a) Three sources radiating at 250 Hz. After the first calculation, 6 of the 25 elements were selected for refinement. Each selected element was divided in four smaller elements, so the refined mesh had 43 elements. After the second calculation (first refinement), only six elements were selected. Each selected element was divided in nine smaller elements, so the final refined mesh had 91 elements. Figure 5.20 shows the strengths of the equivalent sources on the source plane at the radiating frequency and the corresponding spectra, calculated with the three distributions of equivalent sources, the original and after the two refinement steps. With the multigrid approach, it is possible to determine the position and extent of the three sources with accuracy. Interestingly, the spectra obtained for the equivalent sources with the second refinement show a peak at the third harmonic of the radiating frequency, i.e., 750 Hz, but several dBs below the real frequency. The appearance of that fictitious peak is a numerical artifact.

(b) Three sources radiating at 400 Hz. In this case, we consider a higher frequency. The positions of the three sources are not the same as in the previous test. After the first calculation, three elements were selected and each one divided in four smaller elements. The refined mesh had 34 elements. After the second calculation, seven elements were selected for refinement. Each selected element was

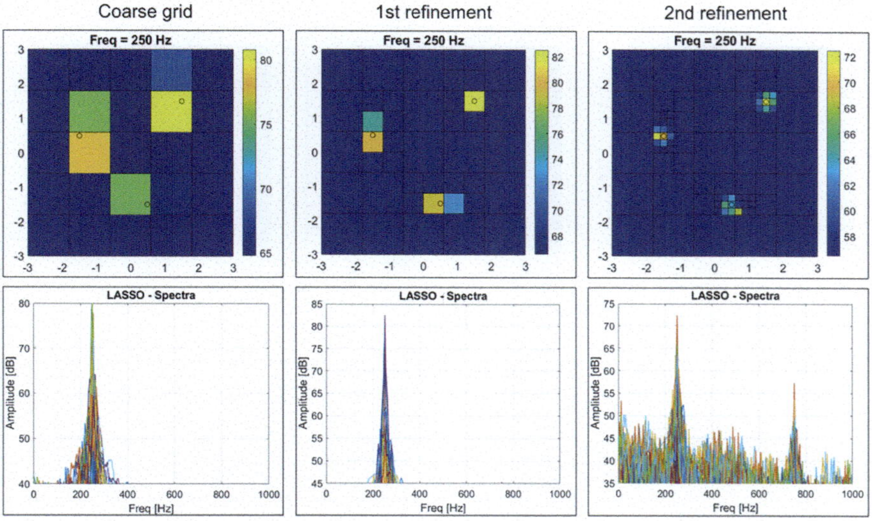

Fig. 5.20 Source localization at 250 Hz applying multigrid approach

divided in nine elements. The final mesh had 90 elements. Figure 5.21 shows the strengths of the equivalent sources on the source plane at the radiating frequency and the corresponding spectra, calculated with the three distributions of equivalent sources, the original and after the two refinement steps. The position and extent of the three sources are obtained with great accuracy. A fictitious peak at 1200 Hz (3rd harmonic) appeared but with an amplitude of 30 dB lower than the maximum value.

(c) Sources radiating at different frequencies. We combine the previous two examples, so now we have a total of six sources radiating at 250 and 400 Hz. After the first calculation, nine elements were selected. Each selected element was divided into four smaller element. The refined mesh had 52 elements. After the second calculation, 14 elements were selected. Each selected element was divided again in four smaller elements. The final mesh had 94 elements. Figure 5.22 shows the amplitudes of the equivalent sources on the plane for the two radiating frequencies and the spectra of all equivalent sources for the three calculations. The method is able to localize accurately the six sources and their radiating frequency.

5.3.2 *Test Case: Source of Broadband Noise*

For the last example, we received from our project partner, the Laboratory for Acoustics/Noise Control (Empa), virtually generated time series of the pressure at 230 microphones arranged in a spiral array due to a set of point sources passing over

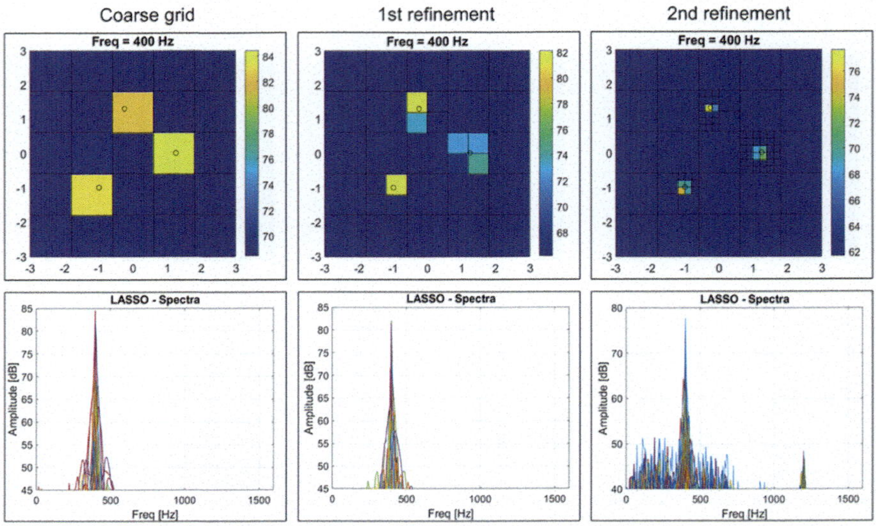

Fig. 5.21 Source localization at 400 Hz applying multigrid approach

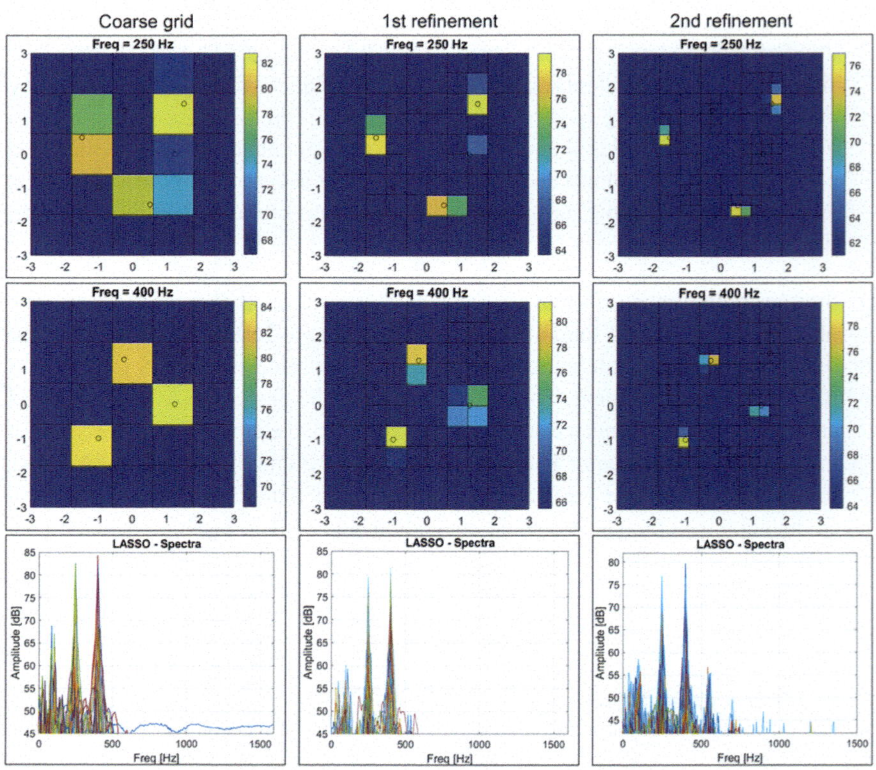

Fig. 5.22 Source localization at 250 and 400 Hz applying multigrid approach

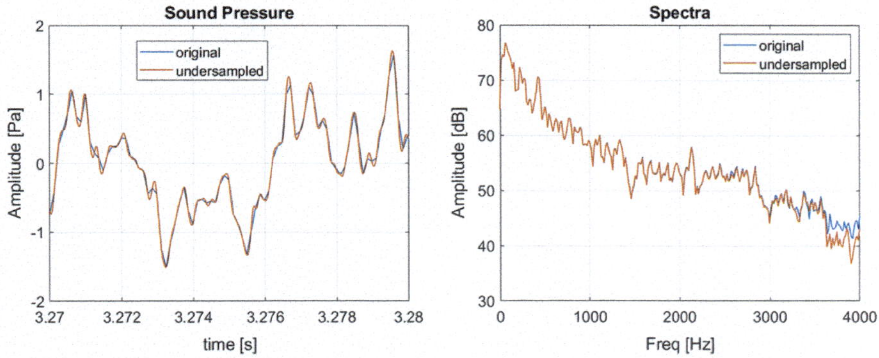

Fig. 5.23 Comparison of original and undersampled signals

at the height of approximately 190 m. The sources radiate broadband noise that emulates the sound produced by a real airplane. The time series has a frequency sampling of 48,192 Hz and a duration of more than five seconds. We were not told the number of point sources or their positions, only that they were located within a square of side $L_e = 43$ m, and our task was to localize them.

The method allows us to consider only a small number of time steps N_{seg}, i.e., we need to extract short signals from the given signal. However, the high-frequency sampling would lead to a very coarse frequency resolution if we perform the Fourier Transform of the short signals. Therefore, the sound pressure had to be undersampled with a frequency sampling of 8 kHz (we picked one of every six time steps). Figure 5.23 shows a comparison of the original and undersampled signals in time and frequency domains. We see that the undersampling affects the signal only beyond 3000 Hz.

The strengths of the equivalent sources were computed in nine overlapping intervals of 512 time steps, and the averages were built. Since the source plane is large, the starting grid has 81 sources distributed uniformly over the plane, and after refinement, only the new element was taken into account. After the first calculation, seven elements were selected for refinement. Each one of these elements was divided in four smaller elements. After the second calculation, nine elements were selected for refinement. Each selected element was divided into nine smaller elements. The refined elements have a source density that corresponds to a 54×54 grid. The final grid has 81 elements, i.e., 81 equivalent sources. The refinement steps are presented in Fig. 5.24.

To evaluate the performance of the method, the results were compared with the results of a conventional "delay-and-sum" beamforming.

Conventional beamforming: While the ESM is an inverse method, the beamforming is a direct one, i.e., no system of equations needs to be solved since each equivalent source is assumed to be independent of the rest. The strength of the i-th equivalent source c_i^{bf} at emission time τ^j is obtained by the delay-and-sum approach defined by the expression [Sij12]

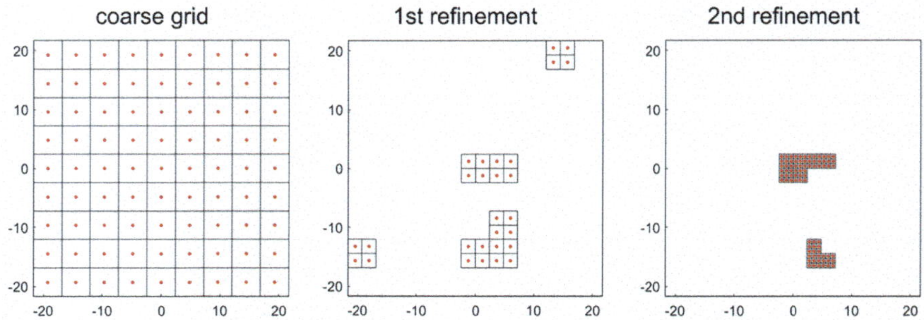

Fig. 5.24 Refinement of the grid for the source localization

$$c_i^{bf}\left(\tau^j\right) = \frac{4\pi}{M}\sum_{m=1}^{M} d_{mi}\left(\tau^j\right)p_m\left(t\left(\tau^j\right)\right) \qquad t\left(\tau^j\right) = \tau^j + d_{mi}\left(\tau^j\right)/c \tag{5.22}$$

where $d_{mi}(\tau)$ is the distance from the m-th microphone to the i-th source at emission time τ, $d_{mi}(\tau) = \vec{r}_m^A - \vec{r}_i^q(\tau)$ and $p_m(t)$ is the pressure at the m-th microphone and reception time t. Usually, $t\left(\tau^j\right)$ does not match any reception time t^k. Therefore, $p_m\left(t\left(\tau^j\right)\right)$ needs to be interpolated. Since the beamforming is a direct method, we are able to consider a bigger number of equivalent sources and extract longer time signals for the averaging than for the ESM. So, we perform nine calculations with a 41×41 source distribution and time signals with 4800 time steps. Figure 5.25 shows the source grid for the beamforming and the microphone arrangement of the spiral array.

Figure 5.26 presents a comparison between beamforming and ESM for different 1/3-octaves. The ESM localizes three sources at all considered frequency bands, while the beamforming distinguishes two sources only beyond 300 Hz and three sources

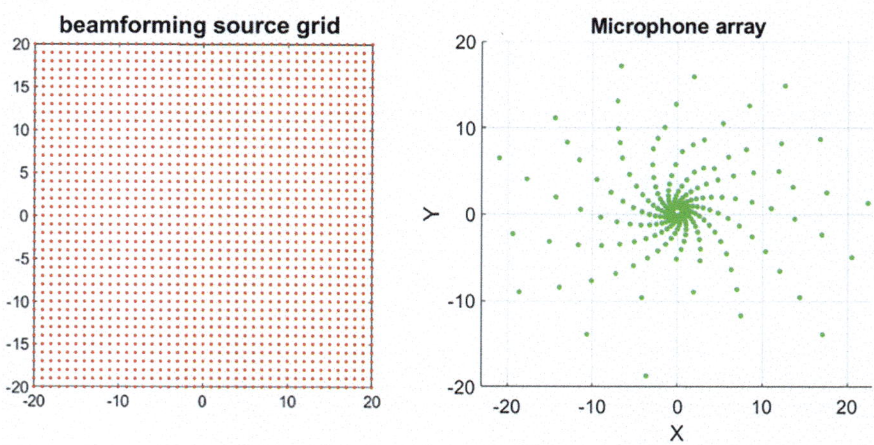

Fig. 5.25 Distribution of equivalent sources for beamforming and microphone positions

only beyond 1300 Hz. The positions of the three sources match between ESM and beamforming.

The multigrid ESM is able to localize the three point sources with good accuracy even at low frequencies where conventional beamforming fails.

The series of calculations performed in this chapter have shown that the ESM-based method in the time domain for source localization that produces the better results is the one that builds a global matrix and solves it with the LASSO regression.

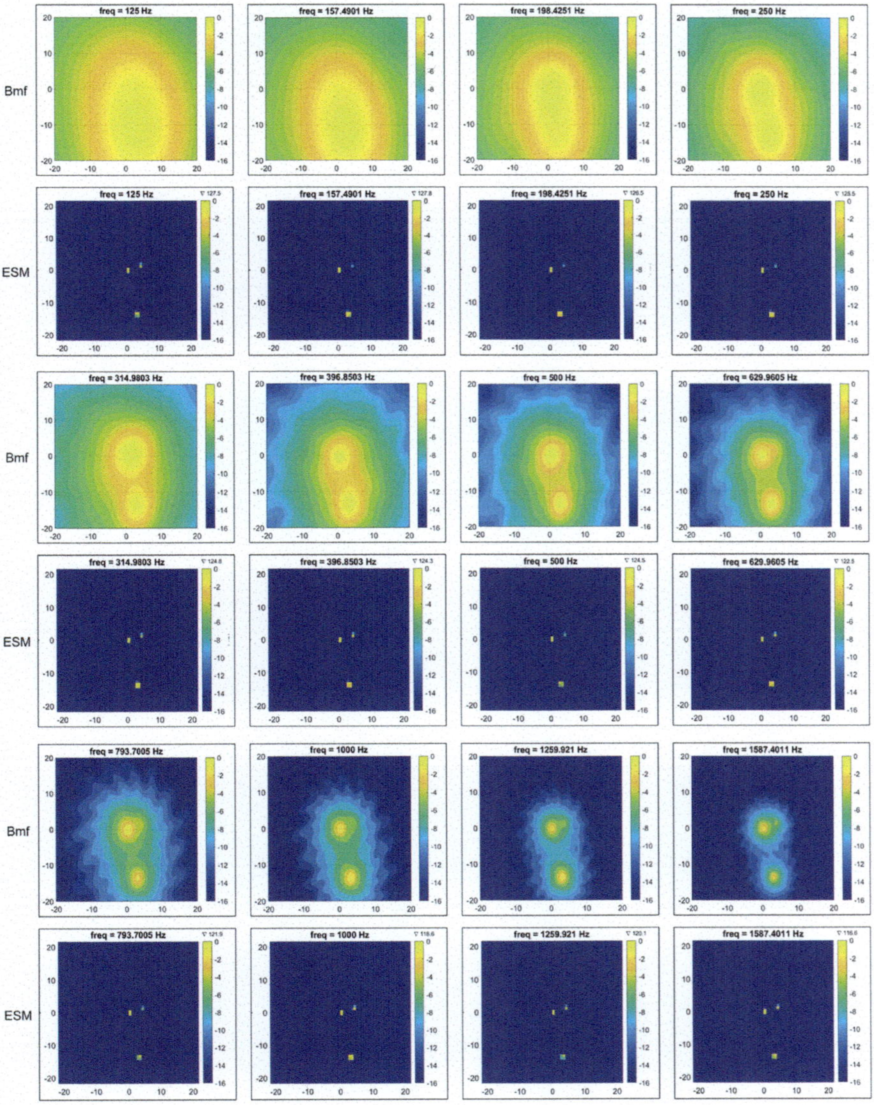

Fig. 5.26 Comparison between beamforming (Bmf) and ESM

The method's drawbacks are the high sensitivity on the position of the equivalent sources and the appearance of fictitious pulses.

References

E. Sarradj, Optimal planar microphone array arrangements, in *Proceedings DAGA 2015*, pp. 220–223.

P. Sijtsma, Acoustic beamforming for the ranking of aircraft noise, NLR-TP-2012-137.

Chapter 6
Final Remarks and Outlook

Abstract In this short final chapter, we show that the superposition of point sources is an essential component of the methods developed here and runs like a common thread through the entire book—from the description of simple moving point sources up to the analysis of moving oscillating bodies and source localization.

Looking at the methods presented in this book for determining sound fields emitted by moving sources, the following common features become apparent: In most cases, the solutions result from the superposition of point sources. One could almost say that these sources are the "*acoustic atoms*" from which fairly arbitrary sound fields can be composed.

For example, the Cartesian convolution integral (2.111), introduced in the frequency domain, can be interpreted as a superposition of stationary point sources that, so to speak, represent the trajectory of the moving source.

This observation is neither surprising nor new, as the so-called "point source radiator synthesis" for stationary sources has been known for a long time ([1], Sect. 2. 6) and is used to assemble complex sound sources from simple elementary radiators.

The sound field reflected from the ground of a half-space can also be modeled by additional mirrored point sources. Therefore, the sound field of moving sources in half-spaces can be composed by a combination of point sources on the trajectory plus point sources on the mirrored trajectory.

The situation is similar for extended vibrating bodies. In the stationary case, the sound field caused by such objects can be treated using the boundary element method, which in turn describes the vibrating body surface using monopoles and dipoles. This procedure also works for moving bodies. Here, too, we are dealing with a double superposition of simple point sources: Firstly, the sources must map the trajectory. Secondly, the surface of the body must be modeled using monopoles and dipoles.

The superposition method could, of course, be extended even further, e.g., to analyze vibrating bodies moving in half-spaces. This would even require a triple superposition of point sources: (1) point sources along the trajectory plus, (2) sources on the body surface plus, (3) point sources on the mirrored body surface to account for

M. Ochmann and R. Piscoya, *Moving Acoustic Sources*,
Springer-AAS Acoustics Series, https://doi.org/10.1007/978-981-96-0871-3_6

reflection. This was not performed in this book and would, therefore, be a rewarding and challenging task for further research.

All these superpositions of point sources lead to integrals, often multiple integrals, over sets of point sources. It must, therefore, always be shown that these integrals converge. In the case of convergence, the challenge is then either to find closed-form expressions for these integrals or to solve them using suitable numerical methods.

The problem of convergence is particularly serious for half-space problems with impedance boundary conditions. With real mirror sources, only a few impedance boundary conditions can be treated, which are often insignificant in practice. Therefore, point sources with singularities located in the complex plane must also be used to ensure that the superposition integrals converge. This was already mentioned in Sect. 2.8 and will be further explored in future work.

Similarly, the principle of the equivalent source method (ESM) is based on the replacement of real sound sources by the superposition of equivalent sound sources, which are usually point sources. Obviously, the ESM, which we have used extensively to analyze moving sources in both the frequency and time domains, is also relying on the superposition of point sources.

Source localization by means of the ESM, as presented in Chap. 5, is a new approach. A comparison with established methods such as beamforming, etc., or even a combination with these methods should open up a further exciting field of research.

References

1. L. Cremer, *Vorlesungen über Technische Akustik (Lectures about technical acoustics)* (Springer, Berlin—New York—Heidelberg, 1975)

Appendices

M. Ochmann and R. Piscoya, *Moving Acoustic Sources*,
Springer-AAS Acoustics Series, https://doi.org/10.1007/978-981-96-0871-3

Appendix A
The Fourier Transform of the 3D Potential Eq. (2.35)

We have to proof Eq. (2.36)

$$F_t[\varphi(t)] = \frac{i\hat{Q}}{4u} e^{i\left(\frac{\omega-\Omega}{u}\right)x} H_0^{(1)}\left(r\sqrt{\left(\frac{\omega}{c}\right)^2 - \left(\frac{\omega-\Omega}{u}\right)^2}\right) = \hat{\varphi}(\omega), \quad \hat{Q} = -Q_1/\rho \tag{2.36}$$

with

$$\varphi(t) = \frac{\hat{Q}}{4\pi R_1} e^{ik_0\gamma^2 R_1} e^{-ik_0\gamma^2(ct-Mx)} = \frac{\hat{Q}}{4\pi} e^{ik_0\gamma^2 Mx} \frac{1}{R_1(t)} e^{ik_0\gamma^2(R_1(t)-ct)} \tag{A.1}$$

and

$$R_1(t) = \sqrt{(x-ut)^2 + \left(y^2+z^2\right)/\gamma^2}, \; r = \sqrt{y^2+z^2}, \;\; k_0 = \Omega/c. \tag{A.2}$$

Thus, we must calculate

$$F_t[\varphi(t)] = \frac{\hat{Q}}{4\pi} e^{ik_0\gamma^2 Mx} \int\limits_{-\infty}^{\infty} \frac{1}{R_1(t)} e^{ik_0\gamma^2(R_1(t)-ct)} e^{i\omega t} dt. \tag{A.3}$$

Using the substitution

$$\tau := ut - x, \;\; d\tau/dt = u, \;\; t = (\tau + x)/u \text{ for fixed } x, \tag{A.4}$$

we obtain

M. Ochmann and R. Piscoya, *Moving Acoustic Sources*, Springer-AAS Acoustics Series, https://doi.org/10.1007/978-981-96-0871-3

$$F_t[\varphi(t)] = \frac{\hat{Q}}{4\pi u} e^{ik_0\gamma^2 Mx} \int_{-\infty}^{\infty} \frac{1}{\tilde{R}_1(\tau)} e^{ik_0\gamma^2\left(\tilde{R}_1(\tau)-\frac{c}{u}(\tau+x)\right)} e^{i\omega\frac{1}{u}(\tau+x)} d\tau \tag{A.5}$$

$$= \frac{\hat{Q}}{4\pi u} e^{i\frac{\Omega}{c}\gamma^2(M-\frac{1}{M})x} e^{i\omega\frac{x}{u}} \int_{-\infty}^{\infty} \frac{1}{\tilde{R}_1(\tau)} e^{ik_0\gamma^2\left(\tilde{R}_1(\tau)-\frac{1}{M}\tau\right)} e^{i\omega\frac{\tau}{u}} d\tau$$

with

$$\tilde{R}_1(\tau) = \sqrt{\tau^2 + (y^2+z^2)/\gamma^2}. \tag{A.6}$$

Since we have

$$M - \frac{1}{M} = \frac{-1}{M\gamma^2} = \frac{-c}{u\gamma^2} \quad \left(\gamma = 1/\sqrt{1-M^2}\right), \text{ we get} \tag{A.7}$$

$$F_t[\varphi(t)] = \frac{\hat{Q}}{4\pi u} e^{i\frac{x}{u}(\omega-\Omega)} \int_{-\infty}^{\infty} \frac{1}{\tilde{R}_1(\tau)} e^{ik_0\gamma^2\left(\tilde{R}_1(\tau)-\frac{1}{M}\tau\right)} e^{i\omega\frac{\tau}{u}} d\tau. \tag{A.8}$$

Comparing (A.8) with the right-hand side of Eq. (2.36), it only remains to show that the integral yields

$$I := \int_{-\infty}^{\infty} \frac{1}{\tilde{R}_1(\tau)} e^{ik_0\gamma^2\left(\tilde{R}_1(\tau)-\frac{1}{M}\tau\right)} e^{i\omega\frac{\tau}{u}} d\tau = \pi i H_0^{(1)}\left(r\sqrt{\left(\frac{\omega}{c}\right)^2 - \left(\frac{\omega-\Omega}{u}\right)^2}\right). \tag{A.9}$$

To proof (A.9), we write the integral in the form

$$I = \int_{-\infty}^{\infty} \frac{e^{i\frac{\Omega}{c}\gamma^2\sqrt{\tau^2+\frac{1}{\gamma^2}(y^2+z^2)}}}{\sqrt{\tau^2+\frac{1}{\gamma^2}(y^2+z^2)}} e^{i\frac{\tau}{u}(\omega-\gamma^2\Omega)} d\tau \tag{A.10}$$

or by using $\alpha = \gamma\tau$

$$I = \int_{-\infty}^{\infty} \frac{e^{i\frac{\Omega}{c}\gamma\sqrt{\alpha^2+y^2+z^2}}}{\sqrt{\alpha^2+y^2+z^2}} e^{-i\left(\frac{1}{u\gamma}(\gamma^2\Omega-\omega)\right)\alpha} d\alpha. \tag{A.11}$$

A slightly modified application of Formula (2.26) immediately yields

$$I = \int_{-\infty}^{\infty} \frac{e^{i\frac{\Omega}{c}\gamma\sqrt{\alpha^2+r^2}}}{\sqrt{\alpha^2+r^2}} e^{i\alpha\left(\frac{1}{u\gamma}(\omega-\gamma^2\Omega)\right)} d\alpha = \pi i H_0^{(1)}\left(r\sqrt{\left(\frac{\Omega}{c}\gamma\right)^2 - \frac{1}{(u\gamma)^2}(\omega-\gamma^2\Omega)^2}\right),$$

with

$$r = \sqrt{y^2+z^2}. \tag{A.12}$$

Now, by rearranging the radicand in (A.12), we obtain by a simple calculation

$$\left(\frac{\Omega}{c}\gamma\right)^2 - \frac{1}{(u\gamma)^2}(\omega-\gamma^2\Omega)^2 = \left(\frac{\omega}{c}\right)^2 - \left(\frac{\omega-\Omega}{u}\right)^2 \tag{A.13}$$

and thus

$$I = \pi i H_0^{(1)}\left(r\sqrt{\left(\frac{\omega}{c}\right)^2 - \left(\frac{\omega-\Omega}{u}\right)^2}\right), \tag{A.14}$$

which is Eq. (A.9) and completes the proof. □

Appendix B
Comparison of the Solution for the Spectrum Obtained by Ollendorff [1] and Our Result (2.20)

Our result for the frequency spectrum is given by

$$\hat{\varphi}(\omega) = \frac{-iQ_1/\rho}{4u} e^{i\left(\frac{\omega-\Omega}{u}\right)x} H_0^{(1)}\left(r\sqrt{\left(\frac{\omega}{c}\right)^2 - \left(\frac{\omega-\Omega}{u}\right)^2}\right) \tag{2.20}$$

In our case, the source moves in a fluid at rest. Ollendorff refers to this situation as the reciprocal Doppler effect (see [1], Sect. II). Ollendorff considers the scalar sound potential $\varphi = \varphi_{Oll}$ defined in his Eq. (II.1.13). This potential φ_{Oll} (a cosine function) and our potential φ of Eq. (2.35) (a complex exponential function) are connected by the relationship

$$\varphi_{Oll} = \frac{1}{2}(\varphi + \varphi^*), \tag{B.1}$$

if our source strength $\hat{Q} = -Q_1/\rho$ is chosen such that $\hat{Q} = F_0'\sqrt{1-M^2}$ (F_0': see Eq. (II.1.5) of Ollendorff). Here, φ^* denotes the complex conjugate of φ. Ollendorff also defines the dimensionless potential

$$\Phi = \frac{4\pi r}{\hat{Q}}\varphi_{Oll}, \; r = \sqrt{y^2 + z^2}, \tag{B.2a}$$

which he splits into a positive and a negative part (see his Eq. (II.1.23))

$$\Phi = \frac{\Phi_+ + \Phi_-}{2} = \frac{1}{2}\left(e^{-i\Omega\frac{x}{u}}h_+ + e^{+i\Omega\frac{x}{u}}h_-\right) = \frac{1}{2}\left(e^{-i\Omega\frac{x}{u}}h_+ + \left(e^{-i\Omega\frac{x}{u}}h_+\right)^*\right). \tag{B.2b}$$

M. Ochmann and R. Piscoya, *Moving Acoustic Sources*,
Springer-AAS Acoustics Series, https://doi.org/10.1007/978-981-96-0871-3

Therefore, it is sufficient—due to (B.1) and (B.2a, B.2b)—to show that the following two spectra agree:

$$F_t\left(\frac{\hat{Q}}{4\pi r}\Phi_+\right) = F_t(\varphi). \tag{B.3}$$

Ollendorff then calculates the Fourier transformation F_ϑ of the velocity potential Φ with respect to the dimensionless time variable $\vartheta = \frac{1}{r}(ut - x)$ in a somewhat unusual way, see his Eqs. (II.1.23 and II.2.4):

$$F_\vartheta(\Phi) = \int_{-\infty}^{\infty} \Phi(\vartheta) e^{i\omega \frac{r}{u}\vartheta} d\vartheta. \tag{B.4}$$

This gives the following correspondence with the normal Fourier transform with respect to t

$$\begin{aligned} F_\vartheta(\Phi) &= \int_{-\infty}^{\infty} \Phi(\vartheta) e^{i\omega \frac{r}{u}\vartheta} d\vartheta = \int_{-\infty}^{\infty} \Phi\left(\frac{1}{r}(ut - x)\right) e^{i\omega \frac{r}{u}\frac{1}{r}(ut-x)} \frac{u}{r} dt \\ &= \frac{u}{r}\int_{-\infty}^{\infty} \tilde{\Phi}(t) e^{-i\omega \frac{x}{u}} e^{i\omega t} dt = \frac{u}{r} e^{-i\omega \frac{x}{u}} F_t(\tilde{\Phi}(t)), \quad \tilde{\Phi}(t) := \Phi\left(\frac{1}{r}(ut - x)\right). \end{aligned} \tag{B.5}$$

On the other hand, it follows from Ollendorff's Eqs. (II.1.23), (II.2.4) and (B.2b) that the Fourier transform is given by

$$F_\vartheta(\Phi_+) = e^{-i\Omega \frac{x}{u}} s_+, \tag{B.6}$$

where

$$s_+ = \pi i H_0^{(1)}\left(r\sqrt{\left(\frac{\omega}{c}\right)^2 - \left(\frac{\Omega - \omega}{u}\right)^2}\right). \tag{B.7}$$

(B.7 will be proven below).
Due to (B.5) and (B.6), it follows that

$$F_t\left(\frac{\hat{Q}}{4\pi r}\Phi_+\right) = \frac{\hat{Q}}{4\pi r}\frac{r}{u} e^{i\omega \frac{x}{u}} F_\vartheta(\Phi_+) = \frac{\hat{Q}}{4\pi u} e^{i(\omega - \Omega)\frac{x}{u}} s_+. \tag{B.8}$$

Using Eq. (B.7), *we thus obtain*

Table B1 Comparison of symbols

Symbol of Ollendorff	Symbol in the present paper
γ	M
Ω_g	$\Omega \frac{r}{u}$
Ω	$\omega \frac{r}{u}$

$$F_t\left(\frac{\hat{Q}}{4\pi r}\Phi_+\right) = \frac{\hat{Q}i}{4u}e^{i\left(\frac{\omega-\Omega}{u}\right)x}H_0^{(1)}\left(r\sqrt{\left(\frac{\omega}{c}\right)^2 - \left(\frac{\Omega-\omega}{u}\right)^2}\right) = F_t(\varphi),$$

which agrees with the spectrum given in Eq. (2.20).

Proof of Eq. (B.7):

We consider formula (II.2.16) of Ollendorf

$$s_+ = \pi i H_0^{(1)}\left(\sqrt{\frac{\gamma^2}{1-\gamma^2}\Omega_g^2 - \left(\Omega\sqrt{1-\gamma^2} - \frac{\Omega_g}{\sqrt{1-\gamma^2}}\right)^2}\right) \tag{B.9}$$

for his complex amplitude density, which is valid for frequencies in the Doppler regime. Table B1 shows the meaning of Ollendorff's notation:

Thus, Eq. (B.9) is in our notation

$$s_+ = \pi i H_0^{(1)}\left(\frac{r}{u}\sqrt{\frac{M^2}{1-M^2}\Omega^2 - \left(\omega\sqrt{1-M^2} - \frac{\Omega}{\sqrt{1-M^2}}\right)^2}\right). \tag{B.10}$$

Now, the radicand can be rearranged as follows:

$$\frac{M^2}{1-M^2}\Omega^2 - \left(\omega\sqrt{1-M^2} - \frac{\Omega}{\sqrt{1-M^2}}\right)^2 = (\omega M)^2 - (\Omega - \omega)^2, \tag{B.11}$$

and therefore, we obtain

$$s_+ = \pi i H_0^{(1)}\left(r\sqrt{\left(\frac{\omega}{c}\right)^2 - \left(\frac{\Omega-\omega}{u}\right)^2}\right).$$

□

Equation (II.2.13) for the "supercritical range of high-frequency phase rates" and Eq. (II.2.19) for the "subcritical range of low frequencies" according to Ollendorff's designation in [1, p. 108] can be obtained from his Eq. (II.2.16) = Eq. (B.9) by choosing for the square root with negative radicand (cf. the discussion at the end of Sect. 2.2.1.3(a))

$$\sqrt{\frac{\gamma^2}{1-\gamma^2}\Omega_g^2 - \left(\Omega\sqrt{1-\gamma^2} - \frac{\Omega_g}{\sqrt{1-\gamma^2}}\right)^2} = i\sqrt{\left(\Omega\sqrt{1-\gamma^2} - \frac{\Omega_g}{\sqrt{1-\gamma^2}}\right)^2 - \frac{\gamma^2}{1-\gamma^2}\Omega_g^2}.$$

Appendix C
2D Green's Function (Eq. 2.138) Written in Exponentials

Following [2], p. 279, Eq. (5.1.30), the 2D Green's function in polar coordinates is given by

$$g_2(r, \phi; r_0, \phi_0) = \frac{i}{4} \sum_{n=-\infty}^{\infty} H_n^{(1)}(kr_>) J_n(kr_<) \cos[n(\phi - \phi_0)] \tag{2.137}$$

or by

$$\begin{aligned} g_2(r, \phi; r_0, \phi_0) &= \frac{i}{4}\frac{1}{2} \sum_{n=-\infty}^{\infty} H_n^{(1)}(kr_>) J_n(kr_<) \left[e^{in(\phi-\phi_0)} + e^{-in(\phi-\phi_0)}\right] \\ &= \frac{i}{4}\frac{1}{2} \sum_{n=-\infty}^{\infty} H_n^{(1)}(kr_>) J_n(kr_<) e^{in(\phi-\phi_0)} + \frac{i}{4}\frac{1}{2} \sum_{n=-\infty}^{\infty} H_n^{(1)}(kr_>) J_n(kr_<) e^{-in(\phi-\phi_0)} \\ &= \frac{i}{4}\frac{1}{2} \sum_{n=-\infty}^{\infty} H_n^{(1)}(kr_>) J_n(kr_<) e^{in(\phi-\phi_0)} + \frac{i}{4}\frac{1}{2} \sum_{m=-\infty}^{\infty} H_{-m}^{(1)}(kr_>) J_{-m}(kr_<) e^{im(\phi-\phi_0)}. \end{aligned} \tag{C.1}$$

Since for real n

$$J_{-n}(z) = (-1)^n J_n(z) \text{ and } H_{-n}^{(1)}(z) = (-1)^n H_n^{(1)}(z) \tag{C.2}$$

(see [3], p.358, Eqs. 9.1.5 and 9.1.6), we obtain

$$H_{-n}^{(1)}(kr_>) J_{-n}(kr_<) = H_n^{(1)}(kr_>) J_n(kr_<) \tag{C.3}$$

and therefore

M. Ochmann and R. Piscoya, *Moving Acoustic Sources*,
Springer-AAS Acoustics Series, https://doi.org/10.1007/978-981-96-0871-3

$$g_2(r, \phi; r_0, \phi_0) = \frac{i}{4} \sum_{n=-\infty}^{\infty} H_n^{(1)}(kr_>) J_n(kr_<) e^{in(\phi-\phi_0)}, \tag{C.4}$$

which is the exponential form of 2D Green's function (Eq. 2.138).

Appendix D
The Limit Value at the Higher Doppler Frequency

We have to prove that the limit process (2.193) is valid:

$$I = \frac{-2/\pi}{\sqrt{\left(\frac{\omega-\Omega}{v_r}\right)^2 - \left(\frac{\omega}{c}\right)^2}} \ln\left[\frac{\omega}{c} \Big/ \left[\left(\frac{\omega-\Omega}{v_r}\right) + \sqrt{\left(\frac{\omega-\Omega}{v_r}\right)^2 - \left(\frac{\omega}{c}\right)^2}\right]\right] \underset{\omega\to\omega_+}{\longrightarrow} \frac{2}{\pi}(1-M_r)\frac{c}{\Omega}. \tag{D.1}$$

For this, we consider the argument of the logarithm

$$\frac{\omega}{c} \Big/ \left[\left(\frac{\omega-\Omega}{v_r}\right) + \sqrt{\left(\frac{\omega-\Omega}{v_r}\right)^2 - \left(\frac{\omega}{c}\right)^2}\right] = 1 + \frac{\frac{\omega}{c} - \left(\frac{\omega-\Omega}{v_r}\right) - \sqrt{\left(\frac{\omega-\Omega}{v_r}\right)^2 - \left(\frac{\omega}{c}\right)^2}}{\left(\frac{\omega-\Omega}{v_r}\right) + \sqrt{\left(\frac{\omega-\Omega}{v_r}\right)^2 - \left(\frac{\omega}{c}\right)^2}}$$
$$=: 1 + z. \tag{D.2}$$

Taking into account Eq. (2.187)

$$\left(\frac{\omega-\Omega}{v_r}\right)^2 - \left(\frac{\omega}{c}\right)^2 = \frac{1-M_r^2}{v_r^2}(\omega-\omega_-)(\omega-\omega_+) \tag{D.3}$$

and

$$\frac{\omega}{c} - \left(\frac{\omega-\Omega}{v_r}\right) = \frac{M_r - 1}{v_r}(\omega-\omega_+), \tag{D.4}$$

we obtain

M. Ochmann and R. Piscoya, *Moving Acoustic Sources*, Springer-AAS Acoustics Series, https://doi.org/10.1007/978-981-96-0871-3

$$
\begin{aligned}
z &= \frac{\frac{\omega}{c} - \left(\frac{\omega-\Omega}{v_r}\right) - \sqrt{\left(\frac{\omega-\Omega}{v_r}\right)^2 - \left(\frac{\omega}{c}\right)^2}}{\left(\frac{\omega-\Omega}{v_r}\right) + \sqrt{\left(\frac{\omega-\Omega}{v_r}\right)^2 - \left(\frac{\omega}{c}\right)^2}} \\
&= \frac{\frac{M_r-1}{v_r}(\omega-\omega_+) - \sqrt{\frac{1-M_r^2}{v_r^2}}\sqrt{(\omega-\omega_-)(\omega-\omega_+)}}{\left(\frac{\omega-\Omega}{v_r}\right) + \sqrt{\frac{1-M_r^2}{v_r^2}}\sqrt{(\omega-\omega_-)(\omega-\omega_+)}}.
\end{aligned}
\tag{D.5}
$$

Since $z \underset{\omega\to\omega_+}{\longrightarrow} 0$, we can estimate the logarithm by the first term of its power series $\ln(1+z) \approx z$. Thus, (D.5) and (D.1) yield

$$
\begin{aligned}
I &= \frac{-2/\pi}{\sqrt{\frac{1-M_r^2}{v_r^2}}\sqrt{(\omega-\omega_-)(\omega-\omega_+)}}\left[\frac{\frac{M_r-1}{v_r}(\omega-\omega_+) - \sqrt{\frac{1-M_r^2}{v_r^2}}\sqrt{(\omega-\omega_-)(\omega-\omega_+)}}{\left(\frac{\omega-\Omega}{v_r}\right) + \sqrt{\frac{1-M_r^2}{v_r^2}}\sqrt{(\omega-\omega_-)(\omega-\omega_+)}}\right] \\
&= \frac{-2/\pi}{\sqrt{\frac{1-M_r^2}{v_r^2}}\sqrt{(\omega-\omega_-)}}\left[\frac{\frac{M_r-1}{v_r}\sqrt{\omega-\omega_+} - \sqrt{\frac{1-M_r^2}{v_r^2}}\sqrt{(\omega-\omega_-)}}{\left(\frac{\omega-\Omega}{v_r}\right) + \sqrt{\frac{1-M_r^2}{v_r^2}}\sqrt{(\omega-\omega_-)(\omega-\omega_+)}}\right] \\
&\underset{\omega\to\omega_+}{\longrightarrow} \frac{2}{\pi}\frac{v_r}{\omega_+-\Omega} = \frac{2}{\pi}(1-M_r)\frac{c}{\Omega}
\end{aligned}
\tag{D.6}
$$

which proves (D.1).

Appendix E
The Superposition Integral Fulfills the Helmholtz Equation in V_+

We have to show that the formulas (IF1) and (IF2) satisfy the Helmholtz equation (HSPa) in the cases under consideration. This applies in particular if the additional part

$$u_{1add}(z) := \overline{u}(z) - u_0(z) = u_0(-z) + 2\Gamma \int_0^\infty e^{\Gamma\tau} u_0(-z-\tau)d\tau \tag{E.1}$$

or alternatively

$$u_{2add}(z) := \overline{u}(z) - u_0(z) = u_0(-z) + 2i\Gamma \int_0^\infty e^{i\Gamma\tau} u_0(-z-i\tau)d\tau \tag{E.2}$$

fulfills the homogeneous Helmholtz equation in V_+.

It is, therefore, sufficient to show that $u_0(-z-\tau)$ or $u_0(-z-i\tau)$ satisfies the homogeneous Helmholtz equation in V_+ for $\tau \geq 0$, assuming that the integral and the Helmholtz operator can be interchanged.

(a) ***2D case***

Equation (2.286) shows that

$$u_0(-z-\tau) = \frac{iQ}{2\kappa u} e^{i(\omega-\Omega)\frac{x}{u}} e^{i\kappa|z+h+\tau|}, \ \kappa = \sqrt{\left(\frac{\omega}{c}\right)^2 - \left(\frac{\omega-\Omega}{u}\right)^2} \tag{E.3}$$

and, therefore, the application of the 2D-Helmholtz operator gives (see Eq. (2.67))

M. Ochmann and R. Piscoya, *Moving Acoustic Sources*, Springer-AAS Acoustics Series, https://doi.org/10.1007/978-981-96-0871-3

$$\left\{\Delta_2 + \left(\frac{\omega}{c}\right)^2\right\} u_0(-z-\tau) = \frac{Q}{u} e^{i(\omega-\Omega)\frac{x}{u}} \left\{\frac{\partial}{\partial z^2} + \kappa^2\right\} \frac{i}{2\kappa} e^{i\kappa|z+h+\tau|}$$
$$= -\frac{Q}{u} e^{i(\omega-\Omega)\frac{x}{u}} \delta(z+h+\tau) = 0 \tag{E.4}$$

for $z > 0$, since $h > 0$ and $\tau \geq 0$.

With regard to the complex representation (E.2), we must consider

$$u_0(-z-i\tau) = \frac{iQ}{2\kappa u} e^{i(\omega-\Omega)\frac{x}{u}} e^{i\kappa\sqrt{(z+h+i\tau)^2}} \tag{E.5}$$

and obtain for $z > 0$

$$\left\{\Delta_2 + \left(\frac{\omega}{c}\right)^2\right\} u_0(-z-i\tau) = \frac{Q}{u} e^{i(\omega-\Omega)\frac{x}{u}} \left\{\frac{\partial}{\partial z^2} + \kappa^2\right\} \frac{i}{2\kappa} e^{i\kappa\sqrt{(z+h+i\tau)^2}}$$
$$= -\frac{Q}{u} e^{i(\omega-\Omega)\frac{x}{u}} \left[\cosh(\kappa\tau)\delta(z+h) + (i/\kappa)\sinh(\kappa\tau)\frac{d}{dz}\delta(z+h)\right] = 0 \tag{E.6}$$

for $z > 0$, where we have used Theorem 1 of [4]. Please note that the complex counterpart to $|z+h+\tau|$ is $\sqrt{(z+h+i\tau)^2}$, but not $|z+h+i\tau|$ (see [4], Eq. (4.5)), where the sign of the complex square root is also explained).

(b) ***3D case***

The proof that $u_0(-z-\tau)$ and $u_0(-z-i\tau)$ satisfy the homogeneous 3D-Helmholtz equation *in* V_+ is completely analogous to the 2D case. From Eqs. (2.312) and (2.313), we obtain $u_0(-z-\tau)$ and $u_0(-z-i\tau)$, respectively. For the latter case, we must use Theorem 2 from [4].

(c) ***1D case***

Equation (2.346) shows that

$$u_0(-z-\tau) = \hat{p}(-z-\tau) = \frac{iQ}{2ku} \int_{z_0}^{z_1} e^{i(\omega-\Omega)\frac{\zeta}{u}} e^{ik|z+\tau+\zeta|} d\zeta, \tag{E.7}$$

and the application of the 1D-Helmholtz operator yields

$$\left\{\frac{\partial}{\partial z^2} + k^2\right\} u_0(-z-\tau) = \frac{Q}{u} \int_{z_0}^{z_1} e^{i(\omega-\Omega)\frac{\zeta}{u}} \left\{\frac{\partial}{\partial z^2} + \kappa^2\right\} e^{ik|z+\tau+\zeta|} d\zeta$$
$$= -\frac{Q}{u} \int_{z_0}^{z_1} e^{i(\omega-\Omega)\frac{\zeta}{u}} \delta(z+h+\zeta) d\zeta = 0 \tag{E.8}$$

for $z > 0$, since $h > 0$ and $\zeta \geq z_0 \geq 0$. Just as in the 2D case, it can be shown that the complex representation with $u_0(-z - i\tau)$ also fulfills the homogeneous 1D-Helmholtz equation by using *Theorem 1 of* [4].

Appendix F
The Boundary Condition Eq. (2.393) for a Compact Sphere

To derive the approximate boundary condition

$$\hat{\varphi}(\vec{y}_1) = \frac{c}{k_0} A e^{i\frac{3}{2}Mk_0 y_{11}} \left[i(k_0 a)^2 - \frac{3}{2} M k_0 y_{11} \right], \quad M = \frac{u}{c}, \quad k_0 = \frac{\Omega}{c}, \tag{2.393}$$

we start with the boundary condition for the velocity potential at *low Mach number,* given by Eq. (2.426)

$$\hat{\varphi}(\vec{y}_1) = -\frac{c}{k_0} A e^{ik_0 \frac{3M}{2} y_{11}} \left[ik_0 a \frac{h_0^{(1)}(k_0 a)}{h_0^{(1)\prime}(k_0 a)} - \frac{3M}{2} (2 - k_0^2 a^2) \frac{h_1^{(1)}(k_0 a)}{h_1^{(1)\prime}(k_0 a)} \frac{y_{11}}{a} \right]. \tag{2.426}$$

According to Eqs. (10.1.25/26) of [3], the spherical Hankel functions satisfy the relationship

$$h_n^{(1)}(z) = -iz^n \left(-\frac{1}{z} \frac{d}{dz} \right)^n \frac{e^{iz}}{z}. \tag{F.1}$$

Hence, we obtain

$$h_0^{(1)}(z) = -i \frac{e^{iz}}{z}, \quad h_0^{(1)\prime}(z) = -ie^{iz} \left(\frac{i}{z} - \frac{1}{z^2} \right)$$

and thus

$$\frac{h_0^{(1)}(z)}{h_0^{(1)\prime}(z)} = \frac{-i\frac{e^{iz}}{z}}{-ie^{iz}\left(\frac{i}{z} - \frac{1}{z^2}\right)} = \frac{z}{iz - 1} \approx -z \tag{F.2}$$

M. Ochmann and R. Piscoya, *Moving Acoustic Sources*, Springer-AAS Acoustics Series, https://doi.org/10.1007/978-981-96-0871-3

for small arguments z. In the same way, we obtain

$$\frac{h_1^{(1)}(z)}{h_1^{(1)\prime}(z)} \approx -\frac{z}{2}. \tag{F.3}$$

Inserting Eqs. (F.2) and (F.3) into Eq. (2.426) gives the following for small arguments $z = k_0 a$

$$\begin{aligned}\hat{\varphi}(\vec{y}_1) &\approx \frac{c}{k_0} A e^{ik_0 \frac{3M}{2} y_{11}} \left[i(k_0 a)^2 - \frac{3M}{2}\left(2 - k_0^2 a^2\right) \frac{k_0 a}{2} \frac{y_{11}}{a} \right] \\ &\approx \frac{c}{k_0} A e^{ik_0 \frac{3M}{2} y_{11}} \left[i(k_0 a)^2 - \frac{3}{2} M k_0 y_{11} \right],\end{aligned}$$

and this is identical to Formula (2.393).

Appendix G
Head Waves Due to a Moving Line Source

1. **Time** t_{HL}

The expression for t_{HL} is obtained by setting

$$t - \tau - t_H = 0 \tag{G.1}$$

with

$$t_H(\tau) = \frac{|x - Mc\tau| \sin\theta_c + z_h \cos\theta_c}{c} \text{ and } t = t_L + \frac{M}{c} x_L \tag{G.2}$$

along with

$$x = x_L + Mct_L, \quad z = \frac{z_L}{\gamma}, \quad h = \frac{h_L}{\gamma}, \quad z_{hL} = z_L + h_L \tag{G.3}$$

Inserting (G.2) and (G.3) into (G.1), we have

$$Mx_L + c(t_L - \tau) = |x_L + Mc(t_L - \tau)| \sin\theta_c + \frac{z_{hL}}{\gamma} \cos\theta_c \tag{G.4}$$

We write the absolute value as

$$|x_L + Mc(t_L - \tau)| = \pm(x_L + Mc(t_L - \tau)) \tag{G.5}$$

By squaring (G.4) and rearranging terms, we find the quadratic equation in $c(t_L - \tau)$

$$c^2(t_L - \tau)^2(1 - M^2 \sin^2\theta_c) + 2M\cos\theta_c\left(x_L \cos\theta_c \pm \frac{z_{hL}\sin\theta_c}{\gamma}\right)c(t_L - \tau) +$$

M. Ochmann and R. Piscoya, *Moving Acoustic Sources*, Springer-AAS Acoustics Series, https://doi.org/10.1007/978-981-96-0871-3

$$x_L^2\left(M^2 - \sin^2\theta_c\right) - \frac{z_{hL}^2\cos^2\theta_c}{\gamma^2} \pm \frac{2x_L z_{hL}\sin\theta_c\cos\theta_c}{\gamma} = 0 \tag{G.6}$$

The solutions of (G.6) are written as

$$t_L - \tau = t_{HL} = \frac{1}{c}\begin{cases} \frac{-M\cos\theta_c\left(x_L\cos\theta_c \pm \frac{z_{hL}\sin\theta_c}{\gamma}\right) - \Lambda}{1 - M^2\sin^2\theta_c} \\ \frac{-M\cos\theta_c\left(x_L\cos\theta_c \pm \frac{z_{hL}\sin\theta_c}{\gamma}\right) + \Lambda}{1 - M^2\sin^2\theta_c} \end{cases} \tag{G.7}$$

where

$$\Lambda^2 = M^2\cos^2\theta_c\left(x_L\cos\theta_c \pm \frac{z_{hL}\sin\theta_c}{\gamma}\right)^2 -$$

$$\left(1 - M^2\sin^2\theta_c\right)\left(x_L^2\left(M^2 - \sin^2\theta_c\right) - \frac{z_{hL}^2\cos^2\theta_c}{\gamma^2} \pm \frac{2x_L z_{hL}\sin\theta_c\cos\theta_c}{\gamma}\right) = 0 \tag{G.8}$$

Fortunately, (G.8) can be simplified to

$$\Lambda^2 = \left(\frac{x_L\sin\theta_c}{\gamma^2} \mp \frac{z_{hL}\cos\theta_c}{\gamma}\right)^2 \tag{G.9}$$

and the solutions of (G.5) are

$$t_{HL} = \frac{1}{c}\begin{cases} \frac{-M\cos\theta_c\left(x_L\cos\theta_c \pm \frac{z_{hL}\sin\theta_c}{\gamma}\right) - \left|\frac{x_L\sin\theta_c}{\gamma^2} \mp \frac{z_{hL}\cos\theta_c}{\gamma}\right|}{1 - M^2\sin^2\theta_c} \\ \frac{-M\cos\theta_c\left(x_L\cos\theta_c \pm \frac{z_{hL}\sin\theta_c}{\gamma}\right) + \left|\frac{x_L\sin\theta_c}{\gamma^2} \mp \frac{z_{hL}\cos\theta_c}{\gamma}\right|}{1 - M^2\sin^2\theta_c} \end{cases} \tag{G.10}$$

Due to the absolute values, there are eight possible combinations in (G.10), but only four are different, namely

$$t_1 = \frac{x_L(\sin\theta_c - M) + \frac{z_{hL}\cos\theta_c}{\gamma}}{1 - M\sin\theta_c}$$

$$t_2 = \frac{x_L(\sin\theta_c - M) - \frac{z_{hL}\cos\theta_c}{\gamma}}{1 - M\sin\theta_c}$$

$$t_3 = \frac{-x_L(\sin\theta_c + M) + \frac{z_{hL}\cos\theta_c}{\gamma}}{1 + M\sin\theta_c}$$

$$t_4 = \frac{-x_L(\sin\theta_c + M) - \frac{z_{hL}\cos\theta_c}{\gamma}}{1 + M\sin\theta_c}$$

In order to find the correct solution, we require that for $M = 0$, the solution should be equal to the solution for the stationary source. Hence, the correct solution will be

$$t_{HL} = \frac{1}{c}\begin{cases} \dfrac{-M\cos\theta_c\left(x_L\cos\theta_c + \frac{z_{hL}\sin\theta_c}{\gamma}\right) - \left(\frac{x_L\sin\theta_c}{\gamma^2} - \frac{z_{hL}\cos\theta_c}{\gamma}\right)}{1-M^2\sin^2\theta_c} \\ \dfrac{-M\cos\theta_c\left(x_L\cos\theta_c - \frac{z_{hL}\sin\theta_c}{\gamma}\right) + \left(\frac{x_L\sin\theta_c}{\gamma^2} + \frac{z_{hL}\cos\theta_c}{\gamma}\right)}{1-M^2\sin^2\theta_c} \end{cases} \quad \text{(G.11)}$$

Similarly to the stationary case, t_{HL} should be continuous. We can determine the intersection of both straight lines in (G.11) which is given by $x_L = -\gamma M z_{hL}\cos\theta_c$. Hence, the correct expression for t_{HL} is

$$t_{HL} = \frac{1}{c}\begin{cases} \dfrac{-x_L(\sin\theta_c + M) + \frac{z_{hL}\cos\theta_c}{\gamma}}{1+M\sin\theta_c}, & x_L < -\gamma M z_{hL}\cos\theta_c \\ \dfrac{x_L(\sin\theta_c - M) + \frac{z_{hL}\cos\theta_c}{\gamma}}{1-M\sin\theta_c}, & x_L > -\gamma M z_{hL}\cos\theta_c \end{cases} \quad \text{(G.12)}$$

2. **Region of Wide-Angle Reflection**

Head waves due to a stationary source appear at angle of incidence greater than θ_c. When the source is moving, the border of the region of wide-angle reflection is obtained by setting $p(t_L - \tau = t_{0sL}) = 1/c_m$. An easier way is however to set

$$t_{HL} = t_{0sL} \quad \text{(G.13)}$$

For the left branch of (G.12), we have

$$-x_L(\sin\theta_c + M) + \frac{z_{hL}\cos\theta_c}{\gamma} = (1 + M\sin\theta_c)\sqrt{x_L^2 + z_{hL}^2} \quad \text{(G.14)}$$

By squaring (F.14) and rearranging terms we write a quadratic equation

$$x_L^2\left((\sin\theta_c + M)^2 - (1 + M\sin\theta_c)^2\right) + z_{hL}^2\left(\frac{\cos^2\theta_c}{\gamma^2} - (1 + M\sin\theta_c)^2\right)$$
$$- 2x_L z_{hL}(\sin\theta_c + M)\frac{\cos\theta_c}{\gamma} = 0 \quad \text{(G.15)}$$

that can be simplified to

$$\left(\frac{x_L\cos\theta_c}{\gamma} + z_{hL}(\sin\theta_c + M)\right)^2 = 0 \quad \text{(G.16)}$$

which gives the straight line limit to the wide-angle reflection zone behind the source

$$x_{HL}^{(-)} = -\gamma z_{hL}\frac{(\sin\theta_c + M)}{\cos\theta_c} \quad \text{(G.17)}$$

For the right branch of (G.12), we have

$$x_L(\sin\theta_c - M) + \frac{z_{hL}\cos\theta_c}{\gamma} = (1 - M\sin\theta_c)\sqrt{x_L^2 + z_{hL}^2} \tag{G.18}$$

By squaring (G.18) and rearranging terms, we write a quadratic equation

$$x_L^2\big((\sin\theta_c - M)^2 - (1 - M\sin\theta_c)^2\big) + z_{hL}^2\left(\frac{\cos^2\theta_c}{\gamma^2} - (1 - M\sin\theta_c)^2\right) + 2x_L z_{hL}(\sin\theta_c - M)\frac{\cos\theta_c}{\gamma} = 0 \tag{G.19}$$

that can be simplified to

$$\left(\frac{x_L\cos\theta_c}{\gamma} - z_{hL}(\sin\theta_c - M)\right)^2 = 0 \tag{G.20}$$

which gives the straight line limit to the wide-angle reflection zone in front of the source

$$x_{HL}^{(+)} = \gamma z_{hL}\frac{(\sin\theta_c - M)}{\cos\theta_c} \tag{G.21}$$

3. **Head Waves Arriving Before Direct Waves**

Head waves arrive at some positions before the direct waves. The limit of the zone where that occurs can be determined by setting

$$t_{HL} = t_{OL} \tag{G.22}$$

For the left branch of (G.12), we have

$$-x_L(\sin\theta_c + M) + \frac{z_{hL}\cos\theta_c}{\gamma} = (1 + M\sin\theta_c)\sqrt{x_L^2 + (z_L - h_L)^2} \tag{G.23}$$

By squaring (G.23) and rearranging terms we write a quadratic equation in x_L

$$x_L^2\big((\sin\theta_c + M)^2 - (1 + M\sin\theta_c)^2\big) - 2x_L z_{hL}(\sin\theta_c + M)\frac{\cos\theta_c}{\gamma} + z_{hL}^2\frac{\cos^2\theta_c}{\gamma^2} - (1 + M\sin\theta_c)^2(z_L - h_L)^2 = 0 \tag{G.24}$$

that can be rewritten as

$$x_L^2\frac{\cos^2\theta_c}{\gamma^2}+2x_Lz_{hL}(\sin\theta_c+M)\frac{\cos\theta_c}{\gamma}+(z_L^2+h_L^2)(\sin\theta_c+M)^2$$
$$+2h_Lz_L\big((\sin\theta_c-M)^2-2\big(1+M^2\sin^2\theta_c\big)\big)=0 \tag{G.25}$$

The solution of (G.25) gives the limit to the region behind the source where the head waves arrive before the direct wave at positions

$$x_L=\frac{-z_{hL}(\sin\theta_c+M)\frac{\cos\theta_c}{\gamma}\pm\Gamma}{\frac{\cos^2\theta_c}{\gamma^2}} \tag{G.26}$$

where

$$\Gamma^2=z_{hL}^2(\sin\theta_c+M)\frac{\cos^2\theta_c}{\gamma^2}-\frac{\cos^2\theta_c}{\gamma^2}\big((z_L^2+h_L^2)(\sin\theta_c+M)^2$$
$$+2h_Lz_L\big((\sin\theta_c-M)^2-2\big(1+M^2\sin^2\theta_c\big)\big)\big)=0 \tag{G.27}$$

(G.27) can be simplified to

$$\Gamma^2=4h_Lz_L(1+M\,\sin\theta_c)^2\frac{\cos^2\theta_c}{\gamma^2} \tag{G.28}$$

Since for $M=0$, (G.26) should give the solution for the stationary source, the correct solution has the minus sign; therefore,

$$x_{H0L}^{(-)}=-\gamma\frac{z_{hL}(\sin\theta_c+M)+2(1+M\,\sin\theta_c)\sqrt{z_Lh_L}}{\cos\theta_c} \tag{G.29}$$

For the right branch of (G.12), we have

$$x_L(\sin\theta_c-M)+\frac{z_{hL}\cos\theta_c}{\gamma}=(1-M\,\sin\theta_c)\sqrt{x_L^2+(z_L-h_L)^2} \tag{G.30}$$

By squaring (G.30) and rearranging terms, we write a quadratic equation in x_L

$$x_L^2\big((\sin\theta_c-M)^2-(1-M\,\sin\theta_c)^2\big)+2x_Lz_{hL}(\sin\theta_c-M)\frac{\cos\theta_c}{\gamma}$$
$$+z_{hL}^2\frac{\cos^2\theta_c}{\gamma^2}-(1-M\,\sin\theta_c)^2(z_L-h_L)^2=0 \tag{G.31}$$

that can be rewritten as

$$x_L^2 \frac{\cos^2\theta_c}{\gamma^2} - 2x_L z_{hL}(\sin\theta_c + M)\frac{\cos\theta_c}{\gamma} + (z_L^2 + h_L^2)(\sin\theta_c - M)^2$$

$$+2h_L z_L\big((\sin\theta_c + M)^2 - 2(1 + M^2\sin^2\theta_c)\big) = 0 \tag{G.32}$$

The solution of (G.32) gives the limit to the region in front of the source where the head waves arrive before the direct wave at positions

$$x_L = \frac{z_{hL}(\sin\theta_c - M)\frac{\cos\theta_c}{\gamma} \pm \Gamma}{\frac{\cos^2\theta_c}{\gamma^2}} \tag{G.33}$$

where

$$\Gamma^2 = z_{hL}^2(\sin\theta_c - M)\frac{\cos^2\theta_c}{\gamma^2} - \frac{\cos^2\theta_c}{\gamma^2}\big((z_L^2 + h_L^2)(\sin\theta_c - M)^2$$

$$+2h_L z_L\big((\sin\theta_c + M)^2 - 2(1 + M^2\sin^2\theta_c)\big)\big) = 0 \tag{G.34}$$

(G.34) can be simplified to

$$\Gamma^2 = 4h_L z_L(1 - M\sin\theta_c)^2\frac{\cos^2\theta_c}{\gamma^2} \tag{G.35}$$

Since for $M = 0$, (G.33) should give the solution for the stationary source, the correct solution has the plus sign; therefore

$$x_{HOL}^{(+)} = \gamma\frac{z_{hL}(\sin\theta_c - M) + 2(1 - M\sin\theta_c)\sqrt{z_L h_L}}{\cos\theta_c} \tag{G.36}$$

References

1. F. Ollendorf, Das Frequenzspektrum des Doppler-Effektes (The frequency spectrum of the Doppler effect). ACUSTICA **50**, 99–108 (1982)
2. G.D. Duffy, *Green's Functions with Applications* (Chapman&Hall/CRC, Boca Raton-London—NewYork—Washington DC, 2001)
3. M. Abramowitz, I.A. Stegun, *Handbook of Mathematical Functions*, 9th edn. (Dover Publications, New York, 1972)
4. M. Ochmann, R. Piscoya, *Theory and Application of Acoustic Sources Using Complex Analysis*, 1st edn. (Springer Nature, Singapore, 2021)

M. Ochmann and R. Piscoya, *Moving Acoustic Sources*,
Springer-AAS Acoustics Series, https://doi.org/10.1007/978-981-96-0871-3

Index

This index mainly relates to Chap. 2 and only contains entries for terms that occur less frequently in the book—such as "Archimedean spiral"—but does not include very frequently occuring terms like "frequency domain" or "Fourier transform", in order to avoid overloading the index.

M. Ochmann and R. Piscoya, *Moving Acoustic Sources*, Springer-AAS Acoustics Series, https://doi.org/10.1007/978-981-96-0871-3

The manufacturer's authorised representative in the EU is Springer Nature Customer Service Centre GmbH, Europaplatz 3, 69115 Heidelberg, Germany. If you have any concerns regarding our products, please contact ProductSafety@springernature.com

Printed and bound by CPI Group (UK) Ltd, Croydon, CR0 4YY
30/09/2026
02225302-0009